Service Assurance for Voice over WiFi and 3G Networks

For a listing of recent titles in the *Artech House Telecommunications Library*, turn to the back of this book.

Service Assurance for Voice over WiFi and 3G Networks

Richard Lau
Ram Khare
William Y. Chang

ARTECH HOUSE
BOSTON | LONDON
artechhouse.com

Library of Congress Cataloging-in-Publication Data
A catalog record for this book is available from the Library of Congress.

British Library Cataloguing in Publication Data
Lau, Richard
Service assurance for voice over WiFi and 3G networks.
—(Artech House telecommunications library)
1. Internet telephony 2. Telecommunication—Quality control
I. Title II. Khare, Ram III. Chang, William Y.
621.3'8212

ISBN-10: 1-59693-000-4

Cover design by Igor Valdman

685 Canton Street
Norwood, MA 02062

International Standard Book Number: 1-59693-000-4

10 9 8 7 6 5 4 3 2 1

To my parents who inspired me most, to my wife, Carrie, and my two children, Ryan and Kenneth, who were totally supportive while I spent many hours of otherwise family time on the writing

— Richard Lau

To my life companion, Amla, and my daughters, Anupriya and Rashmi, and son, Jayant, who supported and kept encouraging me during periods of doubt in completing this book

— Ram Khare

To my wife, Kathy, my son, Robert, and all sentient beings

— William Y. Chang

Contents

Foreword

Communications technology has a profound impact on improving the life we lead as individuals, as members of a larger society, and as participants in the enterprises that drive the economics of the modern world. A shared view among futurists is that we are moving towards an age where we will be able to reach anyone, anywhere, from any place, in our own style, and enjoy services of community, entertainment, convenience, and safety. As an optimist, I also believe that the growth of communication services brings widespread prosperity. To get there, I can identify three important ingredients: conditions for individual researchers and technologists that will make the journey possible; continued evolution and perfection of the underlying technology; and, finally, the creation of sustainable businesses that can economically deliver the services we want. This book is an important contribution to that journey and I would like to address the three points I made in the previous sentence.

The authors of this book are current or former senior staff members of Applied Research (AR) and OSS Engineering Organization at Telcordia Technologies. I commend them for the effort and dedication they have shown in writing the book you are about to read. While much of the underlying research for the book comes from projects either supported by Telcordia or some of our customers, the initiative and hard work to pull the basic material together in comprehensible form is theirs. The industrial strength research, understanding what is practical for service providers, and the ability for dealing with problems of scale are unique to few places. I would like to think that AR is one of them. An essential ingredient is accomplished individuals who over time learn the details and nuances of their craft and show the thought leadership and passion that lead to significant results. The authors of this book have passed that test many times over. It is also important that they are surrounded by peers who have the depth and willingness to challenge their ideas and provide the feedback necessary to arrive at useful conclusions or produce valid concepts. Next is access to customers who provide the context for our research and whose practices and requirements distinguish works such as this from academic exercises on model problems. Without resources, none of this would be possible. Here changes in the way organizations such as AR operate have been painful and difficult. What were once plentiful internal resources for long-range industrial research have practically vanished. Business imperatives have made it necessary to stitch resources and funding from more focused internal projects, commercial programs, and government research, which has proven critical for addressing problems that have a longer horizon. The authors have lived through this transition and, despite the extra pressures to market projects and produce contract deliverables, have completed this book.

The development and perfection of technology still have a long way to go to fulfill the vision of utility and ubiquity for communications services using the underlying network computing, storage, and software building blocks. As the book illustrates, there is a great gulf between the notion of plug-and-play and the delivery of "real," robust, and reliable services. For voice on WiFi and 3G networks, there is just functional performance. While it may be possible on a small network controlled by one user who can dictate all the system parameters and control the volume of traffic, the job becomes progressively more difficult as traffic increases, the network must support heterogeneous elements, and control is in the hands of multiple operators. Moreover, the network may have to share allocation for many different services such as video or data. The idea of quality of service (QoS) is central to this as is the idea of assurance. This is the heart of the book and the technology that the authors address. In a simple way, QoS is about measurable and explicit metrics that describe performance at all levels—from application to network functionality. Assurance is the process of monitoring performance parameters and translating through analysis and prescription the settings and conditions on the end-to-end service network to deliver the desired QoS. The science and engineering for attaining acceptable levels of QoS and assurance are complex and multidisciplinary. The authors have done an outstanding job of codifying what we know and what yet needs to be done.

The last issue I would like to address is business sustainability for service providers. Here there are two issues that stand out, both related to the economics of providing services. The first deals with the science of automation, the second with the science of scale. When the number of customers, network size, and service types is large, the systems that monitor, control, and run the services must do so automatically. Even a small degree of human intervention—which we call fallout—results in a degree of reliability for customers that is not acceptable and costs for an operator that are not sustainable. No matter how desirable the services are, there is a performance threshold that must be crossed to get acceptance and the cost of doing so must be contained. It means simply that the tools used to deliver "assurance" must indeed be very good. Scale possesses demands at two extremes. When the uptake rate for a service is small, the initial cost to provide a service should have a low barrier. It is hard to imagine that a small number of customers can absorb and pay for a great deal of infrastructure when a service is first introduced. At the same time, the service should be excellent or desirable to have a significant uptake. Once a service has attained subscription from a large number of customers, the costs operation should scale predictably at a rate that is linear or less with the number of subscribers. Furthermore, the rate per subscriber should be commensurate with the value of the service and competitive with alternate technologies to be sustainable. The science and engineering for these two issues are the domain subject that the authors address for voice on advanced packetized mobile and fixed networks.

Adam Drobot
Senior Vice President
Applied Research, Telcordia Technologies
Piscataway, New Jersey
August 2005

Preface

The growth of both cellular phone use and Internet access is a spectacular phenomenon that happened at similar periods during the last decade. Interestingly, the evolutions of these two informational services have taken quite independent paths. Moreover, until recently, cellular and IP technologies have been finding acceptance in vastly different marketplaces. However, as both markets are maturing and saturating, providers are discovering that the combination of these key technologies, together with disruptors such as WiFi, can potentially offer a lot more than each can separately deliver.

The availability of these technologies to both wireline and wireless operators suggests the blurring of traditional boundaries between various market segments. Incumbent operators are now presented with major external challenges embodied in rapid technology changes, a growing field of aggressive competitors, and the need to establish intimate and in-depth alliances with nontraditional partners. This means that service providers must find the a winning business strategy that includes creating new services at an unprecedented rate, implementing an aggressive cost reduction plan, and providing superior service quality for attracting new customers and at the same time, minimizing churn.

Facing these challenges, service and network providers can no longer rely on traditional network management tools but must take advantage of a new management model that is closely aligned with their business strategy. Such a model will provide flexibilities for business options, allow win-win partnering relationships, and have a sound theoretical foundation. Naturally, such a model would be service-centric rather than the traditional network-centric model. The introduction of a powerful service model is our aim in this book, to present a unified methodology that can help the wireless service provider to automate overall service management. A desired service model also would contain appropriate tools to assist operators in making the right decisions, and it should also have hooks that allow seamless flows between various operations support systems (OSS).

The idea of constructing a service model to support various operations functions is a relatively new concept in the telecommunications industry. It places the focus where it matters most. The challenge of designing a useful service model is not only to deal with today's known applications but also to be extensible for future service growths. For a customer- and service-centric business, focusing on the user experience is the key to empowering the operators' market and pricing differentiators. Therefore, these needs are well recognized by those service providers, and the incorporation of service-model-based management has been demanded by a number of advanced operators, and more are following the trend.

There are currently many different books that address wireless and Internet technologies. However, the topic of how to manage these technologies from an operator's perspective is less written about. It is even more unusual to find books discussing management from the operations process viewpoint, yet explaining clearly the tight relationships among business, service, technology, and operations. By presenting a service model concept and explaining how such a model can be integrated into the operations flow, we hope to fill such an important gap.

Overall, this book presents a practical view of the service model definition and management issues that are necessary to implement an effective management solution for operating a successful wireless business.

Four major themes are dominant in this book. The first is to define the new generation of service assurance and its business applications in telecommunication value chain. Chapter 1 profiles several service assurance management methods and provides some history and background to demonstrate the evolutionary path of service management and quality control. The service model is introduced as a new approach with the potential to form a significant element in meeting the new business needs of evolving wireless service networks. In Chapter 2, we describe the content, design options, and implementation considerations of a generic wireless service level agreement (SLA) and their direct implications for service assurance. An additional focus of Chapter 2 is to identify the value adders in wireless applications and the needed procedures and service functions to understand their place in OSS operations. Chapter 2 provides a path for an integrated process that can maximize the interests of both clients and providers from the standpoints of financial and service quality; it is important for service managers but can be omitted or just skimmed for buzzwords by engineers.

A second theme is the background of technologies and operations. Chapter 3 provides the necessary background on WiFi and 3G networks and their interworking, network domains and network elements, and roaming, which will help in building service models of VoWiFi/3G services. Chapter 4 puts the emphasis on OSS life cycles and their functionalities in the application of service assurance from the standpoints of solution developers and operators. It describes the depth of OSS architectures to understand their major flows and interactions with other systems. Discussions focus on the data flows and control flows that will provide essential business procedures that will allow proactive and continuous improvement to the services and networks. Chapter 4 is primarily for the benefit of engineers or service managers with a domain-focused background, who are likely to be unfamiliar with the life cycle of an end-to-end operational environment.

A third theme is the description of the scope and research results of the service model. The service model has been partly brought about or accepted by leading operators, but it is a critical need for providing service assurance that has not been fully realized by service providers and resellers. In Chapter 5, much more attention is given to the theoretical and practical views of service model applications in the context of service assurance and managing SLAs in a cross-technology service. This chapter describes how statistical theories such as hypothesis, detection, and correlation can be applied to solving service problems. Also highlighted are service decomposition, service dependence, containment relationship, and QoS alarm propagation, as well as an intelligent "rules engine" embedded inside every service subcomponent to perform impact analysis. Chapter 5 also provides many built-in assuring functions within the service model that can be used to extract knowledge from the key performance indicators (KPIs) and key quality indicators (KQIs). It is important to note that the service model approach is independent of specific technology. This chapter should be useful to service architects and engineers responsible for service definition and service management. Material in the chapter is important background for the subsequent chapters.

A final theme is to present a wireless service as a composite identity of information model (service model) and processes (integrated within OSS) and to describe in detail how a wireless service provider can achieve the highest yields by utilizing the new service assurance method. Chapter 6 features how the emergence of voice over IP (VoIP) technology in wireline networks presents an opportunity for combining packetized voice and data over one convergent network infrastructure and their associated management issues. The overview of the Voice over Wireless LAN (VoWLAN)/3G service architecture and domain definitions provides the necessary background for the service model discussion in the following chapters. Chapter 7 formulates the service model in the context of three business arrangements, including enterprise, hotspot, and dual-mode VoWiFi/3G integration, with the focus on defining the basic building blocks of the service model and laying down the foundation for the discussion of service assurance in Chapter 8. A set of critical KQIs and KPIs is provided at the end of this chapter.

Chapter 8 makes best use of all the information described in prior chapters with respect to packet voice, WiFi, 3G technology, associated KPI/KQI, and statistical data analysis and shows how the service model and technology are stitched together in an operator's flow-through OSS environment.

While the book is aimed at solution developers, system architects, service engineers, and service managers working with VoWiFi service in some form, Chapters 6, 7, and 8 provide a stand-alone assurance solution that is also applicable to VoIP and wireless networks in isolation.

Finally, Chapter 9 presents conclusions and a path forward for the future service assurance. A service-centric, model-based approach for service assurance can deal with the increasingly complex and demanding management of services. We recognize that a new set of assurance functions are required to deal with the demanding deployment support of a complex and emerging technology when SIP-based packet voice and wireless and mobility requirements are taken into consideration. As a path forward, the service model and the emerging service assurance OSS can help in ensuring high service quality, for services in which QoS is critical, which can be a significant differentiator for operators of capacity-scarce networks such as cellular. Lastly, the application of the service-model-based approach is a technology-independent method. We see that our models and ideas have broader applicability outside the areas we have addressed so far.

Acknowledgments

A number of people have provided insight, guidance, content, and encouragement in the writing of this book. We would like to thank Adam Drobot for writing an excellent foreword with well-articulated insights in the topics covered in the book. We also would like to thank Howard Sherry, Arnie Neidhardt, Judith Jerkins, and several anonymous reviewers for reviewing the manuscript, Matthew Chiger for his input on business assurance applications, Bill Schoneman for legal review, and Nim Cheung, Gerald Chu, and Rick Thornton for encouragement and support.

We thank Mike Kelly of TeleManagement Forum for guiding us to the latest TeleManagement Forum developments. Finally, we would like to thank Artech House editors: Mark Walsh and Christine Daniele, acquisition editors, Barbara Lovenvirth, development editor, and Rebecca Allendorf, senior production editor, for their valuable help in the preparation of this book.

Chapter 1

Introduction to Service Assurance

The early 1980s saw tremendous expansion in network deployment. Telecommunications service providers realized the cost benefits and productivity gains that could be created by network technology and began to expand existing services almost as rapidly as new technologies were introduced. Because each new network technology required its own set of experts, the staffing requirements alone for managing complex, heterogeneous networks created a crisis for many companies. By the mid-1990s, some service providers were starting to experience new levels of pain from the lack of interoperability between network technologies. An urgent need arose for automated network management across diverse environments, affecting both day-to-day network operations as well as strategic network growth planning.

Entering the new postdivestiture millennium, many modern service providers were presented with major external challenges embodied in rapid technology changes, a growing field of aggressive competitors, and the need to establish intimate and in-depth alliances with nontraditional partners. Significant among the unavoidable challenges that these operators faced were

- Rationalization of service values and their impacts on revenue assurance;
- Collaborative efforts for new service introductions sensitive to shorter life cycles;
- A shift of management focus from network-centric to service-centric operations;
- Integration of operational and business islands within the service provider.

These drivers required a more detailed and aggressive approach to managing business. A new emphasis on service assurance was required to keep up with the increased pressure of market demands and competition. New, or at least improved, operational processes needed to be developed.

In response to the need for improvement across industry, both the government and private industry extended considerable efforts toward quality improvement. There were significant improvements in the quality methodologies available to businesses. For example, statistical quality control (SQC), which deals with the acceptable quality levels of finished products, and statistical process control (SPC), which deals with the quality of processes, yielded industry-wide quality standards that were developed and applied with significant success. SPC-based tools certainly helped to increase the quality of products exiting the shipping dock of many companies. The application of ISO 9000–related quality guidelines helped improve many company operations and was successfully used to shape company policy in many places. The discipline described in the Capability Maturity Model® (CMM®) [1] guidelines has certainly improved software production operations in many companies, both large and small. Process improvement tools such as Six Sigma and Balanced Scorecard have also contributed to improved operations in many places and have begun making inroads into telecommunications service providers' information technology (IT) operations.

Today, however, service providers that apply these methods may find themselves achieving less than the anticipated results. Absent an integrated approach that views the business as a whole and understands how the performance of each part of the business affects the rest, any attempted quality initiative seems often to end up optimizing results solely within its own sphere of application. The measurable benefits of such an approach often seem small compared with the considerable pain of implementation. The introduction of an integrated service model is our attempt to present a unified methodology that can help the service provider to achieve overall business balance between cost containment and the processes required of a strong quality-oriented and service-oriented business operation.

The idea of constructing a service model to support wireless operators and enterprise operations is a relatively new concept in the telecommunications industry. The challenge of designing a useful service model is not only to deal with today's known applications but also to be extensible for future service growths. A desired service model should also be able to produce appropriate tools that can support the operators in planning and decision making. For a customer- and service-centric business, focusing on user experience is the key to empowering service relationships and pricing differentiators. These drivers are well recognized, and the incorporation of a service-based business model has been demanded by a number of advanced operators, and more are following the trend.

This book presents the fundamental building blocks for service providers to comprehend the service challenges and their corresponding solutions. Guidelines are provided that allow the service providers to improve their operations through collaborative execution of business processes and the creation of appropriate control and feedback mechanisms that will provide continuous enhancement of both new and existing processes.

1.1 SERVICE ASSURANCE

One of the expected behaviors of service management is to help the service provider discover the technologies that enable its business models and processes. This does not mean knowing the details of programming but rather understanding how technology can impact service operations in making them more or less valuable to service customers.

A service provider's strategy at the highest level is to achieve business success. Because telecommunications technologies can radically affect the business model and strategies of the service provider and its enterprise customers, there is a strong need for the service provider to establish solid service assurances that can address the more commonly cited concerns associated with service quality, such as:

- Failure to meet performance expectations;
- Higher than expected costs;
- Poor account management;
- Poorly defined service-level agreements (SLAs).

The service provider must also be able to exercise control over internal processes that have a direct effect on customers' service experience, such as:

- Limited vendor performance knowledge;
- High staff turnover;
- Inability to respond to changing needs and customer concerns.

1.1.1 What Is Service Assurance?

The enterprise customers that have major SLAs (contracts) with their service providers will expect certain quality of service (QoS) guarantees. These customers will be evaluating the percentage of calls blocked, calls dropped, connection delays, provisioning times, and help-desk response times, for instance. They may also demand that their executive management team receive a higher level of performance than other employees or that field personnel be guaranteed reliable service when interacting with the corporate network. To date, wireless carriers have made limited commitments because they have not been able to stand by them completely. New business models for service assurance will have to support commitment-based offers for enterprise or high-value customers.

General requirements for service assurance include a collection of processes and systems that must be scalable, secure, efficient, and capable of seamlessly supporting other functional components [other operations support systems (OSS)]. A service assurance solution must integrate with other OSS and feature a distributed measurement network to test and measure services and a means to collect and analyze the measurements. Key OSS links include:

- Flow-through provisioning;
- SLAs that are an automatic part of service initiation;
- Billing processes that incorporate service outages and SLA violations;
- Trouble ticketing;
- A network management system (NMS) that notifies the provider immediately if service begins to degrade.

In short, service assurance has an intense focus on all aspects of the customer's experience and the associated tools and processes to monitor the performance of the system against measures made from the customer's point of view. The concept is not profound, but the implications for the service provider's operations can be.

1.1.1.1 Building the Measurement Network

Network monitoring [2] is done by receiving network events (status changes) or sampling the network by taking measurements at specific times of the day. The network metrics can be collected from single-system element management systems (EMS) or network management tools. In order to create comprehensive knowledge of network, customer, and service performance reports, many service providers use a labor-intensive, manual method that they offer only to valued customers. Some use service assurance tools to automate this process, making the practice more scalable and providing a broader view of the customers and network.

In addition to real-time service status (events), the collected measurements are used for capacity planning by the network group. Capacity upgrades have been based traditionally on simple utilization measures; now the company is able to include actual application performance in capacity upgrade decisions. To accurately guarantee the customer's experience, the service provider must measure from the customer's site. The approaches that measure service performance only from within the core network have large blind spots. Such systems lack visibility into the access network, which can account for more than two-thirds of the equipment-handling and end-to-end service issues. These traditional network operations center (NOC) management tools do not provide information about which customers lost service or what traffic generates the highest revenue. They may even measure different servers than those the customer is using [e.g., a central domain name system (DNS) server instead of the local server].

Trying to know which customers have service commitments and which transactions have a higher priority is also problematic for the service provider. For instance a mobile service provider must gather information from up to seven parts of the network and compare it to customer and SLA databases. Furthermore, a personal digital assistant (PDA) or mobile phone does not have a specific IP address; thus, it is difficult to prioritize a customer's traffic across the network. Common approaches such as Simple Network Management Protocol (SNMP)–based [3] data-collection can measure core network utilization effectively but fails to measure actual service performance because the measurement appliances are likely to be behind enterprise firewalls. Thus, they are beyond the reach of the SNMP data-collection operations. Instead, an authenticated and encrypted protocol that can be tunneled through enterprise firewalls must be established. This implies a separate arrangement for each customer in the area of its network security which is usually a sensitive subject.

In addition to the measurement appliances, support for embedded agents such as Cisco's Service Assurance (SA) Agent [part of Internetwork Operating System Software (IOS)] [4] allows testing from existing routers. SA Agent and similar solutions are useful for low-end testing, for example, in small branch offices or simple services. Larger virtual private networks (VPNs) and advanced services such as Voice over Internet Protocol (VoIP) will overwhelm the routers typically used for corporate connections and require dedicated, scalable testing devices. Figure 1.1 depicts a high-level data-collection hierarchy.

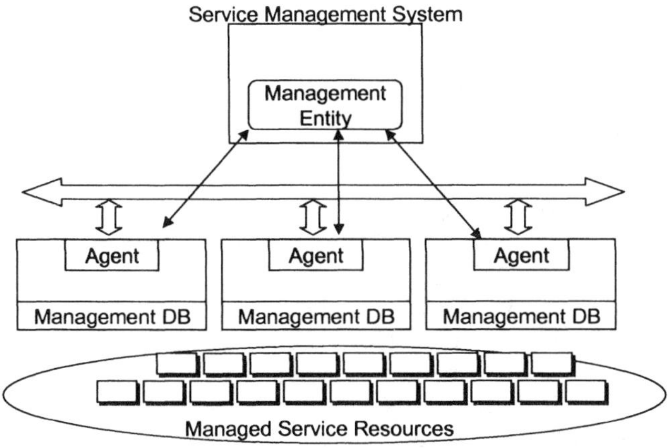

Figure 1.1 Data-collection hierarchy.

1.1.1.2 Operational Scalability

In the network environment, to discover if a service has been delivered and how well, the service provider must pull together data from multiple sources. Efficient management of the measurement network is the key to operational scalability.

Traditionally, efficient management has started with the ability to perform automatic auditing and high-level network status assessment. These measurements allow for the monitoring of equipment performance health and additionally provide a focus on revenue-generating activity and SLAs. These types of service-centric systems manage the configuration of the measurement appliances and embedded software and collect test data in a centralized manner. Metrics are audited automatically against predefined, service-specific, and customer-specific guarantees. Threshold violations are signaled automatically to other OSS components, and high-level network status is available from a Web-based interface. Reports showing service status against weekly or monthly goals or QoS compliances are produced. The same data and reporting interfaces usually correlate results from equipment alarms and other tests in order to support troubleshooting.

The network infrastructure has become complicated, with many more elements playing a part in delivering services. Application components, server and system components, and network and business components are now vital parts of the service delivery chain. Meeting the challenges of enabling broad service deployments of advanced application services and addressing specific customer concerns will come from the underlying service assurance architecture. The service providers that can deploy a scalable service assurance system will be able to meet the demand for relevant, useful service guarantees, which will translate into repeat customers and enhanced revenues. For instance, the service operators need to know if an entire video file was downloaded or if an m-commerce transaction was successful. These measurements are much different from measuring call completion or call blocking and require a broader set of measurements from a wider range of elements.

Integration with other OSS is also the key to scalability. The new OSS components in the service delivery chain, as well as the additional partnerships required to offer content services, have prompted new merits of tight integration in the wireless OSS. The new considerations of carriers are the degree of controllability they can achieve to improve provisioning, ensure delivery, and allow clear SLA specifications. The more tightly integrated network operations infrastructure has given rise to new questions that management must now consider in the evolution of the business operation's processes. For example, how dynamic should the assurance framework be for business process changes and future service options? How tightly coupled should the OSS integration be with respect to the NMS metrics, customer management system trouble tickets, service quality system (SQM) threshold violations, and billing system credit records based on SLAs?

1.1.2 Values of Service Assurance to a Service Provider

As a service provider or a system integrator (system integrator provides design, layout, installation, and consultation of hardware or operational systems), service assurance is critical to the business for the reasons of maintaining and improving the quality of the network service, helping retain customers, and providing the critical difference between winning and losing in today's competitive marketplace. While SLAs can help service providers derive new sources of revenue and gain competitive advantage, they must avoid performance penalties resulting from poor performance. The advantages derived from an efficient, scalable service assurance measurement operation on a large network are numerous for both service providers and systems integrators. Some of the derived benefits include the following.

To service providers:

- Rapidly identify the root cause of problems.
- Improve the network's mean time to repair (MTTR) and mean time between failures (MTBF).
- Eliminate the over-provisioning of bandwidth.
- Improve satisfaction and retention of customers with high-quality and differentiated services.
- Provide end-to-end guarantees with confidence to valued customers regarding their business-critical applications.
- Identify new revenue opportunities by monitoring customer business behaviors.
- Create new sources of revenue by value-added services.

To systems integrators:

- Understand workflow constraints by a clear mapping of service usage to business needs.
- Eliminate unnecessary capital expenditures by understanding the usage patterns.
- Monitor promised versus delivered variances in SLAs.
- Improve the time to troubleshoot performance issues.
- Maximize the availability of mission-critical business applications.
- Prevent fraud through real-time detection of illicit usage.
- Reclaim bandwidth wasted on nonbusiness applications.

1.2 CURRENT SERVICE ASSURANCE PRACTICES

The wireless industry grew at an incredible rate in the late 1990s and early 2000s. Carriers struggled to keep up with subscriber demand and were forced to add cell sites rapidly—at any expense. These were driven by substantial increases in wireless phone users, remote workers, telecommuters, and remote data-communication needs. Large enterprise customers obtained economic justification for mobile operations through productivity gains and competitive advantages derived from improved service to their customers. As a result, the capital efficiency of and overall service quality offered by the majority of carrier network providers were overlooked.

Today, the situation has not yet been improved dramatically. With voice traffic at an all-time high, huge capacity demands have been placed on wireless networks. The world subscriber base passed the 1.5 billion mark in the first week of June 2004 [5]. Nearly two-thirds of all wireless data subscribers are using Short Message Service (SMS) and 19% are using two-way paging [6]. Functions of SMS include sending e-mail, making data inquiries, accessing information, and scheduling reservations. WiFi wireless local area networks (LANs) continue to proliferate with security problems being addressed vigorously and solved to the satisfaction of IT security professionals. Major vendors and enterprise-grade WiFi access points are supporting all IEEE 802.11 standards. Almost all large wireless carriers are planning to support roaming across WiFi and their wireless WANs. Plus, wireless broadband services will become increasingly attractive for under serviced residential and small office, home office (SOHO) customers. Management of mobile assets and wireless networks is becoming increasingly important for enterprise and network carriers.

However, existing wireless networks are still designed and tested with tools that focus primarily on estimating cell site coverage and not tuned to run at optimal capital efficiency; this is because the network inefficiencies had little overall impact on revenue growth historically. The service providers have been so focused on bringing the new services to market and have been less eager to invest in tools that analyze network performance and assure service delivery until the services are successful. As a result, service assurance solutions are rarely rolled out with a new service; instead, most service providers only get a snapshot of network performance, not a complete picture of how poor performance affects customer service. Newer applications, such as location-based services and m-commerce, will not have any service assurance component until they achieve a scale that makes the investment economical.

Another core challenge of monitoring performance in a wireless environment is that most next generation services rely on multiple shared resources. Service delivery depends on existing radio equipment, voice switches, and service nodes, as well as additional elements such as content caches, e-mail, application servers, subscriber identification modules (SIM) cards, and many other components now being added to the wireless network. The solution- and infrastructure-level technologies that can effectively integrate these elements into a unified service view are still maturing, thus may not be immediately available to service providers. Figure 1.2 depicts the current view of service assurance for a telecommunications service provider. We will examine each component in the following subsections.

1.2.1 Fault-Centric View

Since service faults can cause downtime or unacceptable network degradation, fault management is perhaps the most widely implemented OSS. It has the ability to perform the collection and correlation of alarms and other relevant events to provide an accurate view of the health of the service. Because fault management provides service providers with information about the state of the network and allows the operators to detect, log, notify users of, and, to the extent possible, automatically fix network problems, many service providers build their OSS flows around fault management.

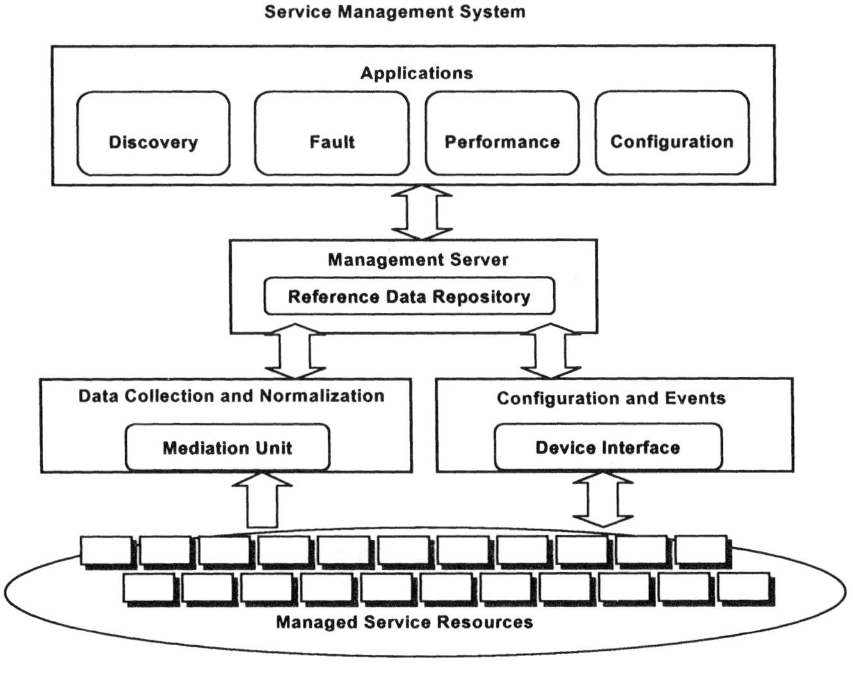

Figure 1.2 Current service assurance view.

Although a technology manager often is the prime executive accountable for network health, to achieve the required focus on customer impact and make decisions on resource allocation, the organization's top executive team (or chief information officer) needs to review the relevant network-health status and measure potential service disruptions regularly. Taking a process view of fault management (Figure 1.3) can help service providers better understand and assess the impact of network faults. We list the fault management cycle as follows:

1. A fault manager [7] collects significant events from the resources, identifies fault events that may lead to the loss of service, and converts them to a common format.
2. The fault manager then analyzes one or more events or system parameters to determine the nature and location of a fault. The diagnosis can be by direct observation of a specific component or by correlating multiple fault events. The correlation is for event number reduction or for increasing the information carried by each event. This step helps the service provider quickly focus on significant information.

The root-cause analysis is an extension of the diagnosis process. It can isolate the problem and prevent unnecessary trouble from being spread throughout the whole OSS.

- *Recovery* is the restoration of the service after a fault is isolated. The recovery process is often multistep, where several actions must be taken in order.
- *Reporting* is the notification or logging of the diagnosis or actions taken to resolve the service problem. The fault manager should map the events into state information about the resources and services.
- *Repair* involves hardware reconfiguration or repair or software changes or upgrades, which can be manual or automatic.

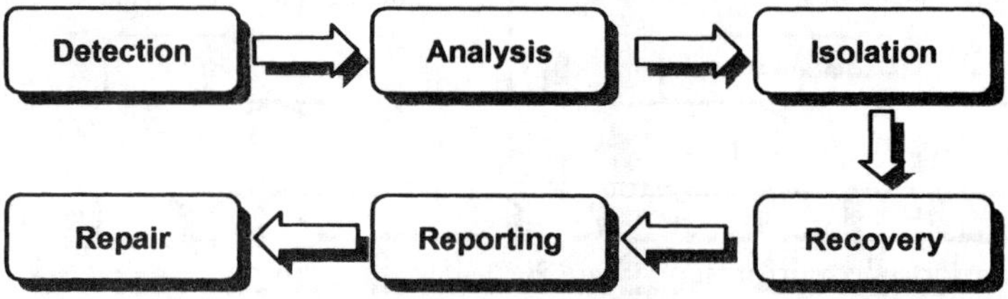

Figure 1.3 Fault-centric service assurance.

The aim of problem management in fault-centric operation is to recover quickly from faults and to look for ways to avoid the faults in the future. Its major component activities are:

- Trouble ticketing tracks the process of fixing a fault, escalating as needed to keep business commitments.
- A problem database maintains information about common problems and how to recover from them.
- Root-cause analysis tries to determine why a particular fault happens and understand what changes can be made to avoid the fault in the future.

1.2.2 Performance-Centric View

The business objective [8–10] of performance management is to measure and make available various aspects of network performance so that the service provider's performance can be maintained at an acceptable level. Examples of performance metrics may include network throughput, user response times, and line utilization.

Performance management's function is to send near real-time alarms when thresholds are exceeded or trend data has indicated a potential service problem is about to occur. If thresholds are defined in SLAs with their customers, the SLA management function monitors the service-level objectives for compliance, if possible notifying the provider before the SLA is violated. The SLA notification allows the service provider to take corrective action before significant penalties are incurred. The SLA management function may also report to customers about compliance levels. Performance management does not report on real-time network events because the system could not deal with the overwhelming amount of performance data in real time. Results are summarized over controlled time intervals and compared to norms. Many providers have chosen the performance-centric view for their service management, especially in the areas of root-cause analysis and troubleshooting.

Performance management involves three main steps:

3. Performance data is gathered on metrics (performance indicators) of interest and stored in a performance database.
4. The performance indicator is analyzed to determine normal (baseline) levels.
5. Appropriate performance thresholds are determined for each important metric [key performance indicator (KPI)] so that exceeding these thresholds indicates a service problem worthy of attention. In the reactive mode, performance alerts are generated and sent to the NMS when performance thresholds are exceeded. In the proactive mode, a "soft"

alert is sent when historical data shows a service trend or behavior that matches a predefined scenario.

Management entities continually monitor performance indicators as depicted in Figure 1.4. The what-if analysis function takes historical data and network engineers' assumptive conditions to derive service results for predicting unknown service behaviors or for trouble identification. For further discussion, we will use VoIP performance management as an example. Because of its packet-data nature, VoIP technology presents new challenges for QoS assurance.

The key to classifying different service levels of VoIP is the use of class of service (CoS) management to monitor and report categories of services based on applications. CoS management adds value to an IP network by prioritizing traffic based on its origin, destination, or nature of application. Without CoS, all the traffic has the same treatment. A fully functioning performance solution should work hand in hand with the inventory systems to recognize devices. It should also automatically discover the CoS structure, the QoS policies (such as policing, queuing, or traffic shaping), and associated parameters. This performance solution uses CoS information to classify the service paths and assign them to different SLA criteria for monitoring. The per-customer traffic performance assessments can then be reported either by application or by a set of IP addresses.

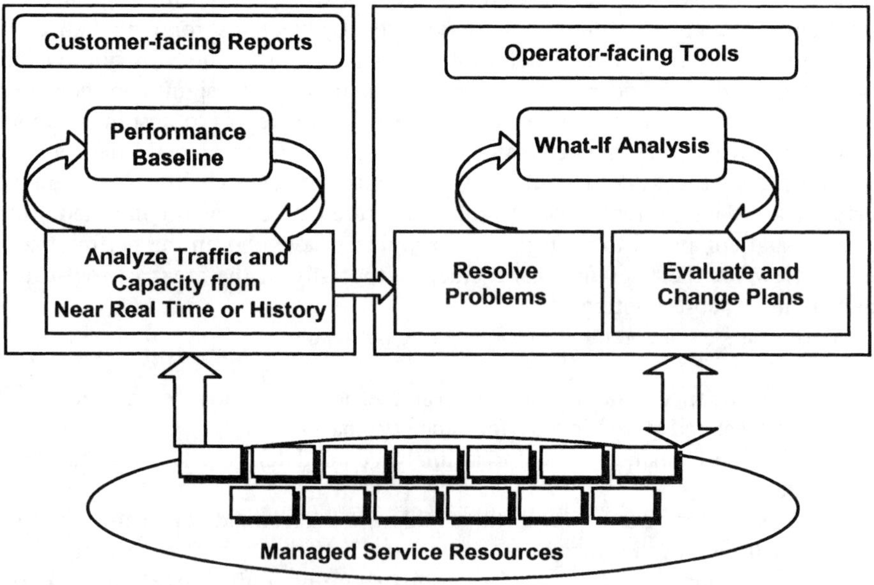

Figure 1.4 Performance-centric service assurance.

1.2.3 Configuration-Centric View

The general goal of configuration management is to monitor network and system configuration information so that various versions of hardware and software elements can perform as intended. It is designed to control and optimize costs, schedules, and resource allocations.

The configuration management relationship provides an integrated way to manage the service provider's assets, or resources, because the management system can have multiple relationships with very complex links. Some surveys have indicated that, in an inefficient company, typical discrepancies or lack of synchronization between the OSS and the actual assets range from 25% to 50%. This is the result of unrecorded equipment changes, service modifications, and inconsistency in inventory presentations in different OSS layers. Inventory discrepancies lead to many downstream inefficiencies, such as stranded equipment assets, stranded bandwidth capacity, increased capital expenditure (without full asset utilization), high failure rates on service activation requests, long lead times for service activation, higher manpower costs for operations and provisioning staff, slower service assurance, difficult back-office integration tasks, and lengthy troubleshooting efforts.

A fair configuration management system (CMS) should be able to "configure" and reconfigure service providers' entire services, not just their network resources. It needs to meet constantly changing customer expectations, increasing product variety, and massive technological change. These products work rather well in the early phases of technology deployment and have attracted billions of dollars in information technology investments, while raising productivity and competitiveness to historic levels.

Configuration management is often viewed as a subset of network management processes. However, these environments are not always capable of delivering the new definition of "full" configuration management functions. Lately, the functional scopes of CMSs are defined as combined hardware and software inventory management—often referred to as asset management—with problem management, event monitoring, software distribution, and security features. Essentially the providers are trying to extend configuration management to the all OSS operations. Nevertheless, this configuration-management-centric view is still limited to its ability to cover all the assurance and financial aspects.

The major benefit of configuration management is its tangible value adding to all stages of service fulfillment. At every stage in the fulfillment, the process is modeled, appropriate closed-loop controls are implemented, and the interface with the next stage is clearly defined. The lack of redundancy and human intervention, as well as improvement in data accuracy, can constitute obvious savings in labor and overhead.

CMSs provide the tools needed to communicate more effectively with the workgroups and the third-party service partners that comprise the supply chain.

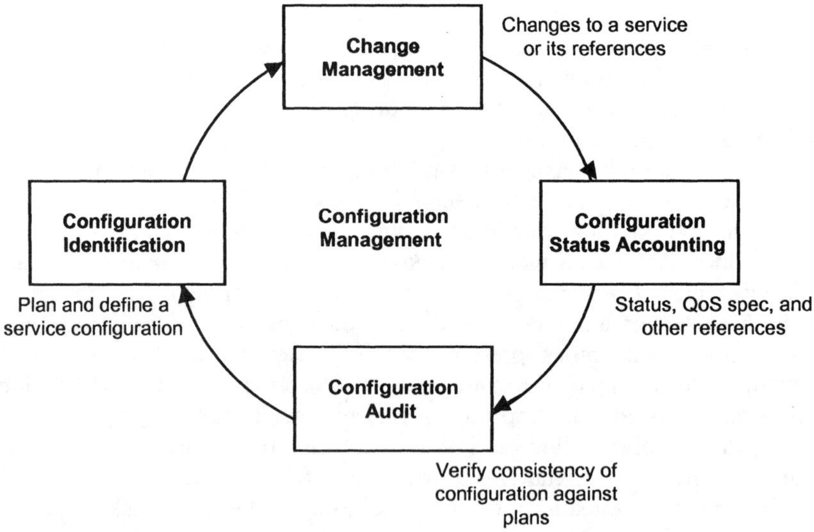

Figure 1.5 Configuration-centric service assurance.

Their purviews have included everything from provisioning to service distribution. The configuration management process addresses the composition of an offering, SLAs defining bundled or single services, and other data and products that support it. Configuration management can be structured into the four integrated processes (shown in Figure 1.5), which provide for complete fulfillment life cycle management.

Business processes are dramatically affected by the integrity of the information used to plan, operate, test, build, and ultimately deliver services. The financial impact of inventory discrepancies on a service provider's operations budget is significant. Eliminating the downstream inefficiencies mentioned earlier goes straight to the bottom line of the typical service provider. This automation could reduce service delivery times, providing the service provider with competitive advantage and differentiation.

1.2.4 Beyond the Basics

As seen earlier, the traditional approach to service management of a telecommunications network has employed OSS that either concentrated on detecting and correcting equipment faults or monitoring the performance of network equipment against expected norms. While these approaches are still valid in next generation networks, they are beginning to lose control over what is needed by the service operators of the evolving telecom environment. A better and more appropriate system cannot adopt only a fault-centric or performance-centric view for network management because either one is lacking the essential ingredients to construct a complete service picture for the providers' business performance. On the other hand, the configuration-centric solution has been oversold for its ability to influence the entire OSS life-cycle management. Some major telecommunications solution providers had bought into the idea that network configuration and provisioning systems should be the core systems that could "glue" the OSS solutions space into a comprehensive service assurance solution. Indeed, well-designed asset management and effective dispatch capability can quickly prove value to service providers, but subtle and profound service behaviors are outside the scope of configuration management. This will be revealed through the new approach to service management that will be addressed in the following sections.

1.3 THE NEW TREND FOR SERVICE ASSURANCE METHODS

Gartner [11] predicts that by 2010, 80% of key business processes will involve the exchange of real-time information involving mobile workers. By 2007, analysts predict there will be nearly 100 million mobile workers with different styles, tastes, and work habits, who will use at least one wireless gadget to make phone calls, manage personal and professional data, or access corporate networks. Alexander Resources predicts that the WLAN market will hit $15 billion by 2007 [12]. The IDC's Europe, Middle East, and Africa (EMEA) WLAN Tracker shows that the EMEA WLAN market increased by 9.2% in the third quarter of 2004 as compared to the second quarter, reaching total end-user revenue of $349.8 million. The residential market continues to be the main driver of WLAN growth; during the third quarter, more than 1 million wireless routers/gateways were shipped, primarily to the residential market [13].

In response to such market demand, service providers are deploying capacity faster than at any other time in history. They must keep up with network capacity and quality requirements at the lowest possible cost to stay competitive. How can a carrier keep up with market demand when prices keep going down and cell sites are getting harder to get past zoning boards? A critical part of the answer lies in the implementation of an appropriate service assurance solution.

Service assurance is the implementation of processes and methodologies that allow service providers to measure and control the quality of services provided from end to end. The goal is to accomplish this at the lowest cost. For instance, the quickest and cheapest way for a wireless service provider to grow its network is to boost the capacity of each individual cell site without degrading call quality.

This means fewer new cell sites must be built, reducing new capacity cost and time to market. For a service provider who has a multibillion-dollar annual capacity-expansion budget, increasing the average capacity of existing and new cell sites by just 10% will result in hundreds of millions of dollars in savings. That said, a successful service assurance solution must emerge to provide growth and performance management that is aligned with new technology evolution.

Until recently, service providers used computer-based propagation models to plan their wireless networks and place cell sites. These models cannot accurately identify or predict capacity-robbing network interference; nor can they incorporate follow-up network behaviors for continuous improvement. Furthermore, there is no real way to predict system performance accurately as a function of increased traffic. Customer complaints about blocked and dropped calls are usually the first sign of trouble, and by then it is often too late to save the customer relationship.

From a mobile operator's perspective, implementing accurate measurements of radio frequency (RF) path loss data (performance), cell site configuration, and switch parameter settings (inventory) at the corporate level is not a simple task. It requires a level of integration in wireless network planning that may not currently be available through traditional data-collection methods.

If this data were available, the operators could more accurately determine how planned network changes would affect overall capacity, call quality, and company profits. If system operators could further incorporate strategic considerations into a broader perspective on network fault management (fault), the overall business impacts (Figure 1.6) could be folded into a framework of key business indicators. If these indicators were combined with customer satisfaction results and new-product introduction metrics, this would present a comprehensive corporate profile that would be very valuable in top management's action planning. This service assurance solution could then be enhanced into a broad business management view that reflected capital costs, profit margins, and market share.

1.3.1 Business Goals

New technologies, service offerings, and legacy systems, compound the challenge for modern service providers. The convergence of voice and data and new intricate value-chain relationships require multiple trading partners to participate in the delivery of value-added services. To optimize revenues and stay profitable, service providers now must establish a systematic service/revenue assurance program with clearly defined objectives, processes, and management information that support this new business environment.

For premium customers (e.g., roaming subscribers) or mission-critical applications (e.g., emergency facilities that use SMS messages for dispatch), mobile service providers are tracking transactions and monitoring network performance closely with surveillance tools. So, when there is a fault or a service interruption, the service provider can use an end-to-end service view to identify the root cause of the problem and quickly restore the service. However, the end-to-end service issue and the idea of becoming service providers rather than just network operators clearly goes beyond basic network management. It also includes integration with other OSS processors such as service fulfillment, billing, and customer network management. Next generation service providers who wish to optimize profits and customer retention need a systems-based approach to revenue assurance that offers real-time business and operational intelligence for their executives to detect revenue slippage across the entire sales and service delivery chain accurately and consistently. Using a front-end graphical tool, these executives can have a clear gauge on revenues, slippage points, magnitudes, and trends and act at the first sign of serious negative trends, rather than wait for subsequent analytical snapshot reports that are most likely postmortems.

We have grouped several important business goals that should be supported by the new generation of service assurance solutions.

Customer front:

- *Customer self-service:* Ability for the customers to conduct their orderings and servicing;
- *Portals:* Ability to differentiate the service provider's own offerings and products from competitors or complementary service providers;

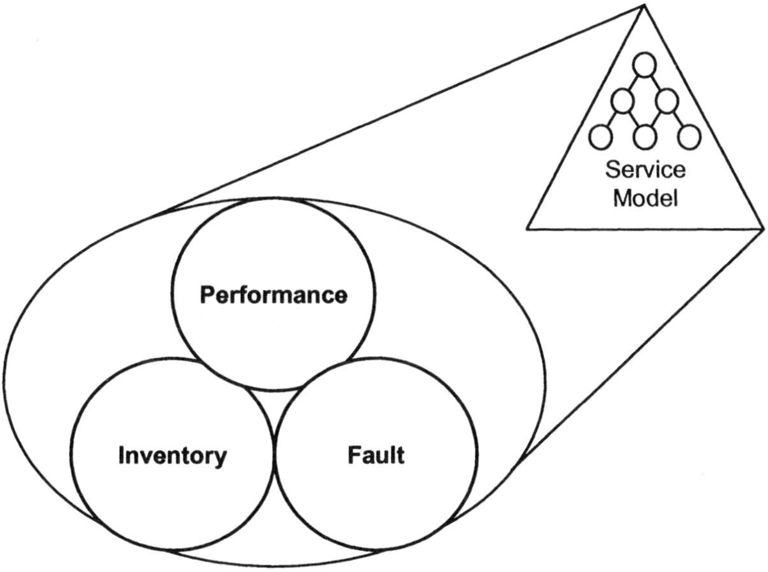

Figure 1.6 Integrated service assurance.

- *Build to order:* Ability for customers to order and configure their specific services.

Service packaging:

- *Aggregation:* Ability to recruit large groups of buyers or sellers to obtain better costs or prices (addresses both demand and supply sides);
- *Flexible bundling:* Ability to bundle closely related, but separate and different, products in combinations that would not be possible on a stand-alone basis.

Sales channel:

- *Producer direct:* Ability for the service provider to sell directly to the end user, bypassing the traditional methods of selling and distributing through distributors;
- *Channel integration:* Ability to integrate diverse channels into a coherent sales and distribution system;
- *Syndication:* Ability to sell services to integrators, who then package them with other products and resell or deliver the package to a third-party.

Marketing capability:

- *Dynamic pricing:* Ability to display different price options based on volumes, attributes, and locations, which may include bidding;
- *One-to-one marketing:* Ability to collect and store enormous amounts of information that can be used to zero in on a prospect with a clearly defined profile, one to one;
- *Marketable knowledge:* Ability to turn internal knowledge into a valuable asset by digitizing it and making it available to the service subscribers.

Service supports:

- *Effective controllability:* Ability to use a generic means to manage business behaviors through different indicators and metrics; business management tools should include statistical features and analytical features allowing the service providers to measure proactively current and future service and business dynamics;

- *Revenue assurance:* Ability to reduce the amount of revenue lost due to leakage in the service delivery life cycle.

1.3.2 Developing a New Focus

Many service providers have been demanding more open, modular systems, while the OSS solution providers have countered with the benefits of tightly coupled systems. Wireless network service providers are requiring that wireless OSS accommodate new services and technology easily. They must be tightly integrated because, as solutions grow more sophisticated, it is critical that the service providers be able to keep the internal openness to share service information among OSS processes.

For instance, content delivery from a primary provider partner will require provisioning on every single transaction, and each time the parameters can be different, depending on the wireless device, the user's location, and the type of content. Multiple points in the business operation touch this service, creating an assembly line–like architecture. Wireless service providers responsible for the entire service supply chain will have to turn content contracts into information that can be provisioned within the infrastructure. Tightly joining the OSS and network in a common framework could lead to new SLAs for business customers and content partners. These more complicated SLA agreements between wireless providers and their content and retail partners may cover completion time, security, connectivity, transaction time, success rates, availability, and throughput. Many of these new SLAs will relate to provisioning rather than performance. Content partners may require a commitment to enable the service within a particular time, to customize the mobile phone interface around the content provider, or to place the content in certain areas on the phone. Some of these SLAs address the look and feel of the content and how it will be accessed and whether it requires revenue sharing.

Service disruptions, billing errors, and unresponsive customer service drive customers away. However, service providers are recognizing that network-performance issues are the critical determinant of customer satisfaction. They realize they must focus on service management and measure end-to-end performance in terms of time, quality, and cost if they are to maintain a high level of customer satisfaction.

Many service providers have come to realize that they cannot find the KQIs of service performance in technical and financial data alone. Their challenge is to obtain the ability to improve network performance continuously by taking appropriate action to minimize network-failure impact on customers. They must also create the necessary information architecture to track financial impacts. Only then can the service providers really begin to recognize the competitive positioning of their services. Table 1.1 illustrates the evolution of the service provider's business model [14].

Some service providers choose to operate their own network or IT infrastructure, while others choose to outsource some or all to other service providers. Whether a service piece is directly operated or outsourced, it is an integral part of the service delivery chain and directly influences the service quality and cost perceived by the end customer. Service providers need to become skilled at making outsourcing decisions and at integrating and managing any outsourced arrangements. With the growth of data and information services, it is becoming evident that the end-customer perception of quality requires service providers to expand traditional measures of quality. They must adopt a more proactive, service-centric management style and enter into more interactive customer-centric relationships.

1.3.3 Fully Integrated Service View

The goal for a fully integrated service is to assure reliable and consistent quality service to the customers. An overarching theme for the service provider is the need to combine service performance data with customer information. Integrated service assurance solutions can assure QoS by directly relating service and network management to business processes.

Table 1.1

Business Model Focus: Changing Expectations

	Past	*Present*	*Future*
Innovation	In-house research Steady improvement Risk averse	Acquisition of new ideas Changing the rules of the game Embracing risk	Culture of innovation Market education Constant delight of the customer
Operational	In-house research Steady improvement Risk averse	Delivery High quality Price	Customized solutions Outsourcing End-to-end process effectiveness
Customer Service	Ease of use Reliability Basic functionality	Excellent support Quality of products Service orientation	Self-service One-to-one marketing Value

Source: [14].

Critical to the success of such a new integrated solution is its capability to provide visibility of the entire managed service, and to draw the appropriate business flow that is affecting the customer experience. Service providers need the ability to create a topological map and service model instance of the infrastructure that includes application and network resources and how they are interconnected physically and logically. Solutions equipped with such a feature can enable the operator's service visualization, which allows quick diagnosis and understanding of the cause of problems and real-time observation on selected service resources. For mobile network monitoring, we divide the technology infrastructure into the following three management layers.

- At the bottom layer, comprising the links, switches, cell towers, and routers, wireless operators deal with fraud, network capacity, and QoS issues with dropped calls and error rates.
- At the middle level, wireless service providers operate heterogeneous networks, including transport bearers such as time division multiple access (TDMA), code division multiple access (CDMA), General Packet Radio Service (GPRS), and Universal Mobile Telecommunications Service (UMTS). Traffic management, service-level availability, transaction speed, and performance are key issues.
- At the top layer, the wireless service providers become part of the global network/service and have to deal with interoperability, interconnection, and termination traffic. The issues at this layer are roaming relationship, value-chain management, and complementary third-party SLAs.

Each layer has a different set of concerns and uses different system interfaces; a centralized service management solution can provide analysis, visualization, and traffic alarms. It correlates the information from the various protocols of the network and creates a view of end-to-end behavior.

For service level assurance, the new system assesses end-to-end performance by aggregating data from all network and system components, links between these components that help set up the initial call or session, and transport bearer data for the user service. As depicted in Figure 1.7, the service assurance process has two directions. The bottom-up flow represents the building blocks formulating business values and assets to support the service provider's revenue generation. The top-down flow represents the means that the service provider uses to ensure the QoS in accordance with committed SLAs.

Using the service model can help the service provider with event correlation, task prioritization, and business impact analysis at all levels.

To support the bottom-up and top-down service assurance process, described above, we need to:

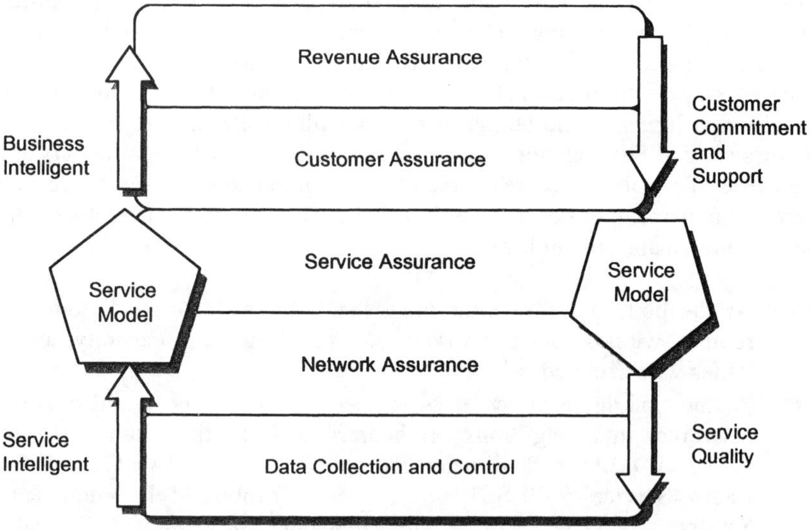

Figure 1.7 Service-model-based service assurance.

- Define the end-to-end service architecture describing all network and service components and their topology and dependencies.
- Specify network and system interfaces and interconnection links that support the service.
- Describe call flows for setup and data transport used by the service.
- Identify aspects of service assurance: what the failure modes are and how they affect the service.

As a starter, the accuracy of equipment inventory in the network is a critical first step for the service provider to avoid stranded assets and misdirected efforts. Any new system must tightly integrate the underlying physical assets of the wireless and wireline networks, enabling real-time, fully synchronized network element and services inventory management across the entire multivendor, multidomain network. Such integration would provide the service provider with a single, accurate, and up-to-date view of its network and service inventory. Furthermore, inventory reconciliation based on the state of the network and network domains can avoid erroneous information for overlying OSS databases and enable error-free order processing.

Ensuring service delivery and understanding network and OSS capabilities are priorities. For end-to-end service delivery, the solution needs to integrate network and service management using a service model that will be described in later chapters. The service model approach enables bottom-up service assessment and impact analysis and top-down troubleshooting with minimal diagnostic testing. As depicted in Figure 1.8, the network characteristics and user load can impact the service performance. Therefore, it is important to isolate problems due to user load and capacity constraints (congested environment) and problems due to conditions in the transport plane (uncongested environment). This can be accomplished by collecting information from the network (e.g., radio elements, network elements, EMS, and gateways), application (e.g., servers and systems), and OSS (involved in the service delivery chain) and by translating this information into representative KPIs for analysis and appropriate action.

In addition to quality delivery, a fully integrated service assurance solution can automatically find the network parameters that contribute to potential performance problems through a correlation process. From the operational perspective, service providers do not need more alarms in the NOC that identify faults on the network, but they do need solutions that relate network data to customer information.

The root-cause diagnostic function collects information from selected infrastructure components periodically and uses this information for reporting on end-to-end service performance. This data can also be utilized to determine corrective actions, if needed, at the domain or network levels.

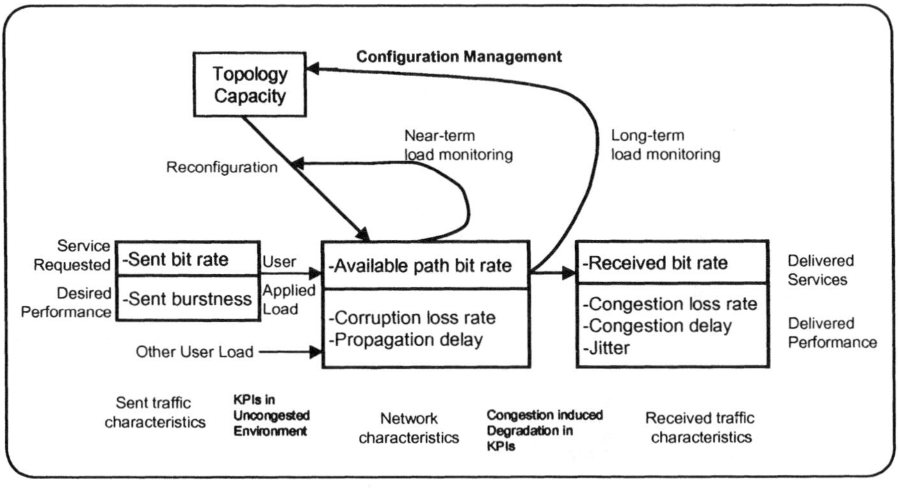

Figure 1.8 Role of network characteristics and conditions in delivered service performance.

Understanding the quality indicators surrounding service delivery and maintenance will be the key to a provider's competitive posture. Furthermore, due to the nature of wireless technology, service providers cannot relate a physical identity (e.g., port) to a specific mobile station; however, it is possible to track the location of their high-revenue customers' stations and improve network performance based on these valuable customers' needs.

Finally, service providers should be able to use their OSS to assist them in improving business efficiency. The integrated service assurance solution, indeed, holds the capability to build their networks cost-effectively or create new services that can improve their new profit margins without additional operating and capital expenses.

1.3.4 Customer Service Aspect

The inundation of new technologies and growing customer demands are complicating the already complex task of mobile network management. The network's increasing intelligence is a blessing in that it allows more specialized new services, but intelligence can also be a management nightmare. One example of the use of more intelligent service equipment is the case where the operator can not only activate a service but also specify the customer's paid-up status and CoS. With millions of customers and tens of thousands of services, management of such combinations is a challenging task.

In today's competitive marketplace, reliability alone cannot satisfy customers anymore. Customers are expecting sophisticated interactions with their service providers, among them being the ability to do flexible one-stop shopping. Customers are demanding the ability to sign up for a provider's offerings and, at the same time, specify products from competitors or complementary service providers. They want the flexibility to construct their own package of bundled services.

Electronic bonding (e-bonding) is the practice of electronically exchanging network-related information between service providers for fault resolution, billing, and provisioning. It is also helpful for meeting customers' service demands. E-bonding is generally part of trouble tickets and billing information, but rapid provisioning is growing into the new customers' requirements. For instance, community-based customers may order service provisioning for high-speed datastreaming at night and regular voice service in the day. The customers may require their primary service provider to perform rapid service provisioning from more than one service provider or, ultimately, may e-bond with these service providers for their own provisioning. Providing customers with this kind of reliability, flexibility, and rapid service provisioning is a virtual market mandate for service providers as they move to become more customer focused.

As mobile services become more mission critical within businesses, and content services begin generating revenue, it is obvious that service users and provider partners will start to demand SLAs that guarantee QoS and exceptional performance. Premium customers need more granular performance details and cross-monitoring because of the complex, bundled-service pricing structures. To reflect reasonable charging policies, these SLAs may surround completion time, security, connectivity, transaction success rates, and many other KQIs. Understanding such real-time knowledge of network, customer, and service performance will be key to the service providers' charging additional dollars.

1.3.5 Operational Aspect

"Better, faster, cheaper" are three magic words that easily sum up the goals of most service business. Service providers are trimming down and rounding up their resources to get fit for the competition. Most of the traditional telecom service providers come from a regulated world with an embedded cost burden of many legacy systems. The new competitors do not have such competition baggage and will be able to provision services faster and less expensively. As the result, almost all major operators have in the past few years undergone "reengineering" efforts to eliminate duplication within their organizations and increase service management efficiencies.

To cut costs, the service providers are collapsing multiple network management centers into megacenters, and many have whittled down their regional network management facilities into one NOC. In doing so, service providers are looking to cut the number of people in remote locations and are also looking at automation. The value of automation comes from the reduction of personnel, as well as the speeding up of the response time to handle service interruption. Another immediate benefit of consolidating network management facilities is to broaden the view of the network. Operators can perform better end-to-end network management after centralizing the management information reports. Automation of management processes implies linkages between islands of mechanization to meet an overall business goal such as reducing order-fulfillment intervals, improving billing accuracy, or meeting SLA targets. A centralized service model can be the effective way to facilitate those linkages by making information about the services and their customers readily available in a standard format to all parts of the service provider's organization.

Although service providers and solution vendors cite a long-term goal of centralizing databases from multiple network OSS, the increasing complexity of applications will require that those applications be distributed across multiple platforms. Service providers are not only looking for a way to communicate among network resources or with other service providers, but they are also looking for mechanisms to scale and distribute the computing load internally (Chapter 4 will provide a more in-depth discussion). The workable trend in the management operations of the future is centralized information but dispersed data gatherers.

The service model management approach enables a service provider to form a hierarchical view of a service, offering a data abstraction of that service so that operators can access that information without having to go through every detail involved. Operators can see a managed service at a high level, then drill down, layer by layer, to recover more detailed information. Service models are the action of standardization that can help service providers tie the performance and billing systems together or tie provisioning systems with accounting systems, for example. It is worthwhile to point out that such expanded system integration would allow the service provider to be much more customer focused.

1.3.6 OSS Maturity

What is the path of OSS maturity? To date, most service problems in a network are identified after the fact, either after users have experienced or complained about troubles or the service providers receive abnormal service alerts. An effective service management solution allows network managers to monitor and manage every detail of network usage, including user, application, and resource usage details, all in real time. It unobtrusively gathers and analyzes service data and provides real-time reporting of information pertaining to the health and security of business applications running on the network. All events in the network are interpreted and reported intelligently; giving network operators the reports needed to improve their services. But are these enough to describe future OSS functionalities?

The TMF has created a road map of OSS development in five levels, listed below and depicted in Figure 1.9. It is a commonly accepted direction for service providers to develop their OSS capabilities with regard to configuration, performance, problem resolution, and billing management. All of these functions will access integrated inventory and common configuration data (level 2). On this basis, OSS advances toward a scaleable, cohesive, unified data model (level 3) will endeavor to include transparent visualization, bottom-up and top-down analysis, end-to-end integrated processes, auto-detection, workflow automation, error detection, trend monitoring, action initiation (level 4), self-correction, and the ability to incorporate new technologies easily (level 5). The more dynamic wireless networks, though, present more challenges than fixed environments. The maturity of OSS environments should be witnessed through capabilities associated with data management, processes, and organization.

- Level 1—Informal: Highly manual, point solutions, multiple databases, difficult to change;
- Level 2—Controlled: Some integrations, data accuracy a problem, difficult to scale;

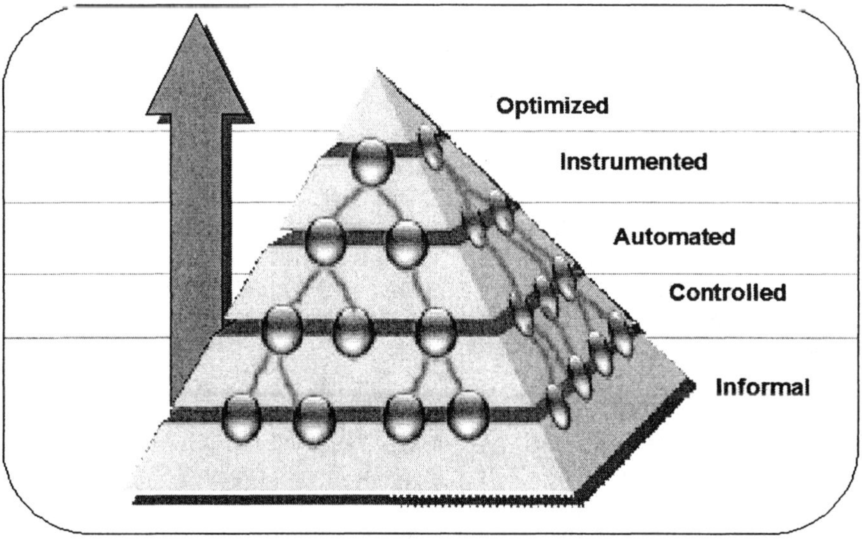

Figure 1.9 OSS maturity. (*From:* [15]. © 2005 AcuMaestro. Reprinted with permission.)

- Level 3—Automated: Largely integrated, cohesive object model, scalable;
- Level 4—Instrumented: Single environment, auto problem detection, and trending, predictable, real-time state quality management, easy to adopt new technologies, and predeployment to meet needs;
- Level 5—Optimized: Zero touch, self-serve, synchronized, continual optimization, constantly evolving.

1.4 GETTING READY FOR THE NEXT STEP

Today, we are on the cusp of a tertiary evolution, where the dramatic increase in communications and computing capabilities is further enhanced by a growing population of industry-specific data and communications standards. The current business systems environment is essentially an electronic labyrinth that consists of a wide variety of systems on disparate platforms, using an even wider variety of applications and communications protocols. Many large corporations are now faced with having hundreds of software systems, each of which carries a cost factor to operate and maintain. The next level of automation requires the combining of these virtual islands of activity to integrate the vast amount of business data intimately and refine that data into pertinent, corporate-wide business intelligence. It is certainly not possible to replace the entire OSS population, but it is possible to connect to the existing systems and extract the relevant subset of information required to provide a complete and current view of overall operations.

The operations and management accountabilities within a business can be thought of as belongings to a kind of layered pyramid. The horizontal layers of the pyramid can be thought of as the various business processes or service operations, from their beginning to their ends (see Figure 1.10). This orientation represents the activities to achieve a result or meet a business target through the contributions of cross-departmental or cross-functional disciplines of the business. An example would be billing or sales department operations or the creation and activation of a new customer connection.

The vertical orientation represents levels of responsibility or accountability for the totality of operations up through a given level. At the top is the chief executive officer (CEO) who is responsible for the entire corporate operation. Functional middle managers each have a spot below the CEO with a narrower span of responsibility.

Corporate middle managers will probably require mostly vertically oriented rollups of operations that display results of the operations up through their level of responsibility; that is, the billing department manager is concerned with billing operations wherever they are needed across the enterprise. He or she may have an interest in other parts of the business but probably only wants to review detailed reports dealing with the results for which he or she is accountable.

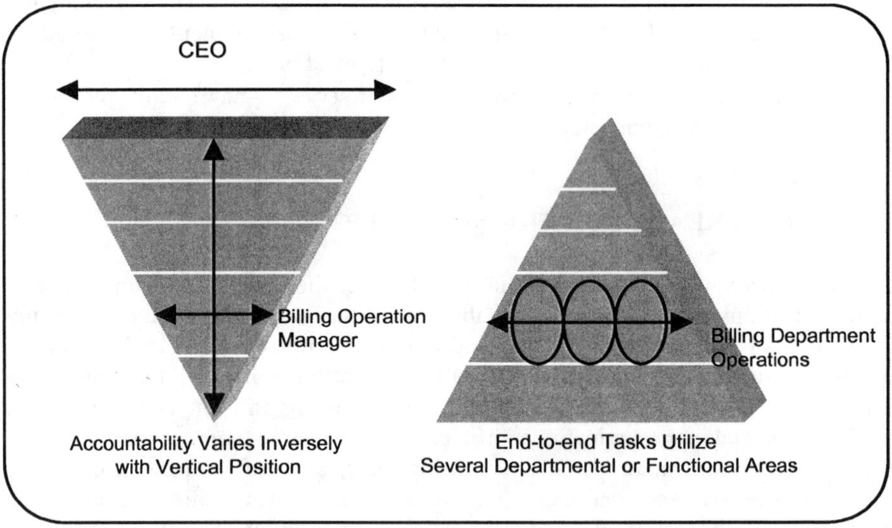

Figure 1.10 The operations and management accountabilities. (*From:* [15]. © 2005 AcuMaestro. Reprinted with permission.)

Operating or functional managers will probably require mostly horizontally oriented rollups that display the various component operations that contribute to the service or function for which they are responsible.

The implementation of the service assurance model lends itself nicely to providing a common business management environment that will support the next generation of business automation. The KPIs and KQIs can be used to provide managers with the horizontal and vertical rollup views of the state of key parts of the business as described above. If the implementation is robust enough, these reports can portray the linkage between operating results and the corporate bottom line.

References

[1] Paulk, M. C., et al., Software Engineering Institute, CMU/SEI-93-TR-24, DTIC Number ADA263403, February 1993.

[2] Sohail, R., "Implementation and Interoperability Experiences with TINA Service Management Specification," BT Labs, 1999.

[3] "RFC 1157—Simple Network Management Protocol (SNMP)," Network Working Group, May 1990.

[4] "Network Monitoring Using Cisco Service Assurance Agent," Cisco Systems, http://www.cisco.com/univercd/cc/td/doc/product/software/ios121/121cgcr/fun_c/fcprt3/fcd301d.htm#1000872.

[5] "Mobile Subscriber Numbers Exceed 1.5 Billion," 3g.co.uk, June 23, 2004, http://www.3g.co.uk/PR/June2004/7947.htm.

[6] "The Mobile Computing Market—The Big Picture—Mobile Computing and Wireless Outlook," http://www.mobileinfo.com/market.htm.

[7] Hesham, H. et al., "A Model Checking Approach to Network Fault Management," CRIM, http://www.crim.ca/rd/publications/ASD_HalHals_SCI04.pdf.

[8] "Performance Reporting Concepts and Definitions—TMF 701," Network Management Forum, November 2001.

[9] "WAN and Internet Service Spending Rises 26%, from $82B in 2002 to $103B in 2006," http://www.bbwexchange.com/news/2002/may/infonetics051402.asp.

[10] "Service Provider to Customer Performance Reporting Business Agreement—NMF 503," Network Management Forum, March 1997.

[11] "Strategic Planning Series—The Mobile Business Value Scenario," Gartner, http://www3.gartner.com/research/spr/attributes/attr_61427_429.pdf.

[12] "Wireless LAN Market Overview," http://www.enterprise.bell.ca/en/resources/uploads/pdf/facts.pdf.

[13] "Residential Market Continues to Drive EMEA WLAN Market in 3Q04," IDC, December 2004, http://www.idc.com/getdoc.jsp?containerId=pr2004_12_06_103519.

[14] "Enhanced Telecom Operation Enhanced Operations Map (eTOM)—The Business Process Framework for the Information and Communications Services Industry, Release 3.0—GB 921," TeleManagement Forum, June 2002.

[15] Chang, W., and M. Chiger, "AcuMaestro White Paper," AcuMaestro Inc., 2005, http://www.acumaestro.com/Solutions.htm.

Chapter 2

An Integrated End-to-End SLA as a Service Provider Value Adder

In this chapter, we will describe the content and design of a generic wireless SLA and how it can function as a value adder for a wireless service provider. More importantly, we will discuss the many SLA design and implementation considerations that can have direct implications for service assurance applications. In light of the current wave of outsourcing activity, service relationships between business partners and between service vendors and customers have become far more complex. Often, many different types of service providers are involved. Particularly, many services are provided by international enterprises with different content and domain interests. Managing the quality of an end-to-end service to an end user is no longer a single process or procedure. It becomes a complicated business practice involving many chained service contracts with different vendors that may be impacted by various national or international regulations.

The focus of this chapter is also the identification of the service provider value-chain in wireless applications. In addition, we will continue our examination of needed procedures and service functions to understand their place in OSS operations. In the following chapters, our discussion of OSS systems will map out how a service's quality and controlling OSS are interrelated and how existing service assurance solutions are no longer viable in the highly competitive wireless market. We will undertake an in-depth analysis of SLA flows in system processes that can help us identify a better, more integrated process that can support the interests of both clients and providers.

Figure 2.1 depicts the relationships of a typical primary service provider (home service provider, in the case of roaming) and its business relationships with other service providers and customers. Most of the business relationships are based on contractual relationships and presented in the form of SLAs [1].

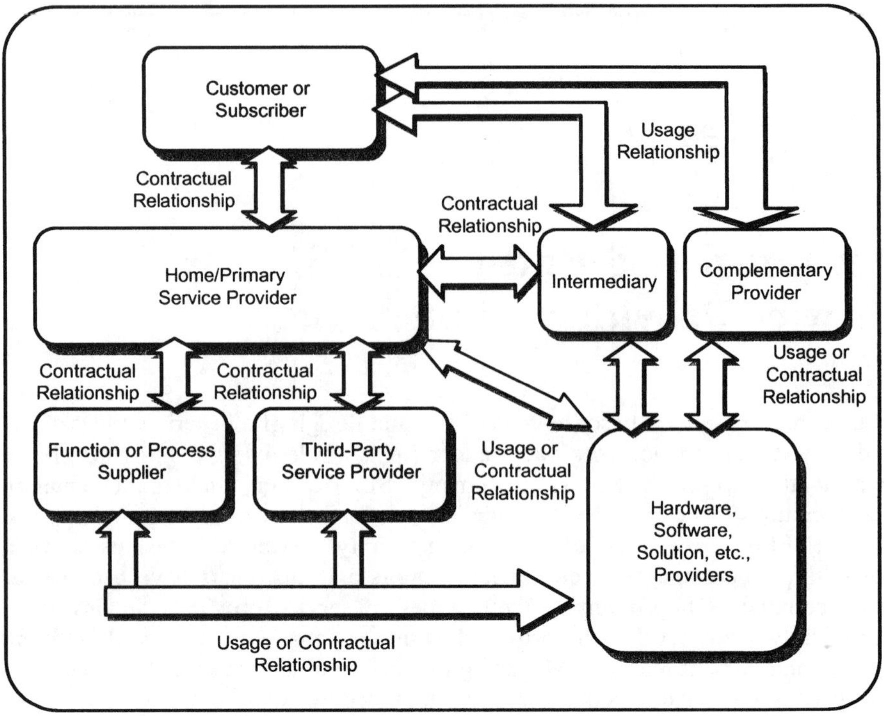

Figure 2.1 Service provider's relationship chart. (*After:* [2].)

2.1 WHAT IS INCLUDED IN AN SLA?

An SLA [3, 4] is a contractual agreement between a service provider and a customer or between cooperating service providers that mandates specific performance levels. For example, a wireless service provider can offer wireless voice service to a corporation with an SLA that specifies that the voice service will be available at least 99.5% of the time in any calendar month or else the customer is given a service credit of one day for each hour of unavailability above the allowed limit. That is, if the SLA (availability) performance target is not met, a specified penalty is assessed. The usefulness of an SLA is not limited to outside services. SLAs can also be useful in inter-organizational service arrangements for operations such as help-desk services, network performance monitoring, or other internal processes. This is becoming a well-accepted and practical business model for service providers (or carriers) to establish responsibility assignments and goal setting that still allows different departments to operate in an independent, yet measurable, manner.

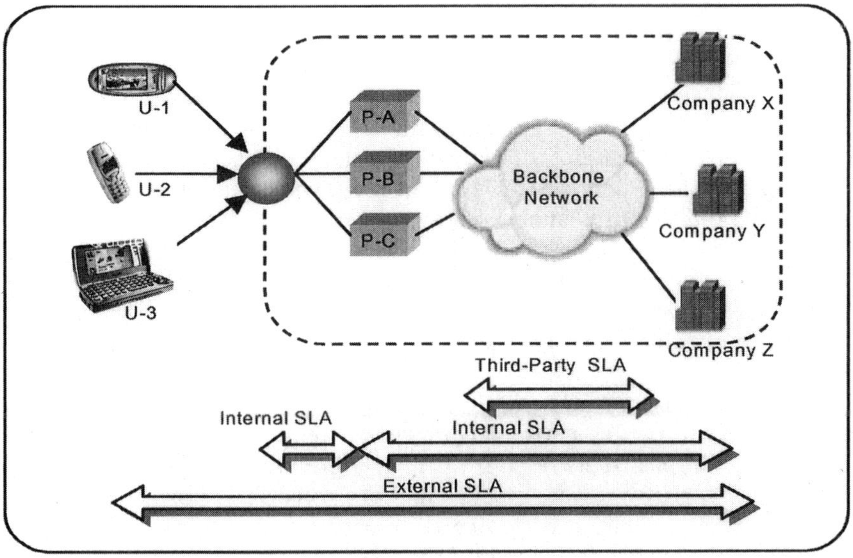

Figure 2.2 Three types of SLAs.

In addition to the above characteristics, SLAs can help the service provider ensure service commitments to their service receivers. The agreements often require review and modification as the result of the changing customer base and associated costs for supporting the wireless products (or services). Changes in usage patterns, volumes, and technology can change a provider's bottom-line numbers as time progress. Normally, an SLA is designed even before a service is offered to the public. An SLA may include content such as a service description, contact details, escalation levels, review schedules, interface processes, parameters, targets, and procedures. The parameters, targets, and operation procedures should be reviewed and evaluated by all stakeholders before the SLA is finalized. Given the dynamic nature of today's business practices, changes to initial service offerings, workflows, and procedures are virtually guaranteed. It is unavoidable that a very flexible service infrastructure is required to support SLA development and enhancement. Therefore, it is easy to understand why SLA design and adoption cycles are beginning to stretch out.

SLAs exist in different forms to fulfill different types of business needs. To help distinguish between these different SLAs, three types are defined in the following sections. While individual SLAs can be created to support any one of these types, SLAs are increasingly becoming more oriented toward end-to-end and structured agreements, with a single SLA made up of internal, external, and third-party layers. Figure 2.2 shows the three relevant relationships in a service provider environment.

2.1.1 The External SLA

Wireless service providers offer SLAs for two types of customers, a general SLA applicable to all service subscribers and a specific SLA negotiated for a single (usually enterprise) customer. For end-subscriber service, the wireless service provider establishes a single SLA for all of the service's customers; such an SLA can cover an offering of a family plan for voice and e-mail services, for example. For enterprise customers, the wireless service provider would negotiate separate terms and conditions in the SLA to cover all of the enterprise's employees utilizing the service. The service quality requirements for these enterprise-version SLAs are normally more stringent and usually command premium service charges.

In the old business paradigm, end subscribers with generic SLAs typically did not have access to the performance reports about their services. Only the enterprise service managers responsible for performing cost analyses would have access to the service provider's performance reports, via passive media such as paper or e-mail. Increasingly, service customers, whether they are individual or enterprise users, are becoming more dependent upon wireless applications. They have begun to be more aware of and concerned with the level of service. As the two-and-a-half generation (2.5G) and the third generation (3G) wireless networks offer more promising bandwidth to application providers, it will encourage making more high-volume and content-rich applications available. This will trigger the service customers' awareness of the service quality agreed to in the SLAs.

An external SLA is depicted in Figure 2.2, which represents two enterprise customers, U-1 and U-3, and a voice subscriber, U-2. Access is provided to the multimedia service through the service providers P-A, P-B, and P-C. These three companies may be channel partners of companies X, Y, and Z.

2.1.2 The Internal SLA

In many cases, service providers have internal organizations that provide service to other parts of the company. In other cases, the service provider may need cross-departmental agreements to provide the end service to their customers or subscribers. These are classified as internal SLAs. In this type of SLA, one department is considered to be the provider, while its customer, who can also be in the same or another organization, is ultimately responsible for providing services to the external customer. It is extremely important for providers to ensure that their internal service objectives can be met, from which the providers in turn offer external guarantees to their customers and subscribers. This reliance of external SLAs on internal SLAs is the key to protecting service quality throughout the entire value chain.

External SLAs carry an economic penalty for nonconformance, so the use of internal SLAs to meet corporate service level objectives is quite reasonable. Direct advantages include managing expectations, boosting productivity, and increasing employee morale. Internal SLAs also provide indirect benefits such as prioritizing work, making interfaces clear for effective processes, and providing motivation to meet company-wide performance targets. In addition, inter-organizational relationships can foster a better foundation for corporate-level teamwork in the pursuit of company excellence.

In Figure 2.2, Company Y may have internal SLAs to track bearer service (X) between the point of presence and the multimedia servers and to track the availability of the back-end application servers.

2.1.3 The Third-Party SLA

In the event of complex media offerings, where two or more service providers collaborate to offer a product for the target customers, the third-party SLA becomes very important. For instance, it may be necessary for two or more service provider's to partner in order to offer a complete end-to-end service. These provider-to-provider service agreements are called third-party SLAs, as portions of the service belong to different providers. In this highly competitive wireless service market, it is impossible for providers to develop all the targeted services in-house within the required time frames and investment targets. A more realistic option to expand the scope of their services is to partner with other providers to produce feature-rich services at low cost. These business relationships facilitate a chain of back-to-back SLAs. Everyone in the value chain is both a customer and supplier, except the very last service provider. The level of service quality for a service chain is tightly coupled with all of the resources in the chain, and all of the associated resources have equal weight in contributing to the ultimate service level. In other words, any existing bottleneck in the chain will affect the entire service; therefore, the SLAs are the glue for the entire value chain.

For example, service provider P-A may offer radio access network and Internet connectivity, while a service provider, Company X, may offer access to a content server. A corporate customer wants to interact with a single service provider for multimedia messaging service with an SLA best suited to its needs. Rather than dealing with multiple service providers, the corporate customer can negotiate an end-to-end SLA with provider P-A. P-A will at the same time arrange another SLA with Company X in order to insure that the performance of the service provider's content server is consistent with the overall end-to-end SLA. This relationship is depicted in Figure 2.2.

When a service provider creates a new offering, they often specify the end-to-end SLA. This SLA might depend upon internal, external, or outsourced SLAs. As additional offerings are created by the service providers, which utilize already agreed services, a parent-child relationship between the associated SLAs is created. This tiered SLA offering can then be included in an order representing the overall SLA. Some considerations for constructing a tiered SLA include:

- An external or internal SLA can include any number of external, internal, or third-party SLAs.
- The third-party SLA type cannot include any other SLAs.

A logical sequence to create a tiered SLA is first to define the third-party SLAs, followed by internal SLAs (which may be associated with previously created internal or third-party SLAs), and then the external SLAs (which also may be associated with any previously created SLA).

2.1.4 Motivation for Using an SLA

From the service provider's perspective, services with SLAs represent higher-quality services that can provide additional revenues. The SLAs also are an additional service attribute that the service providers can modify and tailor to make their services more attractive to their corporate customers. For these reasons, SLAs are very important agreements for both service providers and their customers. From the customer's perspective, SLAs, with their penalties, are an incentive for service provider's to meet the performance targets specified in their agreements. The SLA penalties can take a variety of forms, including monetary penalties, service credits, and even allowing the customer to void the service contract and to seek another service provider. A more comprehensive SLA might cover other performance metrics for wireless messages, such as the percentage of wireless calls that are prematurely dropped. Note that the penalties of any single customer's SLA may not be sufficient to influence a service provider's strategic behavior. However, together the entire set of a service provider's customer SLAs can exert a strong influence on the service provider's long-term actions.

One type of SLA for partnering service providers that we have seen used in practice is the "all in the same boat" type of SLA. For this type of SLA, if there is a service objective that is not met, each service provider is penalized (in proportion to the amount of the overall service that it provides), regardless of whose fault the performance failure is. Although the "all in the same boat" SLA is easy to set up, it can encourage poor performance by a service provider under certain circumstances (e.g., the shared risk may result in lower cost to the provider than the costs required to eliminate the cause of the poor service).

Other ideas for SLAs with partnering service provider's may be drawn from the field of supply chain management where academic researchers have extensively studied the issue of performance contracts between two adjacent members (firms) on a supply chain.

2.2 THE VALUE CHAIN OF WIRELESS APPLICATION SERVICES

In the world of wireless services market share and revenues are under constant pressure from two major factors. The first is fierce competition that drives down market share and significantly reduces profit margins. The second is the evolution of technology at incredible speeds, which reduces the life cycle of service offerings and has major impacts on business plans. As a result of new technology development customers have become increasingly discerning and selective. Wireless service providers are realizing the need to differentiate their products through value-added services. Specific activities, such as the self-service concept, will definitely drive new business opportunities for wireless service providers.

Currently, the industry-standards bodies are pushing new specifications that force open interfaces between businesses and their customers. This pressure has resulted in the unbundling and componentization of service elements and created new opportunities to combine components into new service offerings. The greatest value of componentization is the flexibility to broaden business cases to meet new market demands with reduced costs and time demands. The establishment of new cooperative relationships between the market players characterizes the current stage of wireless application development.

Figure 2.3 [3–6] depicts the business relationships between customers and the service provider and between providers in the form of SLAs, if applicable.

The service subscriber is responsible for the contractual arrangements with the service provider, such as concluding an SLA for service and paying bills. A service subscriber can be a direct customer or a reseller and has a user relationship with the primary service provider. The primary service provider faces the customer as well as the other service providers. It is responsible for the provision of the product to the customer at the service levels agreed to in the SLA, and it bills the subscriber for product usage. It has agreements in the form of SLAs with network providers or value-added service providers, as required.

Network providers play a third-party service provider role, as required, so that the product being supplied by the primary service provider can be delivered.

The value-added service provider may play either a complementary provider role or a third-party service provider role, depending on the product being offered.

Value-added service providers might add content, portal positions, or m-commerce to the mobile service offered by the service provider.

These providers supply products to the customer via the service provider and, so, have a business relationship only with the service provider. The value-added service provider(s) and the user have a usage relationship with the network provider. SLAs would regulate the product as delivered and product usage in the business relationship with the primary service provider.

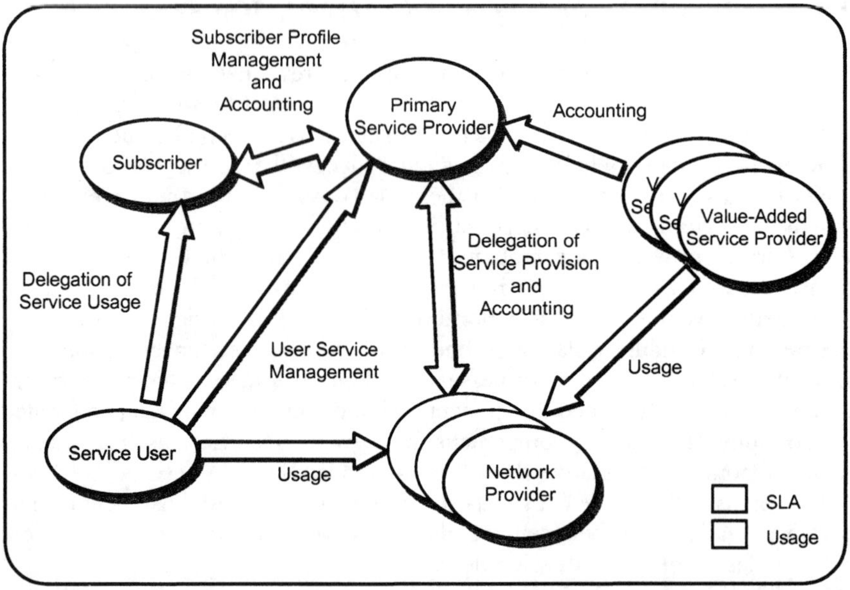

Figure 2.3 Service provider's SLAs with customers and other providers. (*From:* [6]. © 2005 TeleManagement Forum. Reprinted with permission.)

The value chain of wireless application services includes the entities discussed in the following sections.

2.2.1 Customers and End Users

In this book, the term *customer* refers to companies or organizations that sign a contract with a service provider (maybe through a retailer) for one or more services. The customer is the buyer of a wireless application service but may not be the ultimate end user. The relationship with the service provider includes a permanent record of the account and billing data. For example, a business may be the customer of a service provider, but the end users may be employees of the customer (see Figure 2.4).

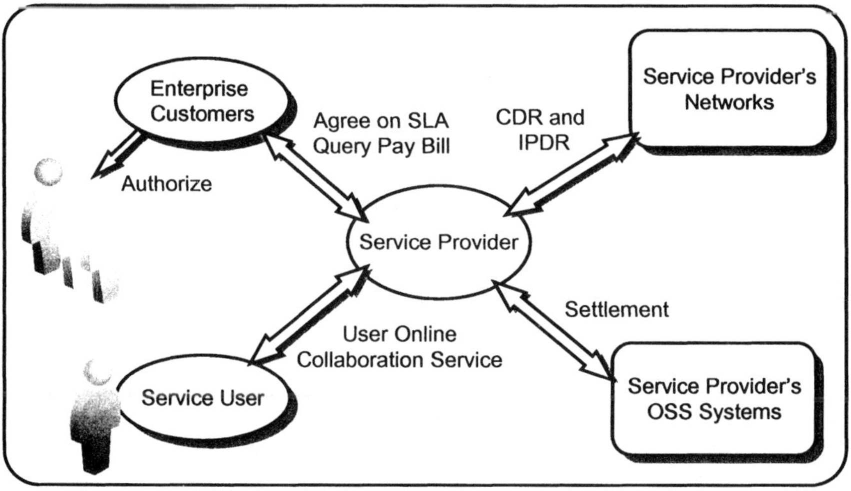

Figure 2.4 Service providers and customers/users.

An *end user* may be a private subscriber or may work for the business customer. The business relationship between these two types of users and the service provider may include prepayment or payment by a third-party provider. The later case is a popular model in various mobile-commerce businesses that use reverse charging.

2.2.2 Service Retailers

A *service retailer* can function as a portal for a wide array of services and can form a customer base for the service provider. It can also function as a value-added reseller who creates limited new services on a customer's demand. The retailer deals with customer queries and billing on the service provider's behalf. A service retailer can also bargain for a better rate and deal with providers' requirements on customers' or users' behalf.

A service retailer has lower entry costs than smaller trading partners (particularly niche service providers and resellers). Service retailers can create opportunities for themselves to become members of service value chains, thus widening the range of players involved and the possible service propositions.

2.2.3 Mobile Virtual Network Operators

The *mobile virtual network operator* (MVNO) is a new business concept invented in Europe. An MVNO markets itself as a private-label wireless network without actually owning spectrum rights and usually does not have its own network infrastructure. In order to accomplish a complete service package, MVNOs have business arrangements with traditional mobile operators to purchase minutes of use (MOU) for sale to their own customers. These operators lease the wireless capacity from traditional operators and then repackage it for a specific vertical industry application. MVNOs generally provide both voice and data services to customers and end users through a prepaid subscription agreement as a means of low-cost market entry.

The MVNO acts as a service retailer and maintains contracts with third-party service providers, while performing the tasks of charging and billing for service usage. Unlike simple resellers, MVNOs typically add value such as brand appeal, distribution channels, and other affinities to the resale of mobile services. MVNOs also have full control over the SIM card, in the sense that the MVNOs own the customers. They also provide customer care, negotiate, and finalize deals between MVNO customers and third-party service providers. Overall, the MVNOs are expected to become drivers of differentiation among operators by providing tailored mobile services to identified target users.

The MVNOs have three sources of revenue: handset and airtime sales, payments for value-added services, and m-commerce sales. Because the MVNOs need to buy airtime and resell it, their margins on basic voice services will be too narrow to draw any considerable revenue. In order to create a profitable business, MVNOs have to establish and sell targeted, value-added services such as providing communications, application, and information services. A successful MVNOs can be beneficial by supplying ample financial resources and sufficient agreements with existing operators to provide a good service coverage area. While MVNOs typically do not have their own infrastructure, there are some sophisticated providers who actually own their mobile switching centers (MSCs) and even service control points (SCPs) to set their standards beyond the limitations of their third-party service providers. Additionally, well-diversified, independent MVNOs can offer a product mix that incumbent mobile operators cannot match. For example, bookstore MVNOs could offer a package of MOUs and books. The goal of offering value-added services is to differentiate services from the incumbent mobile operators, thus allowing for customer attraction and preventing the MVNO from having to compete on the basis of price alone.

2.2.4 Primary Public Land Mobile Network Providers

Public land mobile network (PLMN) is a generic term for a mobile wireless network for mobile stations (MSs) or mobile phones. It is centrally operated and administrated by network service carriers and uses land-based radio frequency transmitters or radio base stations (BSs) as network hubs. Access to the PLMN is significantly different from access to fixed networks. PLMNs can stand alone and interconnect with one another or connect to land-line service users via PSTN gateways. In standard mobile network architecture, the mobile service area is covered by a set of BSs, which are responsible for relaying the calls to and from the MS located in the coverage areas (cells). Physical access in a mobile network is arranged to enable a MS to connect itself anywhere in the network and move about while a call is in progress, as shown in Figure 2.5.

When a mobile user is engaged in conversation, the MS is connected to a BS via a radio link. If the mobile user moves to the coverage area of another BS (another cell), the radio link to the old BS is eventually disconnected. And a new radio link to the new BS is established. This connection moveability requires that the MS be able to maintain continuous radio contact with the network. This can be accomplished by using the MSCs. It allows connections to be switched to another BS as the MS moves. The home location register (HLR) and the visitor location register (VLR) are two special types of databases used to assist as a mobile user moves from one carrier to another (this is called roaming management; see the next subsection).

The primary PLMN providers are responsible for the network layer process operations. The primary PLMN provider plays a crucial role in providing the underlying service frameworks for third-party service providers. The PLMN providers handle queries and establish SLAs on behalf of third-party service providers. This may include the specifications of QoS that mobile users can expect to receive.

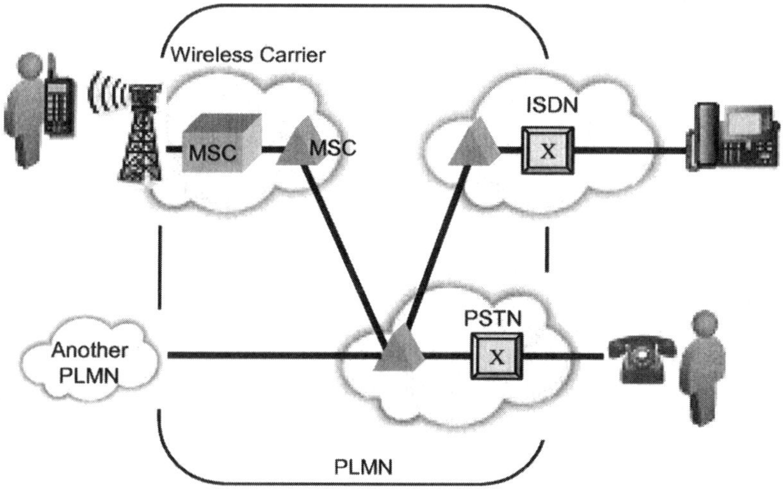

Figure 2.5 PLMN provider.

The primary service providers often piggyback on the services of a third-party provider in order to offer wireless services without the expenses and duties required in originating those services. They also inform third-party service providers of customers' requirements and establish service-provisioning contracts on behalf of the customers.

These service-provisioning contracts can assure conformance to SLAs for the consumption of their respective services. Most importantly, they calculate the charges for the service usage and settle the charges among associated service providers.

Installing traditional telecom cable in rough country (mountainous regions or dense forests, for example) can be extremely labor intensive and costly. Also, updating and expanding the PSTN in a city with an out-of-date cable network can take a long time. In such cases, radio-based solutions are often attractive. They can be based on radio in the local loop or on "fixed cellular" systems. For example, isolated villages in rain-forest regions have been equipped with pay phones connected to the fixed telecommunications network through a PLMN. Fixed cellular systems can be profitable and effective, provided that a mobile network is available and that legislation permits fixed subscribers to use it.

Fixed cellular operation entails a lighter load on the processor in the MSC, compared with regular mobile telephony. One disadvantage is increased transmission costs compared to traditional plain old telephone service (POTS). For example, calls between neighbors must be connected through the PLMN and the MSC, as well as through the local PSTN exchange. Fixed cellular subscribers pay the same tariffs as regular PSTN subscribers.

2.2.5 Third-Party Network and Service Providers

Third-party network and service providers can be organizationally part of the primary service provider or can be independent providers. They offer special features in the mobile service value-chain.

Some third-party network providers handle all of the needs of the end user but are invisible to the end user, who only sees the primary network provider. One example of a third-party network provider is an Internet service provider (ISP); who for a monthly fee provides the mobile user with a software package, username, password, and portal address.

Equipped with the proper hardware devices (e.g., a 2.5G mobile phone) and proper credentials, mobile phone users can log on to the Internet and send or receive e-mails. This type of third-party network provider, in addition to serving individuals, can also serve large enterprises by providing direct connections from the company's networks to the Internet. Other third-party network providers may provide additional services, such as leased lines and Web development.

The third-party service providers offload the burden and infrastructure costs of customer care, service quality management, and billing from the primary service provider. This allows primary service providers to concentrate their resources on their core competencies (service provisioning), while also allowing them to deliver their services at a lower cost.

2.2.6 Content Providers

Content providers can also be classified as third-party entities that manage and distribute software-based services and solutions to customers and end users across a wireless network. A content provider is responsible for the preparation and maintenance of service-related information for its subscribers. The content can be services provided by an organization (e.g., an online consultant), a resource area (software with data), or a discrete publication (data).

As shown in Figure 2.6, a content service provider may be the originator for application content or may simply integrate the services of other providers (who own the actual content). Providing content service to an end user may involve multiple service providers and a number of subcontracted content providers. In essence, content provisioning is a way for companies to deliver their information without having to deal with bearer networking issues. A typical content provider's applications can be broken down into five subcategories:

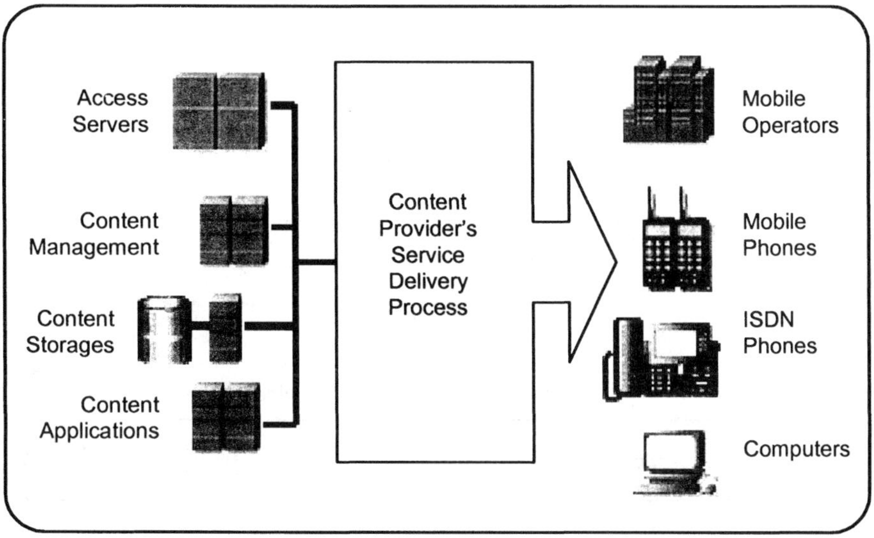

Figure 2.6 A typical content provider's configuration.

- *Enterprise applications:* High-end business applications (e.g., document sharing, calendar sharing, and database access);
- *Volume business applications:* Composite applications for small or medium-sized businesses (e.g., news and tax software packages);
- *Local/regional applications:* Wide variety of application services for regional activities (e.g., baseball schedule and online game);
- *Specialist applications:* Applications for a specific interest (e.g., financial analysis);
- *Vertical market applications:* Support for a specific industry (e.g., trading).

2.2.7 Handset Providers

The *handset provider* is not a critical player in wireless network services. However, the supported features of a handset can make a major contribution to the experience of the end user. It is obvious that the handset is the ultimate user-facing device, and the quality of network services cannot be guaranteed without the handset providers' participation (Figure 2.7). Furthermore, a highly reliable and speedy wireless network service cannot display even, superb features with a poorly performing device. Selecting good handset providers to couple with the wireless service providers' business cases becomes an important business decision.

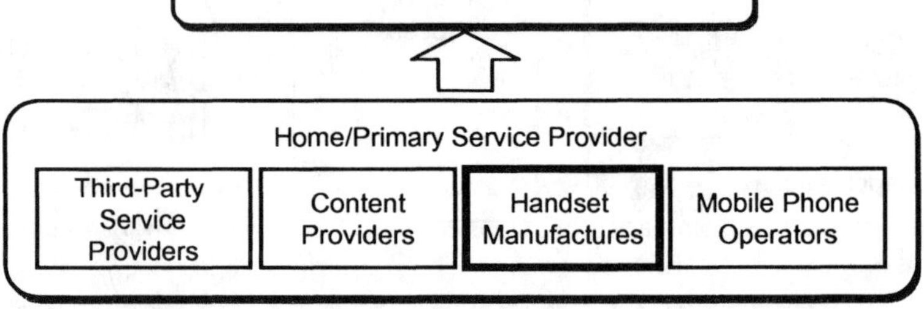

Figure 2.7 Handset providers.

The service agreements that tie the services to customers and users are thus also handset specific. Handset vendors, while trying to promote all of their advanced features, have to be sensitive to the supporting networks and available content. The hardware warranty and customer support have to be delivered hand in hand with the service provider's final SLAs. Such combined solutions that consist of both handset and service commitments can therefore be considered a complete package for end-user SLAs.

2.2.8 Roaming Partners

Roaming means one wireless service provider's subscriber can use the networks or the WLAN services of other providers. The goal is to have a customer receive the same service when traveling in an area supported by another network as the customer receives when in its home service provider's area. In order to accomplish this, the wireless service provider needs to conclude new agreements with *roaming partners* on an ongoing basis.

Through signaling-management systems, the roaming customer's (roamer's) location and service activity are available to the service provider almost instantaneously. Thus, wireless service providers can use these systems to determine the roamer's geographical and service preferences and offer new services continuously in order to support targeted marketing efforts. For instance, service providers may use the location information to provide other location-based services. Furthermore, service providers also can attract roamers and increase revenue by offering new services such as SMS messaging to a new roamer. These types of marketing approaches could include efforts such as welcoming that person into the provider's network and providing information on new services available. There is a strong strategic and tactical business motivation for a service provider to include all possible roaming partners to expand their market coverage and business segment.

Figure 2.8 shows a basic roaming scenario. Here a user moves from coverage area 1 (Loc #1) to 2 (Loc #2). The user is not aware of a different access point being used from area to area. Some access point configurations require security authentication when swapping access points, usually by showing the user a password dialog box. Access points are required to have overlapping wireless areas and the MS automatically swaps to the access point with the best signal.

To allow a subscriber to use service on a nonhome service provider network, the home service provider and the serving service provider need a contractual relationship called a roaming agreement. The roaming agreement can be a direct (bilateral) agreement between two service providers, or it can be established by means of a clearinghouse.

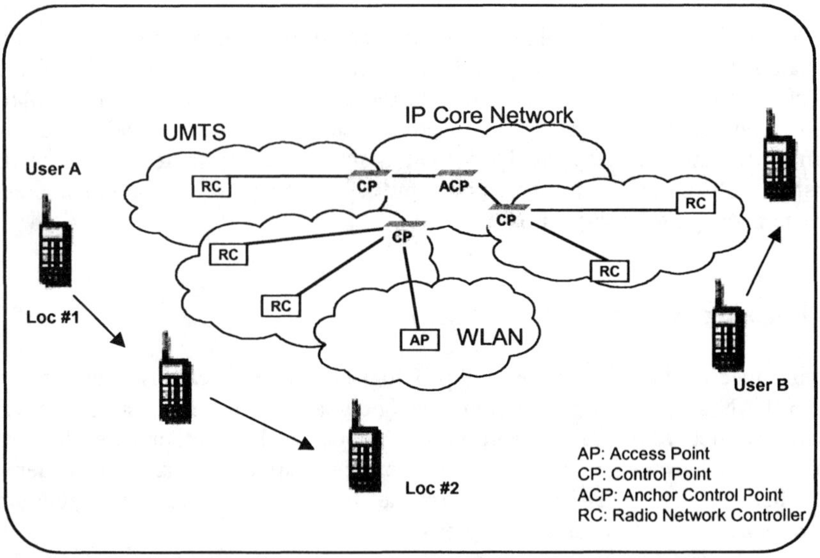

Figure 2.8 A basic roaming scenario.

The home service provider is the customer for the roaming agreement, and the serving service provider is the supplier. Whether as a bilateral contract or a contract with a clearinghouse, the roaming agreement regulates at least the following items:

- Tariff and pricing (see Figure 2.9 for the flows);
- Signaling and traffic interconnection;
- Call detail record (CDR) exchange format and exchange schedule;
- Problem handling.

Some problems may arise when different networks work together. As a result, roaming partners require considerable negotiation and testing of their interfaces before going commercial. In spite of testing, network providers cannot guarantee 100% that the services of their network will be fully available at any time in any of the roaming networks simply because they have so many roaming partners. The potential problems can be caused by a simple software update, for example, or lack of a uniform approach to network address translation. These should all be considered when developing roaming agreements.

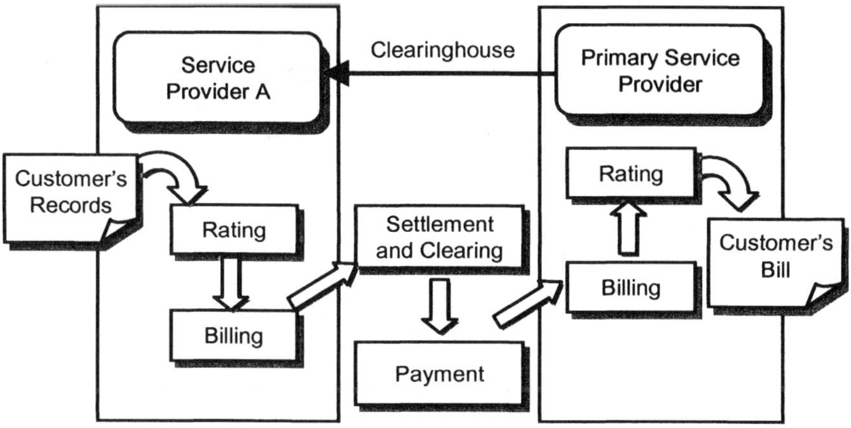

Figure 2.9 User roams onto provider A's network.

2.3 END-TO-END SLA IMPLICATIONS

There are two levels of operations within an end-to-end SLA covering a service. The first level deals with the high-level horizontal business relationships defined across provider domains. The next level represents vertical process relationships in different life phases contributing to the service.

As defined in Section 2.1, an SLA is a form of negotiated agreement between two parties. It is a contract that exists between the service provider and the service receiver covering the scope of services, priorities, responsibilities, and so forth. In addition to the commonly discussed service performance, many processes and procedures that support the offering, such as customer care, billing, and provisioning, can also be within the scope of an SLA. Considering all the aspects of a complete service life cycle, we can draw a two-dimensional relationship matrix based on the associated processes and parties to represent an end-to-end view of an offering. Figure 2.10 shows a simplified service assurance flow and associated accounting management.

Once the end subscriber signs up for a service, the contracts along the full value chain are activated. Each party on the value chain supplies a paying customer, and there is mutual dependency between the end subscriber, integrators, and other parties along the full chain.

Any performance implication that occurs in the chain affects the service receiver, and all providers upstream of the point of failure will be impacted. An effective end-to-end SLA management can be considered in the different perspectives described in the following sections.

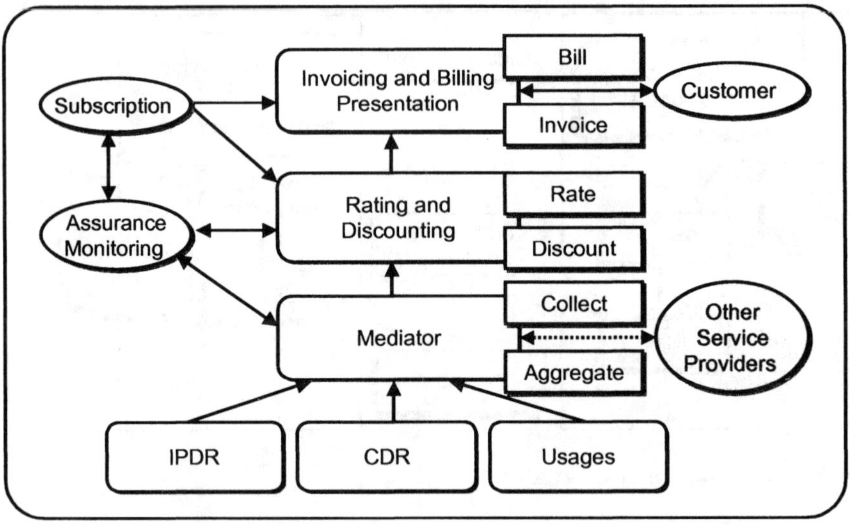

Figure 2.10 Service flows and accounting management. (*After:* [7].)

2.3.1 Customer Perspectives

For a business customer, the SLA should offer describable and measurable indicators to guarantee business values from the purchased wireless services, whether it is networking, content, or both. An effective SLA from the customer's perspective could consider the following:

- Define high-level common terms and definitions for end-to-end service performance in the form of technology-independent QoS metrics, parameters, and reports. Proper baseline references and customizable requirements capable of displaying performance compliance validation for end-to-end service should be part of the SLA.
- Describe the service provider's methods for detecting degraded service performance. Responses to performance events affecting the customer's service, including customer alerts, should be included. Additional reports should include mean time to provision, mean time to identify and resolve malfunctions, service availability, end-to-end throughput, delays, and errors.
- Explain clearly penalty provisions for failure to maintain and deliver the service, including cancellation fees.

- Provide notifications of SLA violations and report any developing capacity problems, changes in usage patterns, and reported delivered performance for comparison with the customer's perceived performance.

2.3.2 Provider Perspectives

Proper methods and procedures should be built in the end-to-end SLA with the following considerations in order to take advantage of the tightly coupled relationships among different service providers:

- The operation should be built upon a dynamic infrastructure allowing flexible operational changes and improving internal measurements and reporting in order to enrich customer relations and differentiate the service provider from its competitors.
- The infrastructure should use a common language and understanding with the customer in characterizing network and operational parameters. The goal is to create provider's internal recognition of the customer's perception of service quality. Through standardized performance-gathering practices across multiple internal domains, the provider can establish common SLA/QoS goals across multiple domains for other third-party providers internally or externally.
- An open and common specification across domains can lead providers to create more knowledgeable customers, who can better express their needs to the service provider, which can reduce the time devoted to the negotiating process.
- This open common specification can also help handset providers and equipment suppliers agree on the mapping of technology-specific parameters and measurement methods to service-specific parameters and measurement methods.

Value-chain management essentially includes the links that have to be established among customers, suppliers, and the internal processes of the service provider. Value-chain considerations therefore influence and may require changes to many of the preexisting internal processes. The changes could be implemented with agreements such as collaboration partner profiles (CPP) and collaboration partner agreements (CPA).

2.3.3 QoS in End-to-End SLAs

QoS is an overall perception of service performance that represents the level of customer satisfaction with the purchased service. QoS measurements are the combined aspects of service support, performance, operation, security, and other factors specific to each service. In this sense, the overall QoS, as delivered to the customer, consists of two major parts:

- *Service-intrinsic criteria:* These criteria, chosen as contributors to the QoS, are typically those that are fundamental to the operation of the service and include both service-specific and technology-specific performance parameters. They are related to the performance of a service provider's service platform.
- *Operational criteria:* These are service- and technology-independent performance parameters but, nevertheless, affect the QoS experienced by the customer. They are related to the performance of an organization.

Each of the measurement criteria, in the form of KPIs, associated with a given service would be individually tracked as part of the SLA. It may be possible to algorithmically combine the chosen criteria into a single QoS value or index intended to provide an overall view as to how close the delivered service is to the contracted service. The techniques involved for integrating these measurements include various aggregation and correlation methods. Weighted metrics are normally part of the formulas to calculate an individual measurement's contribution to the overall QoS as specified in the SLA. Weighting takes into account the significance of each of the criteria to a specific customer's business or use of the service. If a set of metrics or KPIs associated with service characteristics, such as provisioning timing or repair timing, is specified in the end-to-end SLA, then each provider contributing to the service will need to agree to possibly more stringent parameters than a vertically integrated competitor. The KPIs between the service providers are set by negotiation or regulatory imperative and, in general, have to be tighter than that offered end to end.

In the following sections we will review the service life cycle defined in the TMF Telecom Operations Map (TOM) and the enhanced Telecom Operations Map (eTOM) [5].

2.3.4 The Service Order

2.3.4.1 Preorders and Quotes

Preorder and quotation processes are essential to improving inter-provider process throughput and to lowering the levels of manually handled fallout from automated interfaces. Typical considerations of the process for service providers and network operators include the following.

Preorder-related concerns:

- Integration with external (noncarrier) trading partner systems for the following business process purposes: credit bureau/credit check, to qualify customer; credit card/bank, for payment or billing arrangements; local number portability/numbering authority, if applicable; third-party verifier, if applicable;
- Integration with external (carrier) trading partner systems to obtain initial quotation of service availability/service start date; establish service availability at an address, establish equipment/supply availability; establish service availability;
- Reservation of relevant facilities, resources;
- Integration to (carrier) trading partner systems through an intermediate center to configure services provided by the partner.

Quote-related concerns:

- After the preorder confirmation of service, equipment and supply availability, customer qualification, appointments for installation, configuration and price-plan selection, the provider can present the completed quote to customer.

Customer-related concerns:

- Customer accepts quote: Provider converts quote to order and flows the associated tasks to external trading partners, which convert the reserved facilities, resources, and so forth, to be ordered;
- Customer declines quote: Provider updates external trading partners to release formerly reserved facilities, resources, and so forth.

2.3.4.2 Order Realization

Ordering processes are often spread over different organizations and may involve dozens of legacy systems or standard applications. Many offers involve nontraditional external customers. The difficulty with wireless service orders are the complexity of the services offered and the frequent modifications to the service order both before and after its provisioning. Some process difficulties to consider are

- Providing accurate information on the pricing of the product or service because of all the parties involved in the process;
- Tracking all parts of the order among different partners and establishing the status of the order;
- Providing access to relevant information to the partners;
- Preventing delays in provisioning cycles due to the lack of key business information.

2.3.4.3 Order Completion

After a service has been provided, there are processes needed to activate billing (see Section 2.3.8) and confirm the time of completion. As application, content and network service providers explore new value-chain structures, the number of service providers cooperating to provide service to an end customer will increase. In the case of several enterprises forming a closed group and using or providing a service, the charging and billing processes must integrate customer information and distribute the total charges incurred among several service providers.

- *SLA between the primary service provider and customer:* SLAs at this level should specify the terms and conditions of the service that the primary service provider is to provide and customer is to receive. In an SLA, customers and the primary service provider agree on the level of service (QoS) that the primary service provider is to maintain. Customers expect to receive that level of QoS throughout the period during which they use the services.
- *SLA between the primary service provider and a third-party service provider:* SLAs at this level should specify the terms and conditions of the charging and billing service that the primary service provider is to provide to third-party providers and content providers.

Each participating service provider needs to signal jeopardy on a provisioning or repair process to give the cooperating service provider an opportunity to react and take action to deal with any service level agreement they have in place. As the number of partners in the series increase, the KPI targets become tighter.

A primary service provider must be able to apply a combination of charging schemes. Customers can be offered a flat rate, but the costing of a service is often determined by the network technology that a customer uses to access the service. The customer may be charged at a flat rate that is only applicable to a limited usage period. Any usage that goes over a limit could be charged at a higher rate. Therefore, service providers should be able to ascertain: (1) the *features* of a service that are prone to be overused, (2) the *manner* in which they are prone to be overused, and (3) the *time* when they are prone to be overused. A primary service provider must be able to charge for services at different rates on the basis of usage. Service charges may be calculated on the basis of the following broad factors: (1) features of services offered (the parameters of services) that providers use to calculate charges, and (2) customer needs.

2.3.5 Service Fulfillment

Service fulfillment is a manual or automatic process that includes receiving customer orders, installing and configuring the service to order, provisioning the service, and invoicing for the delivered service. The scale and depth of the fulfillment process can vary widely depending upon the complexity of the service offerings and the involvement of partner providers. Fulfillment is accomplished either through a third-party provider—a fulfillment bureau (also called a fulfillment house or fulfillment service)—or internally, with purchased or self-developed solution, hardware, and in-house staff. Each in-house and external solution or bureau has its own capabilities and specialties.

In order to improve their operational efficiency, many service providers have been demanding more open, modular systems, while the OSS providers have countered with the benefits of tightly coupled systems. Wireless OSS integration will be adapted piecemeal to accommodate new services and technologies but will become more tightly integrated as applications grow more sophisticated. Tightly joining the OSS and network could lead to new SLAs for business customers and content partners. The more interesting and complicated SLAs will be between mobile providers and their content and retail partners. Many of the upcoming SLAs will not relate to performance but to provisioning. Content partners may request a commitment to enable the service within a particular time, to customize the mobile phone interface around the content provider, or to place the content in certain areas on the phone. These contracts address the look and feel of the content, how it will be accessed, and whether it requires revenue sharing.

Most wireless infrastructures are built for the consumer; that is, when customers sign up, they are provisioned directly to a network element. Partner contracts are completely different. Multiple points in the business touch this agreement, creating more of an assembly-line architecture. Wireless carriers will have to turn content contracts into information that can be provisioned within the wireless infrastructure. In order to accomplish this, the new partner SLAs should include clear and unambiguous definitions of the following:

- The measurable QoS metrics and parameters that can be guaranteed by the service provider for a specific service in terms that the customer can understand and agree to;
- A service performance measurement method, measurement period, reporting period, and reporting frequency;
- Customer and service provider responsibilities (e.g., maintaining relevant hardware and software); for good practice, the service provider should define thresholds for preventative activities to be initiated before an SLA violation occurs;
- Procedures to be invoked on violation of SLA guarantees;
- Selection of the type of reports associated with the service, specifying each report's contents, format, destination, conditions, and delivery media;
- Service definitions for each service covered by the SLA;
- Process for handling the defined boundary conditions;
- Service cover time (i.e., the limits of service provider support for different times of the day/week/month/year);
- Dispute resolution procedures;
- Customizable fulfillment parameters such as:
    - Parameters to guarantee;
    - Value ranges for the parameters.
- Address service availability, appropriate performance, throughput, and transaction time.

2.3.6 Customer Relationships

Customer relationship management (CRM) is a collection of processes and strategies used to learn more about customers' needs and behaviors. The aim is to develop stronger relationships with them. The goals of CRM are to attract and retain customers successfully and maximize profit by meeting customer expectations and needs. It brings together information about customers, sales, marketing effectiveness, and responsiveness and market trends. A complete CRM solution contains specific methodologies and software that help a company manage customer relationships in an organized way. It usually encompasses business processes in sales, marketing, and service that touch the customer. A successful CRM application helps businesses use technology and human resources to gain insight into the behaviors and values of their customers.

For example, a company might build a database about its customers that describes relationships in sufficient detail so that management, salespeople, service personnel, and even the customers themselves can access account information. The aim might be to match customer needs with product plans and offerings, remind customers of service requirements, or find what products a customer has purchased. In contrast to customer care, CRM tends to be used to deal more specifically with the integration of business operations with each other. The considerations of this process are to deal with the interface between the customer and the service provider and with how the service provider should interact with customer inquiries concerning a service and its SLA.

CRM considerations should include the following:

- Customers should be to able to report troubles or faults, request changes, and make inquiries about their services by telephone, fax, or the Internet and receive notifications in the same variety of ways.
- The service provider should provide a rapid response to contact center inquires concerning service quality levels.
- The customer care contact center should have information available on the status of any service about which the customer could inquire.
- Customer service should have sufficient information to ensure the compatibility of up-selling proposals to the end customer.

2.3.7 Assurance

Service assurance management is a process offered by service providers to support their commitments to service users and customers. This process includes warranties or guarantees in a form of measurements displaying the evidence that the provided services (collectively or individually) satisfy the service agreement requirements. The scope of the assurance management process begins after the service has been provisioned and is being delivered to the customer. It is mainly concerned with the monitoring of service quality levels and the reporting of information to the customer according to the SLA. The process should provide the ability to detect degradation in service performance, alert the service customer, and respond to performance events affecting service agreements.

In current practice, the time-to-market pressure from competition has forced wireless carriers to implement network-monitoring systems for troubleshooting. Most of these systems are not attached to the OSS because of the lengthy system-integration efforts. For instance, a provider may add a new billing system to its process chain after rolling out a new service, but this system will usually not be attached to the network until a later stage. For complex CDMA, GPRS, or 3G networks, there is a critical need for integration among the business support systems (BSS), OSS, and NMS to be able to provide a high-quality customer experience. This is because the next generation services demand that providers create methods to share information more than ever before. Without upfront integration and planning, patch or retrofit solutions could end up being very expensive and time-consuming.

Another obstacle to providing high quality services with many of today's systems is the inability to achieve more than a snapshot of network performance. Most current operations do not show how poor network performance affects individual customer service. Many providers rely on traditional NOC management tools, which do not provide information about which customers have lost service or what traffic generates the highest revenue. For known high-revenue customers, such as roaming subscribers, or mission-critical applications, such as emergency facilities that use SMS messages for dispatch, it is essential that wireless network providers closely track transactions as well as network performance. To discover if the service has been delivered and how well, providers must pull together data from multiple sources. For instance, they need to know if an entire video file was downloaded or if an m-commerce transaction was successful. These measurements are much different from measuring call completion or call blocking and require a broader set of measurements from a wider range of elements.

The network infrastructure has become complicated, with many more elements playing a part in delivering services; application components, server and system components, and network and business components are now a vital part of the service delivery chain. For high-revenue customers, network providers are adding surveillance tools. Even though many of these tools are not universally applicable, providers are recognizing the need to provide fault monitoring and tracking capability to ensure that faults are identified for their high-value services. To create a complete picture of customer service performance, many providers use a labor-intensive, manual method that they offer only to select customers. Providers' staff must access several EMSs and extract statistics that relate to specific customers and export them into a spreadsheet for a summary report.

For a service involving multiple service providers, the customer-facing service provider should be able to account for degradation of service performance resulting from other service providers or network operators. Arrangements should be made with the third parties that will enable the customer-facing service provider to take appropriate action in case of SLA violations caused by third parties. Another core challenge of monitoring performance in a wireless environment is that most next generation services rely on multiple shared resources. Service delivery depends not only on radio equipment, voice switches, and service nodes, but also on additional elements, such as content caches, e-mail, and wireless application protocol (WAP) or application servers, SIM cards, provisioning systems, and many other components required by the wireless network. Large companies that have major contracts with mobile providers will expect service guarantees. These customers will be evaluating the percentage of calls blocked and calls dropped, QoS, provisioning times, and help-desk response times. They may also demand that their executive management team receive a higher level of performance than other employees or that field personnel be guaranteed reliable service when interacting with the corporate network. However, to date, wireless service providers have made no commitments because they have not been able to stand by them. New business models will have to support commitment-based offers for high-revenue enterprise customers.

The behavior of the assurance chain is different from the value-chain of the sales, ordering, and provisioning stages. It is often unclear to the customer which sub-providers are involved. In the sales, ordering, and provisioning stages, single points of contact and project managers mostly succeed in shielding the customer from this detail. Once there is a problem with the service, it is harder to shield the customer from the assurance chain details because the restoration time targets are typically tighter than they are for fulfillment. This may result in providing the customer with inaccurate or incomplete information. The following subsections describe some considerations when dealing with external and internal collaborations.

2.3.7.1 External Collaboration Process

First-line customer service and maintenance personnel are often not highly technical. Without proper processes between the players in the value-chain, problem-resolution issues can lead to customer dissatisfaction. The service provider's internal systems and applications are built to provide knowledge and process support to operational staff. External process steps are typically not supported, not measured properly, and not reported clearly.

Trouble tickets contain the most vital information, and trouble ticketing systems are very often internally integrated with monitoring and reporting systems. There may, however, be information in the ticket that a provider does not want to share. Security, data confidentiality, and government involvement service management issues that can impact directly the exchange of information among the value-chain participants.

2.3.7.2 Internal Collaboration Process

Proactive service assurance is what most network and service providers strive for and what customers expect. The information needs to be directly integrated into the systems used by the NOC and help-desks. Real-time alarm exchange can sometimes be very valuable, if automated event correlation is used by the provider. There is, however, a considerable risk of information overflow. Notification exchanges may reduce the number of interactions and escalations.

Oftentimes, the information is not consistent and complete within current internal systems. Partner management becomes a manual and subjective process disconnected from operational experiences. Lack of clarity and aligned technical standards cause a high percentage of cases where providers do not react to an event or react incorrectly, even though they are proactively monitoring their services and are aware of "something happening here." Electronic collaboration of service management systems must be made very reliable through the use of confirmed protocols, immediate notification, fault tolerance, persistent queues, and the like. It is important for help-desk or NOC employees that someone be identified as having end-to-end responsibility for the service. Clear accountability is a very important element of solutions. Connectivity to these parties needs to be very low cost in order to be efficient.

2.3.7.3 Revenue Factors

Revenues evaporate as orders are delayed, misdirected, held, suspended, uninstalled, mispriced, or just plain lost. A revenue-impacting event is any human or system action (or interaction) in the life cycle of an ordered service—from customer inquiry to provisioning to billing to collection—that could affect revenue intake. Revenue assurance is the identification, monitoring, and management of these revenue-impacting events to maximize revenues.

Traditionally, most revenue-assurance programs have focused on billing and are analytic studies that become outdated the moment they are completed. Some studies examine the differences between third-party invoices and customer bills. In other cases, the analysis compares network element records to billing records, searching for discrepancies between usage and billing. However, billing is merely the tip of the iceberg. The service delivery chain typically has more than a dozen revenue-impacting leakage points. For example, service orders captured by an order-entry system often differ from invoiced services generated by the billing platform, resulting in under billing, late billing, or no billing at all. Temporal events such as product bundling and other discounts that reside in the order-entry system may not make their way to the billing engine. Interactions between the order-entry and provisioning systems or between the provisioning and billing systems are also candidates for leakage, each with its own set of unique revenue-impacting circumstances and events. If the life cycle of the sales and delivery process is not monitored thoroughly as part of the revenue-assurance process, it is highly likely that some leakage points in the service delivery chain will be missed.

It is insufficient to examine a few points along the billing chain, when complex operations and processes, from order entry to provisioning to collections, all play a crucial role in revenue generation and customer service. Revenue assurance must produce systematic, ongoing business and operational intelligence to high-level executives who have the responsibility and authority to manage the business.

2.3.8 Billing and Accounting

Billing services are expected to meet the requirements of the financial and legal regulations of an organization. If charges for a service are prepaid, it is essential to have real-time accounting to prevent users from overusing the service. Usage accounting should be completed within a defined time window. Dynamic credit management can be achieved if external information and internal information, such as usage and payment patterns, can be matched properly. Correlations among usage patterns, payment plans, discounts, and so on, can provide great value to the service provider in identifying customer characteristics and responding more effectively to customer inclinations.

It is highly preferable if billing date, period, statement format, and payment method can be processed in a flexible manner. It is becoming more common for complex billing models to emerge in support of new business models. Examples include the DoCoMo Internet Mode (i-mode) model for mobile content services. In these cases, the availability of billing on behalf of others (BOBO) has been the critical aspect of making the applications successful. BOBO is a form of third-party service that facilitates the establishment of relationships between parties in a value-chain.

The following are some of the main considerations for billing and accounting processes when providing a solution to manage the relationship between the service operator and its partners:

- Provide for a range of business models including revenue sharing, wholesale, sponsorship, advertising, commissioning, and others;
- Provide capabilities to define diversified methods of rating events of all types for the purpose of revenue settlements among partners in a way that will allow the operator to maximize revenues and cash-flow control capabilities;
- Provide a scalable, robust rating and usage calculation engine to support a huge range of call volumes coupled with other content and data-event volumes;
- Provide flexible, user-friendly reporting capabilities, as well as analysis tools for data querying and obtaining better business insight into partner relationships;
- Support special localization features for multiple languages and multiple currencies.

Service-based value chains are magnifying an existing issue in public networks, namely, that of fraudulent use of the network. Value chains create opportunities for service providers further up the value chain (toward the end customer) to:

- Dispute call costs incurred by their customers;
- Defraud service providers further down the value-chains by disputing bills or simply delaying them;
- Neglect to address certain types of fraud when the costs fall elsewhere.

Mostly fraud management is based upon careful analysis of accounting records (CDRs), using algorithms that detect abnormal CDR patterns and trends. The management of fraud-detection processes requires security and confidentiality services and accurate time synchronization between operators.

2.4 SLA DESIGN AND NEGOTIATION

So far, we have described the various value-chain participants, service life cycle, and characteristics of an SLA. Now we will examine the typical service life cycle of an SLA. The TMF has described the SLA life cycle [4] as follows. The TMF divides the SLA life cycle into five phases (see Figure 2.11):

- Product/service development;
- Negotiation and sales;
- Implementation;
- Execution;
- Assessment.

The product/service development process is triggered by the decision to prepare a new SLA. A number of factors can prompt such a decision, including market demand, competitive pressures, or experience with the current network performance and SLAs. Service development encompasses the analysis of customer needs, the analysis of service provider's requirements and constraints, and the preparation of standard SLA templates. In Chapter 5 we will discuss how SLA management can apply to particular technologies.

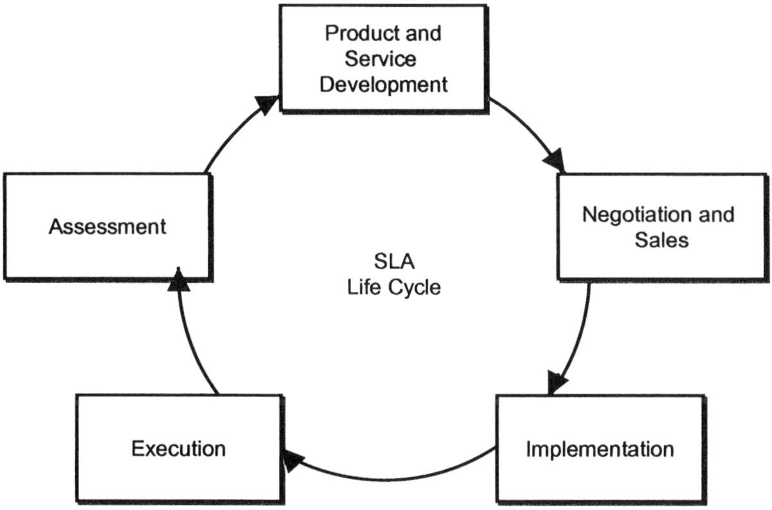

Figure 2.11 SLA life cycle.

2.4.1 Service-Level Definitions

This section will define some of the terms and views used in the development of performance specifications for service implementations.

Grade of service specifies a set of parameter values for engineering the implementation of a specific service and represents the guaranteed QoS that should be obtained for any instance of that service under all normal network conditions. Sometimes these parameter values are divided into performance sets used to market different grades of service—gold, silver, bronze, and so forth.

QoS is the result of the parameter values observed for the actual service performance over a period of time. These terms, however, have been used in a number of different contexts, which often lead to ambiguity or confusion. For example, the guaranteed values of performance parameters are grouped into so-called QoS classes when designing a network to support a range of services.

Performance-level values are calculated, measured, or projected over a period of time during the operation of the provided service. An individual parameter would have a specific performance value (e.g., a bit-error ratio). The collective values for a set of parameters would be a relative indication of the performance level for the service.

QoS parameters contained in a SLAs need to be classified, defined, and measured in a consistent way in order to avoid confusion. In a competitive service environment, customers may want to benchmark what is offered by the service providers, whether they are wireless service providers, third-party service providers, or content service providers. Subscribers generally identify two groups of performance aspects—the administrative or human-related aspects and the network or technical-related aspects. Very often, the human-related aspects (including ease of use) are most important to the user, even though they are not always recognized or measured by the service provider.

Each service design would contain a set of performance parameters for a specific instance of the service. The set of parameter values would then be a relative indication of the quality of the service (e.g., a QoS index). They can be classified as the following sets:

1. The *engineered set* of parameter values is used to design and install the service to achieve an objective performance level.
2. The *guaranteed set* of parameter values specifies the thresholds contained in the SLA between the service provider and customer.
3. The *delivered set* of parameter values is the result of actual measurements over a defined period of time.

The guaranteed parameter values will be specified in the SLA as limits of acceptable service. The engineered set of parameters will vary according to the specific service provider and resources supporting the service. Service providers normally design their engineered set parameters at a higher level than will be specified in the SLA. This provides head room and an opportunity to perform better than the specification, builds loyalty, and generally reduces account churn. In highly competitive situations and where a subprovider is required, there may be little head room. In these cases, there may be little desire to promote active measurement of the parameter values.

Four main factors contribute collectively to the overall QoS perceived by the user of a network service (see Figure 2.12). These are:

1. *Service performance:* Includes both accessibility and availability performance;
2. *Operating performance:* Refers to high quality delivery of expected results;
3. *Service integrity:* Depends on transmission (information transfer) and network performance;
4. *Support performance:* Refers to good customer service and support.

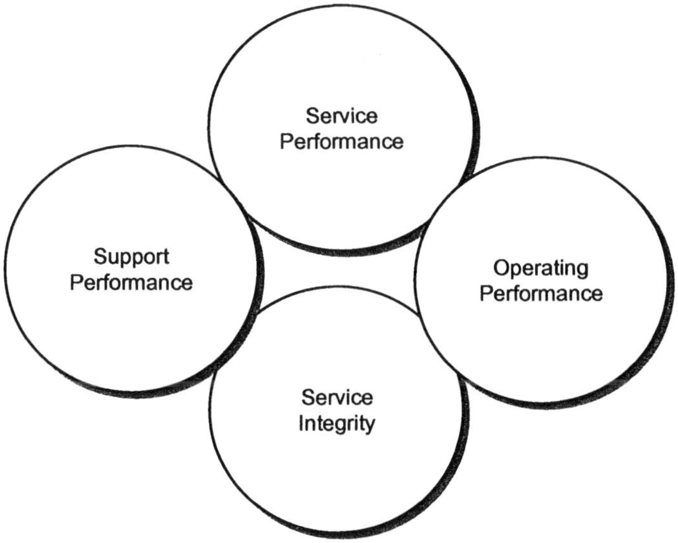

Figure 2.12 QoS factors.

It is important to distinguish between performance *events* and performance *parameters*. Events are instantaneous phenomena that occur within the network supporting the service or its environment that affects the QoS delivered. Performance parameters are derived by events over a measurement interval and provide a defined metric that can be reported.

The parameters may be time-related measures that can be expressed in terms of mean values, of ratios, which are estimators of probabilities, or of event rates or intensities contributing to reliability performance. An overarching theme of service management companies is the operator's need to combine network-performance data with mobile customer information. The service providers do not need more alarms in the NOC that isolate faults on the network, they claim, but they do need tools that match network data with customer information.

SLAs enable the service provider to offer premium services that can increase revenues. With individual SLAs, the service provider can also offer a product that is customized to each subscriber. This way, service provider's can avoid the commoditization of their services and hopefully protect and increase their revenue streams. The goal of increasing or protecting revenues must be balanced by the physical and cost constraints that can arise from network operation, customer care operation, performance measurement, or other considerations.

The SLA can be used to facilitate an effective partnership between service providers and their customers. SLA negotiations can teach the service provider about the concerns and objectives of its customer and teach the customer more about the operation of the service provider's network. This communication between provider and customer can lead to new service-level objectives that are more meaningful to the customer but are still technically and economically feasible for the service provider. It is very important to be very creative in determining these new service-level objectives.

2.4.2 Product/Service Development

Product/service development functions support service planning and development activities in forecasting, defining, and constructing a catalog of available service offerings. Service offerings are created in terms of functionality, characteristics, and the related SLA templates.

The end user only views the QoS as received. Care must be taken to correlate the measures with the "feeling" for performance observed in user terms, which are usually oriented to higher-layer protocol performance rather than to the supporting transmission network technology. The starting point for the user is the SLA specification. The SLA parameters must be easily converted into engineering terms for provisioning the service with adequate head room. Additionally, the SLA specification must be converted into terms that may be passed to the sub-providers and that may be easily verified at the service-provider-to-service-provider interface.

Application support for product/service development helps the service provider determine what service to offer, what CoSs to offer, what parameters to measure for each CoS, and what standard parameter values to guarantee for each CoS. It is also important to identify the impacts of the introduction of a new service on the operation of existing services.

The product/service development process can be started by several events: (1) both internal and external triggers that indicate it is time to develop another SLA; (2) market demand; (3) competitive pressures; (4) internal indications of service conditions; and (5) extreme experiences with current SLAs. The phase of SLA service development covers:

- Identification of customer needs;
- Identification of appropriate service characteristics:
  - What parameters?
  - What levels of service?
  - What values?
- Identification of network capabilities;
- Preparation of standard SLA templates.

Each service description should identify the relevant SLA parameters and indicate whether SLA parameter values can be selected individually or in a bundled fashion. The next step is to identify the metrics and define the baseline requirements that will measure the effectiveness of the parameters such as response time, performance, or availability, which may be covered by the SLA. For any service, the metric used to measure it should be one of the KQIs of service quality. The metric should also be realistic.

The service provider's perspective is different from the customer's. Internally at least, issues related to revenue generation and continuity, differentiated services, and network maintenance are high on their agendas. These issues might figure in the internal SLAs between departments or between service providers.

Some services may contain both technology-specific and service-specific parameters, while some may contain only one or the other. Two examples follow to illustrate:

1. *Technology-specific QoS parameters* are those related to network technology supporting the service, particularly where the service offered is a network bearer service. Some of the technology-specific parameters may not be relevant to a service end user but need to be considered internally by the service provider or between service providers. At the bottom layer, from the switches and cell towers, network providers are dealing with fraud, network capacity, and QoS issues such as dropped calls. At the second level, network providers are running heterogeneous networks, having started with TDMA or CDMA, and layering GPRS. Here, traffic management and performance become difficult. The third level is where wireless providers become part of the global network and have to deal with interoperability, interconnection, and termination traffic. Each layer has a different set of concerns and uses different signaling interfaces.
2. *Service-specific QoS parameters* are those typically related to the application carried by the network and service-specific or application-specific technology parameters such as reliability and availability of computer servers, databases, and so forth.

The key to this step is finding quantifiable factors that are easily measured and analyzed. This can be very difficult, especially when dealing with network performance from a third-party provider. The primary providers may not have any control over many variables and environmental factors that affect performance, availability, and ultimately the success of their SLAs. Therefore, these variables should be carefully addressed in the interprovider SLA. If the primary provider's method of measuring the service fails to take these factors into account, it may have difficulty conforming to the SLA provisions.

2.4.3 Negotiation and Sales

In any value chain comprising a number of trading partners, the main challenge is to establish an effective end-to-end set of processes that deliver a seamless service that is indistinguishable from the same service as provided by a single supplier. The catalog of service offerings is used during the negotiation and sales process to negotiate service options, CoSs, and potentially the values of SLA parameters with the customer. During sales and ordering, the service provider captures the customer-specific information for a particular service offering and verifies the order's feasibility. This requires the verification of available network resources and the capability to meet and assure the requested level of service. In a general negotiation and sales process, the following phases are used as consideration:

- Selection of the values of SLA parameters applicable to a specific service instance;
- Costs incurred by the customer when signing the SLA;
- Costs incurred by the service provider when an SLA violation occurs;
- Definition of reports associated with the service (note that the time when a report may be generated is dependent on the periods related to the relevant SLA parameters, e.g., availability over a unit of time, such as one day, week, month, quarter, or year).

Measurement of QoS by a service provider at the network level may not be the same as user perception of QoS. This leads to the question, is QoS relative or absolute? In fact, there are various types of performance metrics whose performance targets can be specified for a telecommunications service. These different types include:

- Customer care response metrics (e.g., time to answer a customer call, time until a trouble ticket is escalated up to management);
- Operational metrics (e.g., mean time to repair);
- Network metrics for landline services (e.g., availability, blocked calls, delay, and loss);
- Additional metrics for wireless services (e.g., abnormally dropped calls).

As wireless service becomes more mission critical within businesses and content services begin generating revenue, these customers and partners will demand SLAs that guarantee QoS and exceptional performance. High-end business customers will need more granular performance details and cross-monitoring. These SLAs may surround completion time, security, connectivity, and transaction success rates. Finally, providers should use their OSS to help them decrease operating and capital expenses and increase revenue. If applied intelligently, the network provider's OSS can help it to build its networks cost-effectively by planning for capacity. The goal should be to create new services or new profit margins, while incurring only marginal additional expense.

2.5 SLA IMPLEMENTATION AND EXECUTION

One challenge for service providers in SLA implementation is to identify customer-sensitive measures and particularly the values that satisfy the customer. These may vary from one target group (market segment) to another. Armed with this knowledge, the providers in the value-chain can take measures to maximize customer values and expand business opportunities.

2.5.1 Implementation

Once an order is confirmed, the service provider must provision the instance of the requested service and request or reserve the required capacity in the network. This may involve deployment of new network or service resources or just the configuration of existing equipment. These network resources need to be configured to support the required levels of service quality specified in the SLA.

Service implementation is the phase where the service is enabled and the individual customer instance is put into production. For this analysis, the implementation phase is considered to include network provisioning, which may be placed into another phase by some companies. This phase will always include the order and instantiation of the individual customer service per the contract. There are three aspects to service implementation:

- Configuration of the network to support the service in general (network provisioning);
- Configuration of the network to enable a specific instance of a service according to the SLA for a customer (service configuration);
- Service activation.

Once requirements, metrics, and incentives are defined and in place, monitoring capabilities need to be implemented to ensure SLA compliance. Service assurance OSS should monitor the network and application performance from both the end-user and internal operations perspectives at all layers of the open system interconnection (OSI) model. It is desirable that the service assurance OSS provide a scalable, Web-based monitoring capability that provides scheduled, on-demand, or real-time reports of service-level compliance for associated service resources and end-to-end application performance.

Depending upon the actual service implementation, there may not be critical need for third-party products to monitor SLA compliance. The provider may be able to gather the information necessary for analysis easily in a simple Excel spread sheet or database. For example, recording the initial request date and completion date for a new employee setup request can easily be analyzed for SLA compliance. Additionally, if an online request-tracking system is deployed, the measurement process may be as simple as running a database query.

2.5.2 Execution

The network data management process is supported by a number of processes, each one collecting different types of raw performance data from different sources. For example, the network performance data collector collects raw network performance data from network elements or EMS systems. The process performance data collector collects raw operational process performance data from systems such as trouble ticketing and provisioning. The application and server performance data collector collects raw service performance data from other applications.

Raw process data and service configuration details are aggregated to provide service-level quality metrics from the technical, service, and process performance data. QoS analysis and reporting and SLA proactive monitoring use the service performance data to compare service performance against the set of operating targets and SLA guarantees.

The customer SLA reporting function correlates QoS data with customer service descriptive data and SLA thresholds in order to produce SLA reports detailing service performance and quality. The execution phase covers all normal operations of the service covered by an SLA:

- Normal in-service execution and monitoring;
- Real-time reporting and service quality validation;
- Real-time SLA violation handling.

The goal is to find a way to monitor effectively and efficiently the metrics defined in an SLA. If the increased revenue or performance gained by the SLA is completely offset by additional costs incurred to monitor compliance, then the SLA is not doing the company any good. Depending upon the terms of the SLA, QoS violations may result in a variety of actions, including compensation, future discounts, rebates, alternate service routing, and so forth. Customers do not really want refunds via SLAs; they want quality assurance for the services provided. Nevertheless, in a commercially competitive environment, customers will expect rebates if they are provided for in the contract. The conditions under which these occur are a commercial issue between the service provider and the customer.

QoS violation may also be customer induced, such as a violation of the contracted traffic profile parameters. SLA monitors should include warnings for the service provider that violations of contracted performance have occurred.

These levels are set according to the individual service provider's network capability and traffic engineering management strategy. Some network providers choose to overprovision network resources and accept underutilization to avoid congestion. Others choose to operate the network close to capacity limits and, therefore, need to watch network behavior more closely. Not all SLAs described in this subsection are external. Internal SLAs can be used as a tool to help improve customer and employee satisfaction. Providing clearly defined expectations and a measurable metric to evaluate compliance is a key step in establishing an effective agreement.

Because of the rapid pace of technology development, response times and other user expectations change almost monthly. Therefore, all SLAs should be reviewed at least annually. SLAs must be updated periodically to reflect these changes. Without this step, SLAs will quickly become useless and out of date.

2.5.3 Assessment

Assessment takes place in two time frames. The first is scheduled during a single customer SLA contract period, where the assessment is related to the customer's QoS. The second time frame is related to the service provider's overall quality goals, objectives, and risk management. These two assessment and review activities have differing uses within the service provider:

- Customer SLA periodic review elements:
    - Quality of customer service;
    - Satisfaction of customer with service quality;
    - Improvement potential;
    - Changing customer requirements.
- Internal SLA business review elements:
    - Overall service quality across all customers;
    - Realignment of service goals;
    - Realignment of service operations;
    - Identification of service support problems;
    - Creation of different SLA levels.

A successful QoS assessment process can be rewarded in many ways. For the service provider, the benefits of a complete, well-designed, and carefully executed process can include:

- Faster introduction of new services;
- Increased number of services in (a manageable) portfolio;
- Reduced service development costs;
- Reduced system integration costs;

- Reduced service operations costs;
- Manageable service levels;
- Simplified mechanisms for service provisioning and information sharing with business partners like network and content providers;
- New revenue streams from niche services.

By the same token, the benefits for the service subscribers can include the following abilities:

- To select and order the services offered by the provider domain;
- To manage the necessary information about services and contractual relationships between the domains;
- To access and configure a wireless service according to their wishes (customization of end-user preferences).

2.5.3.1 Format and Content for Reports

Types of Reports

Performance reporting covers two types of reports:

- QoS/SLA reports;
- Traffic data/utilization reports.

The QoS report provides overall assessment of service quality according to the SLA parameters. The traffic data report provides usage measurement information for the service. Information is provided in the following format using text, graphs, or tables:

- Summary information;
- Trends;
- Exceptions;
- Comparisons to agreed-upon limits;
- Snapshots.

Common Report Header

All reports should have a common report header providing the following information:

- Customer information (customer name, customer ID, customer contact name, customer contact information);

- Service provider information (service provider name, service provider contact name, service provider contact information);
- Service information (service identifier, service type code, service descriptions ID, service profile descriptions, service access points);
- Report information (report type, reporting period, boundary conditions, suspect flag);
- SLA information (in case of a QoS report, contracted/guaranteed SLA value).

2.6 SLA DEVELOPMENT AND MANAGEMENT FLOWS

Even though the customer versus service provider versus third-party provider relationships can be very complex, the information flows are similar. Figure 2.13 depicts the essential information flows for creating and supporting an SLA as part of a service provisioning agreement. Information flows that are not affected or not changed by standard agreement management flows [8] have been omitted to improve readability.

Fulfillment: Steps 1–11

1. Service request information from a customer (or a user) for a new service agreement or an update to an existing agreement. Service agreement requirements are sent to sales. If the required service capabilities are standard offers or arrangements for the serving service provider, sales OSS defines, negotiates, and updates tariffs between the service provider and serving service provider, resulting in a service order.
2. Service request information sent to order handling OSS. Order handling maintains the internal workflow of the serving service provider required to assign and configure all needed resources to support the new or updated service order.
3. Sales OSS passes new or changed service-requirements information to the service planning and development OSS for analysis and business-case decision. The order is reviewed to determine if:
   - Existing serving service provider service capabilities are not sufficient.
   - It is a new service relationship for the service provider.
   - There are new requirements not currently offered by the serving service provider.

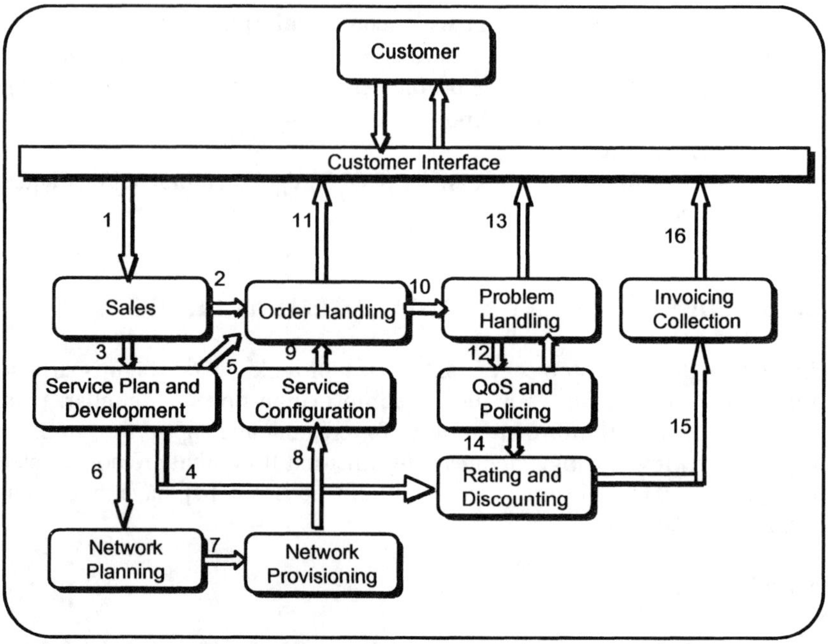

Figure 2.13 Service management information flow. (*From:* [8]. © 2005 TeleManagement Forum. Reprinted with permission.)

If appropriate, service planning and development systems can adjust future design, development, deployment, and implementation based on the information. Service planning and development involves or passes on the following steps.

4. New service or option information is passed to rating and discounting OSS for development or update.
5. Service changes or new service option information is passed to order handling OSS for development or update of order handling OSS and systems.
6. Network requirements are sent to network planning and development OSS, if the implementation of the new or updated service requires network changes. There may be problem handling OSS or customer QoS OSS requirements to support the new or updated service or service options. Interaction is required from service development for almost all other OSS functions to assess and support the impacts of new requirements. An appropriate OSS needs to be involved with development to receive updates to support the new service or option.

7. Network configuration OSS change information required by the new or updated service requirements is passed to the network provisioning OSS.
8. The network provisioning OSS communicates to the service configuration OSS assignment and configuration readiness information. This can be as simple as a test and turn-up or a complex network validation test.
9. Service configuration OSS reports completion for the network and service configuration to order handling OSS. Order handling assigns the most suitable service profile and assigns the tariff plan and passes information to rating and discounting OSS.
10. Order handling registers the SLA terms for the new service relationship and provides them to problem handling and customer QoS OSS.
11. Finally, order handling reports completion of the establishment or update of the service relationship to the customer.

Assurance: Steps 12–15

12. Customer QoS management compiles service performance reports negotiated in the service agreement and provides them to the customer through the customer-facing systems.
13. If customer QoS management detects a violation of a negotiated SLA, it provides violation information to the problem handling OSS.
14. Information about trouble occurring during service execution will be exchanged in notifications and reports.
15. Detection information for SLA violations is reported to the rating and discounting OSS.

Billing: Steps 16–17

16. Rating and discounting generates billing records and sends them to the invoicing and collections OSS.
17. These billing records are collected in a service accounting file. Invoicing and collections processes records into invoices that are provided to the customer.

2.7 CONCLUSION

An SLA itself cannot bring about service assurance. When new technology is being installed and processes need to change, service providers have to set up that new business feature. Throughout this book, we will be looking at various aspects of service assurance. The principle of this chapter has been to give an overview of SLA and the service provider's intimated engagement in the scope, decision, and process of service offers in the value-chain life cycle. Only through a fair level of involvement can a service provider have any realistic chance of measuring return from the increasing investments made in new technologies and services. By now, most service providers have an established service-level management in place for their internal, external, and third-party providers. But how this is actually done for different technologies makes for a fascinating comparison, with different returns for different applications.

References

[1] *Technical Report Value-Chain Issues Facing the ICT Industry—TR 128,* TeleManagement Forum, November 2002.

[2] *SLA Management Handbook, Vol. 2, Concepts and Principles—GB 917,* TeleManagement Forum, April 2004.

[3] "Service Level Management for Wireless IP Interface Implementation Specification—TMF 837," TeleManagement Forum, April 2002.

[4] *SLA Management Handbook, Vol. 1, Executive Overview—GB 917-1,* TeleManagement Forum, July 2004.

[5] "Enhanced Telecom Operation Enhanced Operations Map (eTOM)—The Business Process Framework for the Information and Communications Services Industry, Addendum F: Process Flow Examples Release 4.0—GB 921F," TeleManagement Forum, March 2004.

[6] "Enhanced Telecom Operation Enhanced Operations Map (eTOM)—The Business Process Framework for the Information and Communications Services Industry, Release 3.0—GB 921," TeleManagement Forum, June 2002.

[7] Conor, R., et al., "White Paper: Federated Accounting Management of Service Usage in a Business-to-Business Environment," *FORM Consortium,* August 2002.

[8] "TOM Application Note: Mobile Services: Performance Management and Mobile Network Fraud and Roaming Agreement Management—GB 910B," TeleManagement Forum, September 2000.

Chapter 3

WiFi and 3G Network Technologies

The previous two chapters described how wireless service providers can add value to their services by providing end-to-end service levels to their customers. In general, the wireless services of interest here are the basic voice and data services and, in particular, VoIP using WiFi and 3G networks. These VoIP services use the same underlying network infrastructure as used by the wireless data network, plus added wireless-VoIP-specific components. The wireless data network itself is an overlay network over a wireless circuit-switched network.

This chapter reviews the basic radio technologies of WiFi or wireless LAN (WLAN) and 3G mobile networks. Here the focus is on networking and basic voice and data services capabilities and their architectures, using both standalone WiFi, and 3G and in integrated WiFi and 3G network environments. We briefly describe basic circuit-switched voice service and an example of data service WAP Internet access over GPRS. We also describe integrated services between WiFi hotspots and 3G networks and consider user roaming and terminal mobility.

This chapter provides the necessary background on WiFi and 3G networking and their interworking, network domains, network elements, and roaming, which will help explain further details in building service models and voice over WiFi and 3G services networking and architectures. In Chapter 5, the WAP over GPRS service is used to explain the fundamentals of building service models. Chapter 6 describes VoWiFi/3G services networking and architectures built on the WiFi and 3G networking architectures.

3.1 INTRODUCTION

The term *WiFi* stands for "wireless fidelity" and is proclaimed by the WiFi Alliance. The WiFi Alliance is a nonprofit international association that certifies interoperability of products based on IEEE 802.11 WLAN specifications. Initially, WiFi was a trademark, but it is now used as a general term for WLANs. Also, initially it referred to 802.11b networks, but now it refers to any 802.11 network, whether 802.11a, 802.11b, 802.11g, or dual-band. WiFi networks are also called hotspots. In this book, thc terms 802.11b, WLAN, and WiFi are used interchangeably.

The Third Generation Partnership Program (3GPP) develops 3G standards for General Systems for Mobile Communications (GSM)–based systems in a European Telecommunications Standards Institute (ETSI)–related program for replacement of GSM systems. It develops specifications that are then approved as ETSI standards. The 3GPP specifies and maintains GSM, GPRS, and various releases of UMTS wideband code division multiple access (WCDMA) standards. After the 3GPP was established in Europe, the 3GPP2 was set up as counterpart of the WDCMA camp supporting the American National Standards Institute/Telecommunications/Industry Association/Electronic Industry Alliance (ANSI/TIA/EIA) Interim Standard (IS)–41 and IS-95–based CDMA2000 specifications. The 3G mobile networks considered here are based on standards specified by 3GPP or 3GPP2 organizations. Mobile networks based on both these technologies are now widely deployed, and their user base is still growing.

The recent extensive growth of deployed WiFi hotspots worldwide has led to a demand to integrate them with 3G mobile networks. The goal of this integration is to develop integrated mobile data network services, capable of supporting data services with high data rates when users enter hotspot coverage and with lower data rates outside hotspot coverage by 3G public networks.

As WiFi deployment grows, wireless operators and enterprises are looking into opportunities to leverage their wireless infrastructures. The main opportunity is in integration of WiFi hotspots and 3GPP networks for both data and voice services over the same network support infrastructure. Such integration also helps provide more services, such as presence and location-based services.

3.2 WIFI NETWORKING

In this section, we describe WiFi standards and technologies, networking topologies, systems architectures in enterprise and public hotspot environments, and roaming and mobility.

3.2.1 WiFi Standards and Technologies

The IEEE 802 committee sets standards for local area networks (LANs) and metropolitan area networks (MANs). For WLAN or WiFi, the IEEE 802.11 committee has defined several standards that include 802.11a in the 5-GHz band and 802.11b and 802.11g in the 2.4-GHz band.

802.11b (also called WiFi) is the clear winner in public hotspot and enterprise WiFi and the home networking space. Since WiFi uses the license-free 2.4-GHz industrial, scientific, and medical (ISM) band, it also has some significant disadvantages.

One of the more significant disadvantages of the WiFi environment is that its frequency band is crowded and subject to interference from other networking technologies. Other disadvantages include lack of interoperability with voice devices, and no QoS provisions for multimedia content.

Both 802.11a and 802.11b were approved in 1999; however, lower-cost 802.11b products became available earlier and gained widespread acceptance. With the approval of the 802.11g standard in 2003, the use of 802.11g is now growing and that of 802.11b is declining. In the near future, even the use of 802.11g may be overtaken by the newer 802.11n. Although our focus in this book is on 802.11b, the approach proposed here is applicable to all other 802.11 networks as well. While the 802.11a, b, g, and n standards provide RF and media access control (MAC) layer specifications, the 802.11i, e, and f standards deal with only MAC-layer features. Brief descriptions of the various 802.11 systems follow.

3.2.1.1 802.11b Systems

This is the prominent 2.4-GHz WiFi standard. The 802.11b radio interface is mainly specified for the office environment, where distances and the number of users are small. A brief survey of 802.11b systems follows.

- The 802.11b uses spread-spectrum transmission operating in the ISM band of 2.4–2.4835 GHz. The physical (PHY) layer can use both frequency-hopping spread spectrum (FHSS) and direct sequence spread spectrum (DSSS) transmission techniques. For higher speeds mostly DSSS transmission is used.
- 802.11 networks are carrier sense multiple access/collision avoidance (CSMA/CA) networks that employ the request to send/clear to send (RTS/CTS) protocol for RF transmission.
- The 802.11b standard has defined 14 channels, but in the United States and Canada, only 1 to 11 channels are used. Among these, three nonoverlapping channels used are 1, 6, and 11. In the EU (except Spain and France), 1 to 13 channels are used, among these, three nonoverlapping channels used are 1, 6, and 12. In France, only 10 to 13 channels are used. In Spain, only 10 to 11 channels are used. In Japan, all 1 to 14 channels are used.
- Transmit power regulations are as follows: While most APs are 100 mW (20 dBm), antennas vary. Federal Communications Commission (FCC) part 15 requirements allow up to 1W with 6-dBi antenna (36 dBm) total power.
- Data rates and modulation/coding schemes are:
  - 1 Mbps: Differential binary phase shift keying (DBPSK).
  - 2 Mbps: Differential quadrature phase shift keying (DQPSK).

  - 5.5 Mbps: Complementary code keying (CCK).
  - 11 Mbps: CCK.
- Sensitivity is BER of 1E-5.
- Preamble and header are always carried at 1 Mbps.

3.2.1.2 802.11g Systems

The WiFi standard 802.11g is an extension of 802.11b. It broadens 802.11b's data rates to 54 Mbps within the 2.4-GHz band using orthogonal frequency division multiplexing (OFDM) technology. Because of backward compatibility, an 802.11b radio card will interface directly with an 802.11g (and vice versa) at 11 Mbps or lower, depending on range. The newer 802.11b APs are upgradeable to 802.11g compliance via firmware upgrade.

The 802.11g range at 54 Mbps is less than for the existing 802.11b APs operating at 11 Mbps. As a result, upgrading existing 802.11b APs that currently provide 11 Mbps throughout in all areas will require moving the 802.11g APs closer together and including additional ones to accommodate higher data rates.

Similar to 802.11b, 802.11g operates in the 2.4-GHz band, and the transmitted signal uses about 30-MHz, which is one-third of the band. This limits the number of nonoverlapping 802.11g channels to three, which is the same as 802.11b.

The service assurance issues with both 802.11g and 802.11b remain the same. To eliminate interference, 802.11g devices also have an RTS/CTS mechanism that acts as a traffic cop between 802.11b and 802.11g transmissions, a process that slows overall data transfer.

3.2.1.3 802.11a Systems

The 802.11a standard and FCC spectrum regulatory status is now firmly in place. A big difference with 802.11a is that it operates in the 5-GHz frequency band with 12 separate, nonoverlapping channels. As a result, one can have up to 12 APs set to different channels in the same area without their interfering with one other. This makes AP channel assignment much easier and significantly increases the throughput the WLAN can deliver within a given area. In addition, RF interference for 802.11a systems, as compared to 802.11b and 802.11g, is much less likely because of the less-crowded 5-GHz band and availability of 12 nonoverlapping channels.

Similar to 802.11g, 802.11a delivers up to 54 Mbps, with extensions to even higher data rates possible by combining channels. Due to higher frequency, however, range (around 80 feet) is less than lower-frequency systems (that is, 802.11b and 802.11g). This increases the cost of the overall system because it needs a greater number of APs, but the shorter range enables a much greater capacity in smaller areas due to the higher degree of channel reuse.

One problem with 802.11a is that it is not compatible with 802.11b or 802.11g networks. In other words, a user equipped with 802.11b or 802.11g radio cards will not be able to interface directly with an 802.11a AP. To address this interoperability issue, multimode network interface cards (NICs) are becoming the norm. These multimode card functions are now built in to the newer laptops.

3.2.1.4 802.11n Systems

As use of WiFi has grown more popular, users have been looking for better coverage, which is currently limited in practice to between 30 and 50 feet, and higher throughputs. The 802.11n standard is specifying the next generation of 802.11 networks with a target throughput of at least 108 Mbps. These multiple input, multiple output (MIMO)–based products are backward compatible with older 802.11b and 802.11g equipment. In these systems, increased data throughput is achieved by use of additional transmitter and receiver antennas, spatial multiplexing, and coding schemes, and increased range is achieved by exploiting the spatial diversity.

Currently, there are two variants of 802.11n under consideration. These are WWiSE and TGn Sync, and a final version is expected to be completed by the IEEE by the end of 2006 or early 2007. The WWiSE proposal supports both 20-MHZ and 40-MHz channels, operating in the 5-GHz band and with four MIMO antennas, whereas the TGn Sync proposal supports 10-MHz, 20-MHz, and 40-MHz channels operating in the 5-GHz band, and optionally in the 2.4-GHz band with two or four antennas. For backward compatibility with 802.11b, the 2.4-GHz band may be available as an option in the final version. This new 802.11n will meet the growing need for more data-intensive applications, as well as aggregating traffic from multiple APs or cells together. Although 802.11n is not yet approved, the prestandards products, for example, Belkin's Pre-N wireless router, claiming 800% wider coverage and 600% faster speed are already on the market and are attracting user attention.

3.2.2 Some Selected WiFi-Related Standards

A brief summary of some selected WiFi-related standards and their distinguishing features of interest from a service assurance perspective follow.

3.2.2.1 QoS Improvements: 802.11e

The 802.11e standard specifies QoS mechanisms to support all 802.11 PHY interfaces, streaming traffic, and audiovisual (AV) applications. Whereas the base

802.11 standard treated all traffic classes on a first-come, first-served basis with best-effort traffic delivery, the 802.11e extension to the standard includes support for prioritized access to different classes of data traffic, as well as admission-control mechanisms to regulate access to the wireless medium. This will improve performance for applications such as VoIP and multimedia services over WiFi.

An understanding of 802.11e is important for service assurance of WiFi networks. The 802.11e specifications provide two new mechanisms for resolving contention, enabling QoS. These are enhanced distributed channel access (EDCA) and hybrid controlled channel access (HCCA).

EDCA improves on distributed coordination function (DCF), the default media access protocol used in 802.11a, b, and g, by giving higher priority traffic an advantage during contention. Using EDCA, an AP can support more simultaneous VoIP calls than DCF, for a given quality. However, EDCA does not give deterministic QoS for every application.

HCCA uses a hybrid mechanism, merging the features of DCF and a polling mechanism to avoid contention with DCF and EDCA nodes, and it can preempt the network for QoS traffic.

3.2.2.2 Inter-AP Roaming: 802.11f

The 802.11f standard specifies the Inter Access Point Protocol (IAPP) for roaming compatibility across APs from different vendors.

3.2.2.3 Security Framework and Improvements

Recently, several developments in security framework and standards that increase WiFi security have taken place. A brief summary, of particular interest for the security conscious enterprise user, follows.

Security Framework: 802.1x

The 802.1x standard is a comprehensive security framework for all IEEE 802 LANs, including wireless. It includes authentication [Extensible Authentication Protocol (EAP) and RADIUS] and key management. Note: It is not part of the 802.11 standards.

Security Enhancements: 802.11i

The 802.11i standard deals with the wireless-specific security functions that work with 802.1x. It defines strong authentication and access control mechanisms, leveraging RADIUS and 802.1x. The standard also defines 802.11 key management using 802.1x and support for stronger encryption and data confidentiality, using the Temporal Key Integrity Protocol (TKIP) and Advanced

Encryption Standard (AES), as well as stronger message integrity checking. The AES provides enough security to meet requirements for the Federal Information Processing Standards (FIPS) 140-2 specification described in the following section. 802.11i will make 802.11 wireless networks more secure and is expected to lead to broader adoption in enterprise settings.

Federal Information Processing Standards

The National Institute of Standards and Technology (NIST) has published FIPS 140-2, which defines the U.S. government's security requirements that must be satisfied by a cryptographic module used in a security system protecting sensitive, but unclassified (SBU) information within IT systems. It has also been adopted by the Canadian government's Communication Security Establishment (CSE) and is likely to be adopted by the financial community.

The security requirements cover areas related to the secure design and implementation of a cryptographic module. These areas include basic design and documentation, module interfaces, authorized roles and services, physical security, software security, operating system security, key management, cryptographic algorithms, electromagnetic interference/electromagnetic compatibility (EMI/EMC), and self-testing.

FIPS defines 11 categories of security requirements and identifies four levels of security, from Level 1 (lowest) to Level 4 (highest) for each category. These levels are intended to cover the wide range of potential applications and environments in which cryptographic modules may be deployed. The U.S. and Canadian governments usually call for FIPS 140 Level 1 or Level 2 certification.

Although FIPS 140-2 is not part of the 802.11 standards, the IEEE 802.11i standards' language is being considered for revision to address FIPS 140 cryptographic modules. Now several vendors are offering FIPS 140–compliant WiFi products.

3.2.2.4 Network Management Enhancements: 802.11k

The 802.11k standard specifies enhancements to the current 802.11 standards, to provide radio resource measurement (RRM) of WLANs that will allow uniform measurement of radio information across different manufacturer platforms. By having standardized, repeatable measurements, system designers can use radio environment information to make better decisions about frequency use and transmission power levels. This will lead to WiFi networks that are easier to monitor and manage and that can make more efficient use of the available spectrum, thus help in WiFi service assurance.

3.2.2.5 Roaming Enhancements: 802.11r

The 802.11r standard specifies enhancements to provide fast hand-over capabilities to support roaming among WiFi networks. The 802.11r Task Group is working on reducing the handoff latency when client devices transition between APs or cells in an extended service set (ESS), which includes APs in the same network. Faster handoffs will be critical to meeting the real-time requirements of delay-sensitive applications, such as voice, especially in mobile settings where client devices can be expected to roam frequently. This standard will simplify the deployment of Session Initiation Protocol (SIP)–based VoWiFi portable phones.

3.2.2.6 Mesh Networking Support: 802.11s

The 802.11s Task Group is working on an infrastructure mesh standard for meshed WLANs (MWLANs). This will allow WiFi APs or cells from multiple manufacturers to self-configure into multihop wireless topologies. Example of use scenarios for mesh networks include interconnectivity for devices in the digital home, unwired campuses, and community-area networks, sometimes called hotzones.

3.2.2.7 Regulatory Enhancements: 802.11d, h, and j

These standards extend the PHY and MAC layers to allow 802.11 to operate in the regulatory domains of other countries, which in the 5-GHz band vary from country to country. The International Telecommunication Union (ITU) recommended a harmonized set of rules to allow unlicensed transmitters in this band to coexist with primary-use devices, such as military radar systems in Europe.

- *802.11d:* The 802.11d specification is suited for systems that want to provide global roaming. It will allow APs to communicate information on the allowed radio channels with acceptable power levels for user devices. The 802.11 standards cannot legally operate in some countries; the purpose of 802.11d is to add features and restrictions to allow WiFi to operate within the rules of these countries.
- *802.11h:* The 802.11h standard extends PHY and MAC specifications to make 802.11 consistent with the 5-GHz standard in Europe. It includes two techniques of dynamic frequency selection (DFS), and transmit power control (TPC), which in the 5-GHz band mitigates interference by sharing spectrum among current users and avoids interference between adjacent WiFi BSSs by spreading usage across the band in a uniform manner. These rules allow unlicensed transmitters to employ new versions of "listen before talk" to adjust the transmit power and

intelligently select the operating channel to use the available spectrum more efficiently and to avoid causing harmful interference. It is expected that 802.11h may replace HiperLAN/2, the European wireless standard developed by ESTI, for 54-Mbps data rates operating at 5 GHz.

- *802.11j:* 802.11j extended the MAC and PHY layers to allow for operation in the 4.9-GHz and 5-GHz bands in Japan. The operation of 802.11j in the 4.9-GHz band is unrelated to the 4.9-GHz public safety spectrum allocated in the United States.

3.2.3 WiFi and Ethernet: 802.11 and 802.3

In an enterprise environment, WiFi networks are adjunct networks to their wired LANs. Although both WiFi networks and Ethernet use the same logical link control (LLC) and there is no difference for upper-layer protocols, WiFi networks are not Ethernets. Key differences between these two are:

- The 802.3 (Ethernet) uses CSMA/CD (Collision Detect) scheme.
- The 802.11 (WiFi) uses (CSMA/CA) and DCF and RTS/CTS methods.
- If two clients sense that the channel is idle at the same time, then they send at the same time, and collision can be detected at the sender in Ethernet. However, this is not possible in 802.11 WiFi because WiFi allows either transmit (TX) or receive (RX) (that is, no simultaneous RX/TX).
- Use of wireless media in WiFi, instead of broadcast cables in Ethernet, introduces problems of power management, frequency management, security, bandwidth or capacity, and radio interference.

The above-mentioned problems impact service assurance and are covered in Section 3.2.7.

3.2.4 WiMax: 802.16 and 802.20

The Worldwide Interoperability for Microwave Access (WiMax) Forum was formed to popularize products based on the IEEE 802.16 wireless metropolitan area network (WMAN) standard. These WMANs have much longer range than WLANs, as much as several kilometers and more, as opposed to the nominal 100m of WLANs.

The IEEE 802.16 Working Group has defined 802.16e for mobile WiMax, which is targeted for use in metropolitan and regional networks. These networks offer support for roaming and support applications requiring both low-latency data and real-time voice services. 802.16e has a target shared-data rate of about 70 Mbps, and it will operate in the 2–6-GHz licensed band, with typical channel bandwidths ranging from 1.5 to 20 MHz. Typical client devices will rely on

Personal Computer Memory Card International Association (PCMCIA)–based PC card technology.

Another competing standard in the works is the IEEE 802.20 mobile broadband wireless access standard. It is expected to support data rates of greater than 100 Mbps at ranges of 15 km or more. 802.20 will operate in licensed bands below 3.5 GHz (500 MHz to 3.5 GHz). It will incorporate global mobility and roaming support between base stations. The standard will support real-time traffic with the low latency of about 20 ms or less required for VoIP. Typical channel bandwidth will be less than 5 MHz.

WiMax networks are not extensions of WiFi networks; they differ in many aspects. The WiFi network is a LAN designed to be used indoors at close ranges to distribute Internet access to a few computers at hotspots, the office, or home. The WiMax network is a wireless replacement for a wired broadband connection. WiMax uses highly directional antennas, whereas WiFi uses omni-directional antennas. These two standards also use different approaches for QoS and security. WiMax 802.16d is suitable for fixed point to multipoint networks, and WiMax 802.16e is suitable for mobile both peer-to-peer and ad hoc networks. Our interest in WiMax networks is their use as back-haul networks for WiFi hotspots.

3.2.5 WiFi Networking Topologies

Figure 3.1 depicts the WiFi networking modes in common use and supported by 802.11 MAC layer implementations.

3.2.5.1 Ad Hoc or Peer-to-Peer

The ad hoc [also referred to as independent basic service set (IBSS) topology] mode is a peer-to-peer network [Figure 3.1(a)] in which no dedicated system is required to assume the role of a gateway router. Several wireless nodes will communicate directly with one another in a mesh or partial-mesh topology. Typical instances of such an ad hoc implementation do not connect to a larger network and cover only a limited area. If a client in an ad hoc network wishes to communicate outside the peer-to-peer cell, a member must operate as a gateway and perform routing. Configuring a WiFi network in ad hoc mode establishes a network where wireless infrastructure does not exist or where long-term services are not required, such as a trade show or collaboration by coworkers at a remote location.

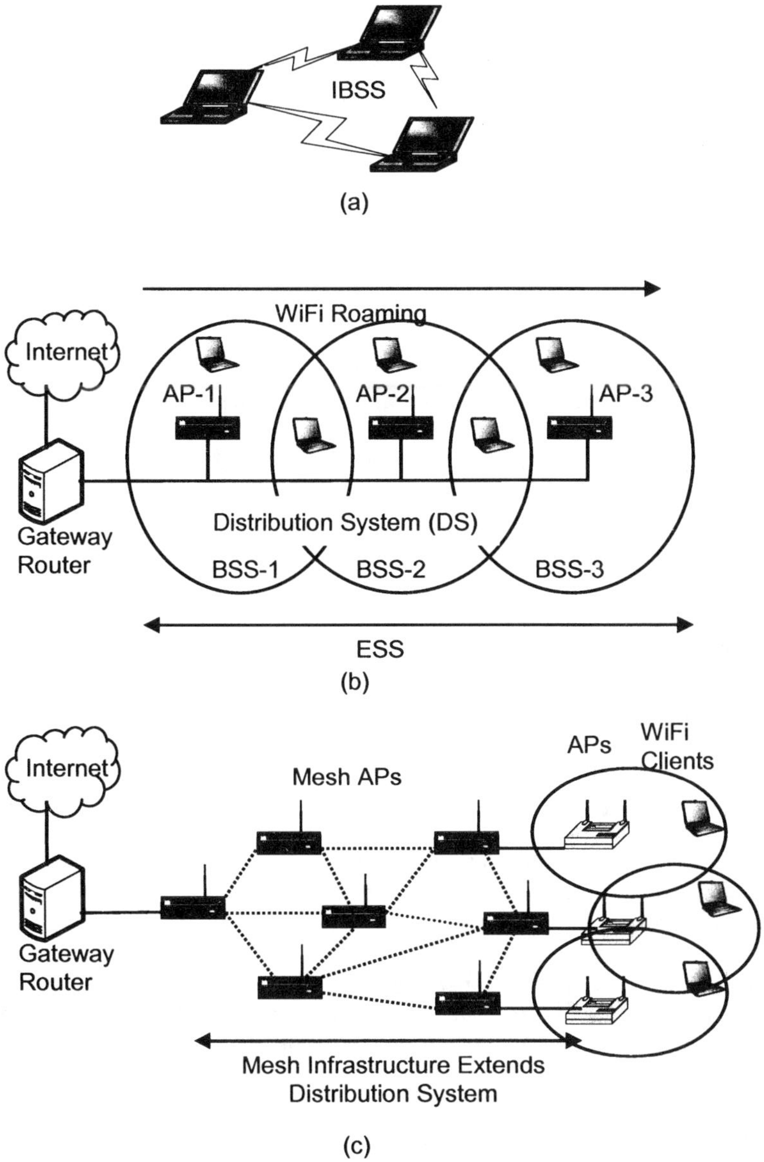

Figure 3.1 WiFi networking configurations: (a) peer-to-peer ad hoc mode; (b) infrastructure mode; (c) infrastructure mesh mode.

3.2.5.2 Infrastructure

Use of the infrastructure mode [Figure 3.1(b)] requires the installation of at least one wireless AP (also referred to as a base station), which is connected to the wired network infrastructure, called the distribution system (DS), and a set of wireless nodes or computers. This most basic configuration is referred to as a BSS topology in the 802.11 standard, which is composed of several wireless stations that are under the control of the same MAC function. Communication between wireless nodes, wireless computers, and the wired network is by the AP. Wireless computers conduct all communications through the AP, unlike ad hoc peer-to-peer communications.

An AP acts as a bridge between the wired and wireless networks. The device consists of a radio, a wired network interface, and bridging software. It thus acts as the base station for the wireless network, aggregating access for multiple wireless stations onto the wired network.

Before being able to communicate data, wireless clients and APs must establish a relationship or an association. Only after an association is established can the two wireless stations exchange data.

All APs transmit a beacon management frame at fixed intervals. To associate with an AP and join a BSS, a client listens for beacon messages to identify the APs within range. The client selection of which BSS to join is carried out in a vendor-independent manner. A client may also send a probe request management frame to find an AP associated with the desired service set identifier (SSID).

Multiple BSS (or APs) that share the same ESS ID and DS form an ESS. Cells in an ESS may overlap, or there may be gaps. The WiFi networks with multiple APs use the same channel or different channels to boost aggregate throughput. Full overlap provides for both an increase in aggregate bandwidth and redundancy. If a single AP fails, other APs are there to take over the load without any dropouts. Each AP and mobile unit (MU) has a unique MAC and IP address. The MU [also called mobile node (MN), mobile station (MS), or mobile client (MC)] can roam from AP to AP. In 3G terminology, an MU is called user equipment (UE) or handset.

Although the WiFi standard defines how a wireless computer communicates with an AP, it does not define how roaming should be conducted and supported within an ESS topology network, in particular when a roaming user crosses a router boundary between subnets. In theory, it is possible to implement dynamic host configuration protocol (DHCP) across the network and force users to release and renew their IP address as they migrate from one subnet to another, but this is not a desirable solution. Roaming between APs is largely reliant on vendor-specific implementations and management. These roaming aspects are covered in Sections 3.3 and 3.4.

Most corporate WLANs operate in infrastructure mode and access the wired network for connections to printers and file servers. The public hotspots provide WiFi service, free or for a fee, from a wide variety of public meeting areas, including coffee shops and airport lounges, and also operate in infrastructure mode.

3.2.5.3 Infrastructure Mesh

The infrastructure mesh mode [Figure 3.1(c)] combines some features of ad hoc mode with infrastructure mode. It is a router network without the cabling between nodes and supports the inherent rerouting for fault tolerance that such networks deliver. The infrastructure mesh is built of peer nodes that are not required to be cabled to a wired port like traditional WiFi APs. Rather, each simply plugs into a power supply. It automatically self-configures and communicates with other nodes over the air to determine the most efficient multihop transmission path. Today, most mesh implantations are vendor specific, until the 802.11s standard specifications for mesh networking are completed.

The infrastructure mesh uses dual-radio mesh APs, which have two radios operating on different frequencies in two different mesh topologies. One radio supports user access, while the other provides back-haul. A typical configuration uses 2.4-GHz WiFi for local access and 5-GHz band wireless for backhaul. The access capacity is not impacted by the forwarding traffic since it is done with a separate radio on a separate RF channel.

In an enterprise environment, infrastructure wireless meshes are important since these cut out the need for costly, wired back-haul to be provisioned to every node.

3.2.6 WiFi Systems Architecture

In this section we describe WiFi network architectures for enterprises and hotspot operators.

3.2.6.1 Enterprise WiFi Architecture for Internet Access

Figure 3.2 illustrates a generic WiFi architecture for enterprises. The main components of this architecture are:

- One or more WiFi hotspots, consisting or one or more APs, at various enterprise locations;
- Hotspots connected by a back-haul network to the enterprise's Managed IP (M-IP) network, and utility server farm.

As depicted in Figure 3.2, the back-haul network that connects the access router (AR) or DSL/cable modem to the ISP's IP network or each hotspot to the corporate server farm or M-IP network consists of one of the four types of links based on standard transport technologies:

- T1/T3 time-division multiplexing (TDM) network;
- Digital subscriber line/asynchronous transfer mode (DSL/ATM) network;
- Hybrid fiber coax (HFC)/cable Data Over Cable Service Interface Specifications (DOCSIS) network;
- WiFi or WiMax infrastructure mesh network.

We will not go into details about these technologies here, but from a service assurance perspective, we need to know the SLA demark points and types of performance measures available at these demark points.

3.2.6.2 WiFi Hotspot Architecture

A public hotspot is a location equipped with a WiFi network for access to the public Internet. It can either be free or fee-based; it can also entail a prepaid or postpaid access fee. Enterprise WiFi networks can be considered private hotspots. There are three types of WiFi hotspot service providers.

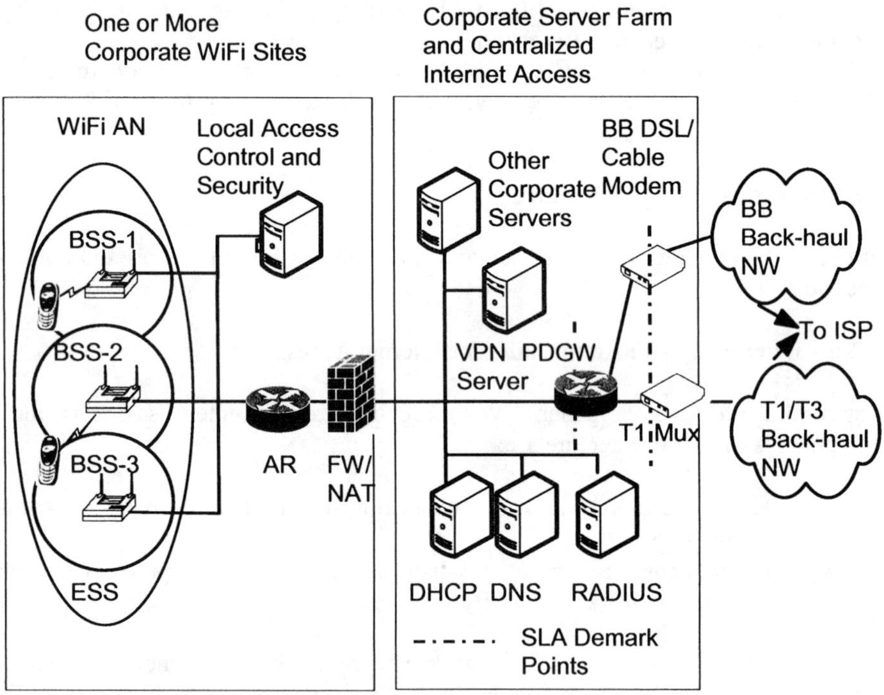

Figure 3.2 Generic WiFi enterprise architecture.

- The wireless Internet service providers (WISPs) which operate hotspots but do not "own" end customers and do not bill end users;
- WiFi service providers, which do not operate hotspots but have a large customer base and bill the end users;
- Both WISPs and service providers.

Figure 3.3 depicts a generic WiFi hotspot infrastructure architecture. This architecture is similar to that described earlier for enterprises. The main difference is that in public hotspots, firewall, and network address translation (NAT), and VPN servers are not used. However, a portal server is included for user registration.

Note that one service provider can support the hotspots of multiple hotspot operators, even using the same AP with multiple SSIDs and logically separate user traffic using virtual LANs (VLANs). We will not go into further detail here, but logically and functionally, these cases are the equivalent of multiple instances of the architecture depicted in Figure 3.3.

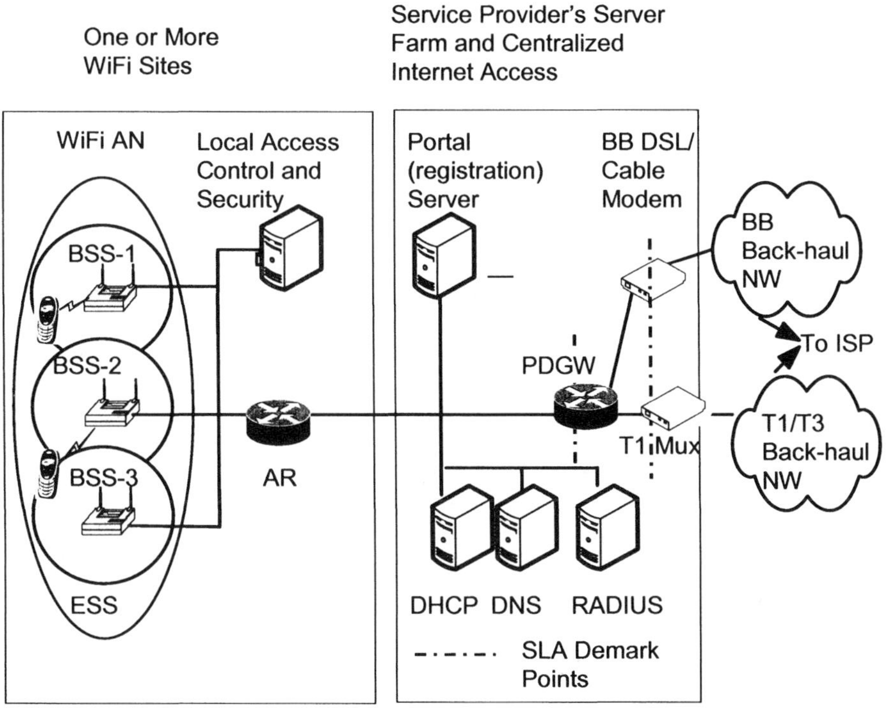

Figure 3.3 Generic WiFi hotspot architecture.

3.2.7 WiFi Management, Performance, and Security Issues

With a growing number of WiFi hotspots within an enterprise and of hotspots operated by hotspot service providers, WiFi management, performance, and security issues are becoming important.

Security and performance are critical in enterprise environments. In this section, we describe the issues in these three areas.

3.2.7.1 WiFi Management Operations

The following management operations are needed to establish a new WiFi wireless link:

- *Scanning:* Scanning is the process of identifying existing WiFi networks. The scanning procedure uses many parameters that include:
    - BSS type: This specifies whether to scan for IBSSs (for ad hoc networks), infrastructure BSSs, or all networks.
    - BSS identification (BSSID): The BSSID determines whether or not an MS scans for a specific network or for any network that is willing to allow it to join.
    - SSID: This SSID is the network name assigned to all BSS in an ESS. MSs wishing to find any network should set this SSID to the broadcast SSID.

  Scanning may be active or passive. Active scanning uses the transmission probe request frame to identify networks. It also uses a list of channels, which an MS will listen on for the existence of a network; a probe delay, which is used before the procedure to probe a channel; and the minimum and maximum amount of time that the scan works with any particular channel. Passive scanning saves battery power by listening for beacon frames, which are sent out by the BS, usually 10 times a second. These frames advertise the presence of the BS, its SSID, and its capabilities, such as security support. At the end of a scan, a report is generated, which lists all BSSs found. The details about each BSS enable an MS to join a selected BSS from the list.
- *Joining:* After scanning, an MS joins a specific BSS. Choosing which BSS to join is an implementation-specific decision that may require user intervention. Usually, the signal strength from the different BSSs determines the joining decision. Joining is not sufficient to enable network access; both authentication and associated are also needed.

- *Authentication:* The function of managing authentication between WiFi clients and networks is part of security and is described in detail in Section 3.2.8.
- *Association:* After authentication, an MS associates with a BS to gain full network access. An MS sends an association request to the BS and receives an association response with an association ID. After this, two-way traffic between the MS and BS can take place. Association is restricted to infrastructure networks, and the join process procedure is sufficient for ad hoc networks.

Besides association, reassociation, which is the process of changing an association from an old BS to a new BS in an ESS, is required to support mobility within an ESS. The reassociation procedure is the same as association. It is initiated when the MS detects that another BS has a stronger signal. To support reassociation, on the distribution backbone network, the old and new BSs may interact with each other to move frames.

3.2.7.2 WiFi Performance Issues

As 802.11b WiFi deployments and number of users grow in enterprise environments, besides security and scalability of network management, the impact of interference on performance and service assurance is becoming important. Common to any wireless system, the performance of an 802.11 system is affected by several external wireless-specific circumstances and design objectives used in WiFi deployment.

Throughput is one of the measures of a system's performance, which can be measured at MS by a user client, or in an aggregate form at the AP. As with wired LAN systems, throughput and errors in WiFi are product dependent and setup dependent, which in turn depends on WiFi design and deployment considerations. Factors that affect throughput include the number of users, propagation factors such as range and multipath, the WLAN system used, and the latency and bottlenecks on the wired portions of the LAN. Frame errors largely depend on interference present in the RF environment. A brief review of the additional factors that affect WiFi performance follows:

- *Frequency management:* Hotspot operators wishing to use RF devices in a given area must cooperate if they are to avoid interference problems. If they operate on the same frequencies, at the same time, and in the same area, their transmissions will interfere with each other's receivers. Each user, in effect, prevents other simultaneous, nearby uses of a portion of the spectrum while transmitting—thus, the need for frequency management.

- *Power management:* While most APs are 100 mW (20 dBm), antennas vary. FCC part 15 requirements allow up to 1W with 6-dBi antenna (36 dBm) total power. Hotspots, which are outdoors, may push this limit and penetrate indoors, in turn causing interference. Power management is one approach to control overlap in coverage areas and mitigating interference.
- *Impact of AP failure on coverage and throughput:* An AP failure can affect coverage for users but is avoided by sufficient capacity planning. When an AP deployment is correctly planned, the failure of a single AP reduces capacity, but not connectivity, for the affected coverage area.
- *Impact of interference:* The site survey process will identify any existing interference patterns and define the preferred locations of WiFi APs. This will ensure that all locations that should have wireless connectivity allow coverage patterns of the APs to overlap, creating a robust WiFi. As interference patterns are identified, it may be necessary to move or remove the source of the interference; this is part of the frequency management process. Frequency management is also vital to the long-term success of a public access zone.
- *Security implementations:* Implementation of security features, such as choice of encryption, impacts AP throughput. Performance drop when using encryption depends on the vendor and AP model; it may be anywhere from 10% to 25% and may go as high as 50%. Further details of security aspects are described in the following section.

3.2.7.3 WiFi Security Issues

Earlier WiFi security efforts have focused on encryption and authentication, with users essentially getting two choices for securing WiFi networks. They can use Internet protocol security (IPsec)–based VPNs or build security architectures around WiFi-specific security standards. While effective, using an IPsec VPN for wireless security carries all the complications of wired IPsec, such as configuration complexity and the requirement of client-side code. The native WiFi security support is expected to win for enterprise WiFi networks, while road warriors on a public WiFi hotspots network will need a VPN to tunnel into the corporate network.

Over the last few years, WiFi network security has evolved from wired equivalent protection (WEP)–based, poor security to WiFi protected access (WPA)–based, improved security, to 802.11i-based robust security. The IEEE's 802.11i includes all elements of the WPA standard while upgrading to stronger encryption.

Besides encryption, we need authentication. The 802.1x authentication standard used in WPA and 802.11i relies on the IETF Extensible Authentication Protocol (EAP), an extension of Protocol for Carrying Authentication for Network

Access (PANA) ping request (PPR). For proper authentication, the client and APs must use the same EAP version.

Figure 3.4 depicts how 801.x supports WiFi network security and the messages exchanged between WiFi client, AP, and RADIUS server.

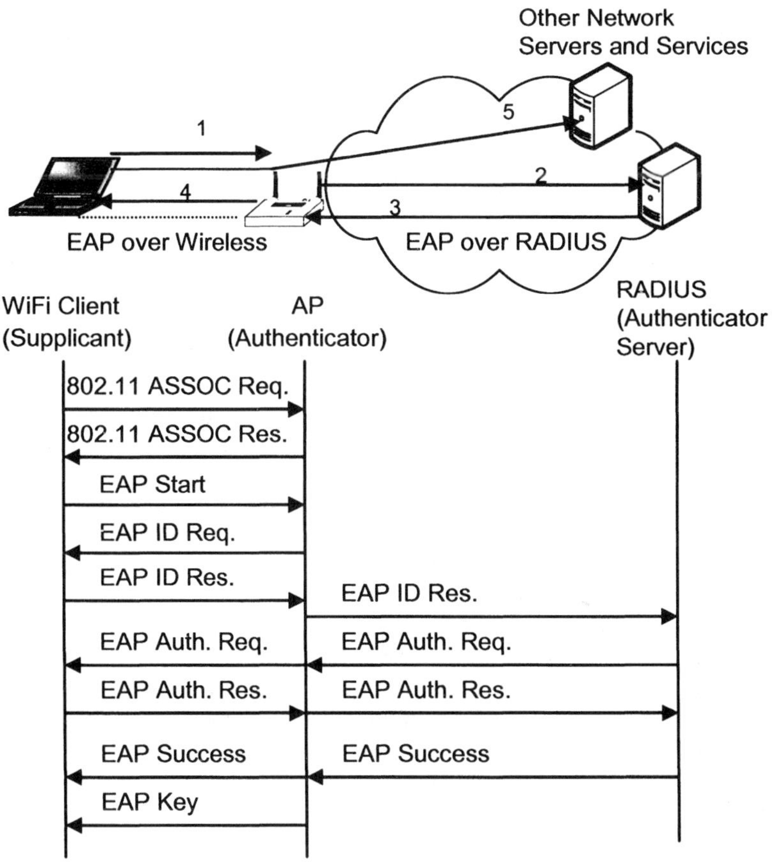

Figure 3.4 The 802.1x WiFi network security.

The key steps are:

1. WiFi client requests access to services. AP prevents wired network access. WiFi client (called supplicant) must have 801.x client software. Initially, AP starts in unauthenticated mode.
2. Encrypted credentials are sent to authentication server.
3. Authentication server validates user and grants access rights.
4. AP port is enabled, and dynamic WEP keys are assigned to client.
5. The WiFi client can now access general network services securely.

Since the 802.1x implementations are not yet widely available, enterprises are using vendor-specific solutions such as Bluesocket, Reefedge, and others.

3.2.8 WiFi Roaming and Mobility

For WiFi public hotspots, roaming refers to the ability to use many WISPs, while maintaining a business relationship with only one. This is user roaming. In an enterprise WiFi context, roaming refers to the ability to use an enterprise's different hotspot sites. The need for WiFi roaming is increasing as the number of hotspots and their users grow. Data from mobile operators' voice service revenue show that roaming services provide higher average revenue per user (ARPU) compared to nonroaming service revenues. Also, roaming enables potential for attracting a larger customer base.

Sometimes, roaming also refers to the ability to move a terminal from one AP coverage area to another without interruption in service or loss in connectivity. This is terminal mobility within an ESS, which is internal to a particular service provider and is usually handled by clients' reassociation from one old AP to the new AP, if there is enough overlap in coverage of those APs.

Here, our concern will be on user roaming between WiFi hotspots rather than terminal mobility. Although Web-based authentication can be used in a single-domain environment, in general, in a multidomain environment, an authentication, authorization, accounting (AAA)–based authorization is necessary, to decide who the user is, if this user is allowed to ask for this service, and how much this user should be charged.

As depicted in Figure 3.5, an intermediary broker is used for interdomain WLAN service registration (authentication and authorization) and as a clearinghouse for account settlements.

The logical components in this architecture are:

- The UE represents the user's equipment (typically a laptop computer, cell phone, or PDA) that is used to access the WiFi network. As depicted earlier in Figure 3.3, within the WiFi hotspot, the AP terminates the air interface to and from the UE. The access router provides access to the

Internet, and the local access controller verifies authorization, enforces access control for authenticated users, and segregates traffic of nonauthenticated (guest) users.

- The visited network AAA server (AAA-V) serves as an AAA proxy for roaming customers.
- The home provider AAA Server (AAA-H) serves as the RADIUS server authenticating the UE user. The home provider and visited network operator AAA servers also take part in transactions involving the reconciliation of billing and settlement records, either mutually or by an intermediate settlement entity.
- The roaming intermediary (INT) represents AAA and billing intermediaries. Such functions include AAA aggregation, wholesale hotspot service aggregation, AAA brokers, and charging, billing, and settlement clearinghouses.

3.3 3G NETWORKING

The term 3G is an umbrella term covering a range of wireless network technologies, including WCDMA, CDMA2000, and others. In this section, we will describe basic network architectures for both 3GPP-based UMTS/WCDMA and 3GPP2-based CDMA2000 networks. For these 3G networks, we describe systems architectures, basic call flows for voice and data services, roaming, and mobility. The intent is not to describe all the details but to focus on the network elements and interaction between those elements whose performance is critical to providing service assurance.

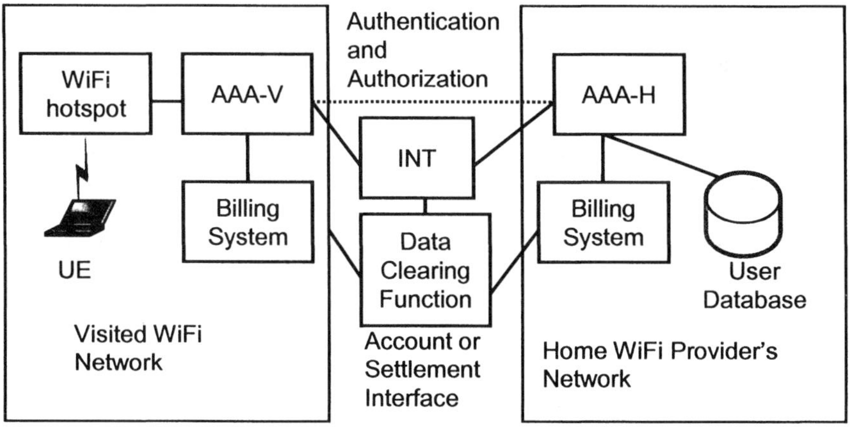

Figure 3.5 Authentication and authorization for inter-WiFi hotspot roaming. (*After:* [1].)

3.3.1 3GPP-Based 3G Networks

The 3GPP-based 3G network architecture grew from the 2G general system for mobile (GSM) architecture, which was originally designed as public land mobile network (PLMN) for circuit-switched (CS) voice services and not for data services. To provide data services, 3GPP developed the 2.5G GPRS network as an overlay packet-switch (PS) domain. The GPRS was later enhanced in several UMTS releases, starting with Release 99, Releases 4, 5, 6, and beyond, which include 3G all-IP network, which supports VoIP and other multimedia services. Table 3.1 summarizes the main features of various 3GPP specification sets, issued as different releases.

Note that the constituent parts of the 3GPP system need to evolve over time, reflecting different economic and technology life cycles. For example introduction of universal terrestrial radio access network (UTRAN) requires a large investment in a mature technology with a slow evolution, whereas introduction of new capabilities of IP multimedia subsystem (IMS) requires relatively low investment, but rapid short-term evolution.

Table 3.1

Summary of Key Features in 3GPP Releases

3GPP Releases	*Key Features*
Release '99 March 2000	A major Radio Access Network (RAN) release. Introduced basic capabilities of UTRAN, WCDMA, and new Core Network-Access Network interface (lu-CS), open service architecture (OSA), and extended Serving GPRS Support Node (SGSN) functions to Radio Network Controller (RNC).
Release 4 March 2001	Features in Release '99 plus a minor release. Added UTRAN access with some QoS enhancements, evolved CS domain from Mobile Switching Center (MSC) to softswitch-based MSC servers and Media Gateways based on IP protocols, and also added location service enhancements and multimedia messaging (MMS), Wireless application protocol (WAP), Mobile Execution Environment (MExE).
Release 5 June 2002	Features in Release 4 plus a major core-network upgrades release. Specified IMS, made IPv6 mandatory for IMS, defined IP UTRAN but did not eliminate ATM, and included several enhancements to WCDMA, MMS, and location services (LCS).
Release 6 March 2005	Features in Release 5 plus further enhancements to IMS, IMS and Internet SIP interworking, 3GPP and 3GPP2 IMS harmonization, WLAN-UMTS integration with UMTS and IMS, further enhancement to LCS and instant messaging (IM) services. Introduced multimedia broadcast and multicast service (MBMS) and digital rights management (DRM), enhanced MExE, virtual home environment (VHE), and OSA.
Release 7 and beyond	Further enhancements to IMS, and WLAN-UMTS integration for handover support and integration with legacy voice, and for MIMO antennas.

3.3.1.1 3GPP GPRS Systems Architecture

Figure 3.6 depicts a simplified view of such a 3GPP GPRS systems architecture. A brief description of the key network domains and components of the GPRS system follows.

Air Interface and Mobile Station

The MS is made up of two key components, the mobile equipment (ME) and the SIM. The ME contains generic radio and processing functions to access the network, human interface, or interface to other terminal equipment. SIM is a smart card, which can be removed from the phone and contains the user profile, such as phone number, barring, PIN number, and confidentiality-related information. This separation of radio functions and subscriber information and intelligence allows a more flexible service environment.

The air interface between the MS and base transceiver station (BTS) is called Um.

Radio Access Network

The Radio access network (RAN) [also called base station subsystem (BSS)] has two components; the BTS, and the base station controller (BSC).

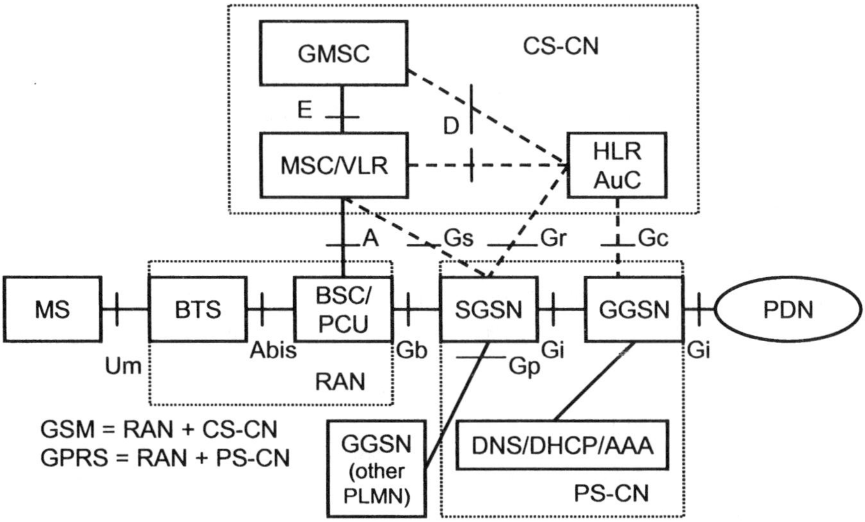

Figure 3.6 3GPP GPRS systems architecture. (*After:* [2].)

The BTS, or simply base station (BS), is responsible for the radio transmission and reception from antennas to the radio-interface-specific signal processing, and it can handle several radio carriers at a time. The purpose is to modulate, amplify, filter, and transmit the downlink signals (and perform the reverse on the uplink).

The BSC is responsible for the radio interface management, allocation and release of radio channels, and handover management of several BTSs. The BSC handles handoffs, cell rankings, locating MSs, power control, channel allocation, frequency/code allocations, coding, and limited switching. It is the network part of the air interface. Sometime the BTS and BSC together are called the BSS.

Underlying network links that connect the BTS and BSC are T1/E1 circuits. The network that connects RAN to packet-switched core network (PS-CN) [BSC to serving GPRS support node (SGSN)] consists of frame relay (FR) links. The RAN has two interfaces, which need to be configured, dimensioned, and monitored for service assurance. Three RAN interfaces of our interest are:

- The A interface between the BSC and the core network (MSC);
- The lu packet-switched data from the BSC/PCU to SGSN;
- The Abis interface between the BTS and the BSC.

Circuit-Switched Core Network

The circuit-switched core network (CS-CN) [also called the mobile switching subsystem (MSS) or the network switching subsystem (NSS)] consists of the mobile services switching center (MSC) [sometimes also called mobile switch (MS), or mobile telecommunications switching office (MTSO)] which is the heart of the circuit-switched domain. It performs the basic switching function, coordinates the setup of calls to and from GSM users, manages communications between GSM and other telecommunications networks, and collects billing-related call details (charging and statistics). It is similar to a fixed-network switching node, but it handles mobility and takes part in radio resource management.

A gateway MSC (GMSC) passes CS-traffic between fixed and mobile networks. The GMSC is an MSC that is able to find the corresponding HLR based on the called number. The GMSC and MSC/HLR may physically be one unit. It interfaces with the HLR, public switched transport network (PSTN), other public land mobile networks (PLMNs), and other networks such as packet-data networks. The MSC and GMSC support both transport and control plane functions.

The underlying network that connects MSC and GMSC comprises TDM links, and the network that connects MSC/GMSC and HLR/VLR comprises signaling system number 7 (SS7) links. In 3GPP Release 98, HLR, AuC, and EIR databases are also considered part of CS-CN.

The HLR is a central, permanent location and management database that holds subscriber information relevant to the provision of telecommunications services [International Mobile Station Identity (IMSI), user profile], some information related to the location information of the subscriber (mainly under which MSC/VLR the user can be found), such as MS roaming number, VLR address, MSC address, and the local MS identity) for routing and charging of calls towards MSC. There is usually one HLR per PLMN. Each GMSC has an HLR, which usually resides with the GMSC.

The VLR is a temporary location and management database of MSC and is utilized to handle mobility and roaming. It temporarily stores subscription data for those subscribers currently located in the service area of the corresponding MSC and holds data about their current location area. It interacts with the HLR in obtaining the subscriber data when needed. The VLR includes all users currently located in the system, including roamers and nonroamers. The MSC updates the VLR with HLR information. A VLR may be associated with one or several MSCs and usually resides with the MSC. The HLR and VLR work together to allow both local operation and roaming outside the local service area.

The authentication center (AuC) is a database that maintains and manages security, authentication and encryption, and related information for each subscriber. The AuC is linked to the HLR. If the mobile identification number (MIN) or IMSI from the MS does not match the AuC, the AuC will inform the HLR to block the call (thereby preventing fraud).

The equipment identity register (EIR) is a database that keeps track of all mobile stations (often implemented with AuC) and their identities [electronic serial number (ESN) or international mobile equipment identity number (IMEI)] in order to prevent the use of stolen or faulty equipments, and maintains security related information about the mobile equipment (separate from subscribers). Stolen or faulty mobiles are black listed in the EIR.

Packet-Switched Core Network

GPRS modifies the RAN part [BSC is updated to include the packet control function (PCF) to support packet-switched data] of the GSM architecture, and adds an overlay packet-switched core network (PS-CN). The components of the PS-CN are two GPRS support nodes (GSN), the SGSN and the gateway GSN (GGSN).

The SGSN is the first IP-aware point of contact for the UE. Functionally, SGSN connects RAN to GGSN and interfaces with the HLR/AuC to obtain user subscription data and to authenticate users, and it interfaces with MSC/VLR to handle UE's combined GSM/GPRS network attach process. The SGSN performs mobility management, encryption, and charging functions, and also supports SMS over GPRS.

GGSN is the gateway that connects to a packet-data network (PDN) such as the Internet or corporate intranet. It is a router that supports tunneling, routing, and accounting functions and also supports DNS, DHCP and AAA client functions. The GGSN also maintains a packet-data protocol (PDP) context of the attached UE to the corresponding SGSN. The setting up of the PDP context also allocates an IP address to UEs visible to ISP's at the GGSN.

Also, for roaming support, border routers, in private IP networks connecting SGSN and GGSN, are used to connect with other PLMNs or a GPRS roaming exchange (GRX).

The underlying networks that connect PS-CN elements, as well as inter-PLMN links, are part of private managed IP (M-IP) networks. The links between PS-CN elements and the HLR are SS7 links.

The PS-CN also includes IP utility servers, such as DNS, DHCP, and AAA, to support data services.

3.3.1.2 3GPP UMTS Systems Architecture

From a service assurance perspective, we need not consider UMTS architectures release by release; rather, we can focus together on Release '99 through Release 5 architecture, since after Release 5, most of the developments have been enhancements to network features and services that do not impact our service assurance approach. Figure 3.7 depicts a simplified view of such 3GPP UMTS systems architecture. A brief description of the key network domains and components of the UMTS system follows.

Air Interface and User Equipment

The UE (the UMTS term for MS) is made up of two key components, the UE and the user identity module (UIM). As with the GSM MS, the UE contains generic radio and processing functions to access the network, human interface, or interface to other terminal equipment. Now CDMA phones have also adopted the use of UIM-like smart cards under the name of the removable user identity module (R-UIM).

The UMTS UEs support two UMTS terrestrial radio access (UTRA) air interfaces called WCDMA and time division synchronous CDMA (TD-SCDMA). The air interface between the UE and NodeB is called Uu.

Radio Access Network

UMTS introduced new terminology and upgraded GPRS RAN. The RAN has two components: NodeB (new name for BTS), and radio node controller (RNC) (new name for BSC).

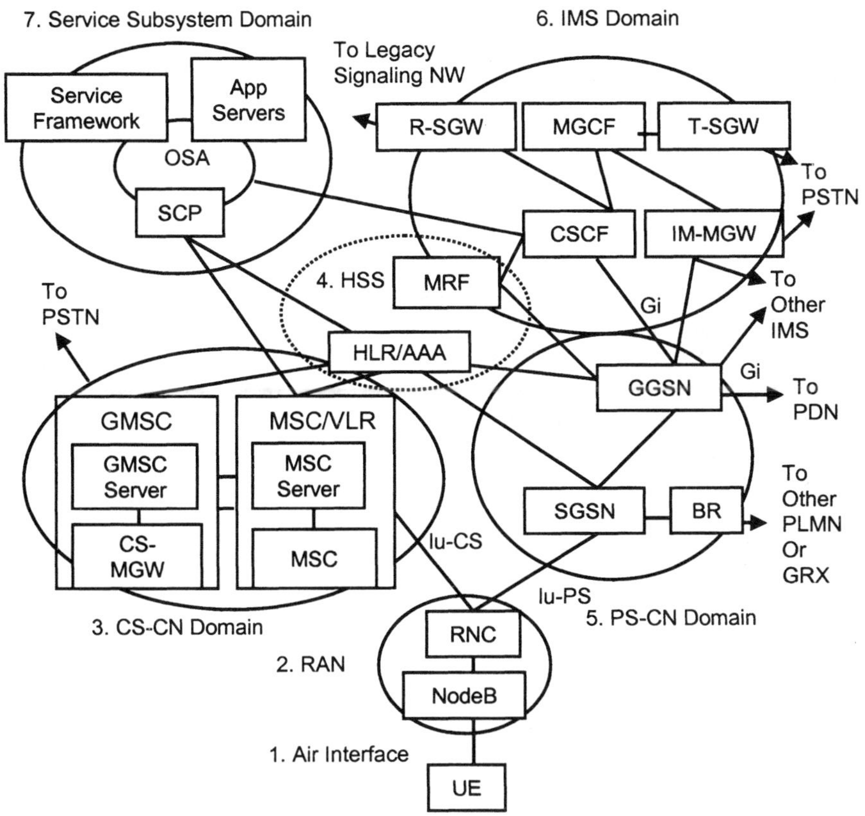

Figure 3.7 3GPP UMTS systems architecture. (*After:* [2].)

Similar to BTS, NodeB is responsible for radio transmission and reception from antennas to the radio-interface-specific signal processing, and it can handle several radio carriers at a time. The purpose is to modulate, amplify, filter, and transmit the downlink signals (and perform the reverse on the uplink).

The RNC is responsible for radio interface management, allocations and release of radio channels, and hand-over management of several NodeBs. The RNC handles handoffs, cell rankings, locating UEs, power control, channel allocation, frequency/code allocations, coding, and limited switching. It is the network part of the air interface. Sometime NodeB and RNC together are called the radio network system (RNS).

Underlying network links that connect NodeB and RNC are ATM private virtual circuits (PVCs); these links in an all-IP network will be IP over ATM. The RAN has several interfaces, which need to be configured, dimensioned, and monitored for service assurance. Five RAN interfaces of interest are:

- Iu: Interface between the RNC and the core network (MSC or SGSN);
- Iu-cs: Iu circuit switched (voice from/to MSC);
- Iu-ps: Iu packet switched (data from/to SGSN);
- Iub: Interface between the RNC and NodeB;
- Iur: Interface between two RNCs.

Circuit-Switched Core Network

In UMTS, CS-CN components are enhanced to support new Iu interfaces; otherwise, the components remain the same as in GPRS CS-CN.

In GPRS, the MSC and GMSC support both transport and control plane functions. In 3GPP UMTS Release 5, these transport and control functions of the MSC have been logically split into two components, MSC server and circuit-switched media gateway (CS-MGW).

As mentioned earlier, the interface between the MSC and RNC is lu-CS, and the links are ATM PVCs.

Home Subscriber System

In UMTS, the HLR is enhanced and now called the home subscriber system (HSS). It consists of HLR, VLR, AuC, and EIR databases, described earlier, and it also includes AAA functions.

The underlying support network for HLR, VLR, AuC, and EIR is SS7 with GSM mobile application part (MAP) protocol; however IP networks support the newer HSS elements such as AAA and home agent (HA).

Packet-Switched Core Network

In UMTS, the components of the PS-CN are the same as in GPRS. However, the SGSN is upgraded to support new ATM interfaces.

3GPP: IP Multimedia Subsystem Architecture

The IP multimedia subsystem (IMS) based on Internet concepts is independent of CS and PS networks and uses PS transport for signaling and bearer traffic and the existing radio access infrastructure to provide multimedia services. It uses SIP as the primary signaling protocol.

To provide new VoIP and multimedia capabilities, the 3GPP enhanced the basic GPRS architecture for Internet data services and specified an access-independent IMS domain that defined several new network components. A brief description of key components of IMS domain, grouped in two categories, as depicted in Figure 3.8, follows:

1. *Call session control functions (CSCF):* The CSCF has taken most of the MSC functionality in the IMS architecture, and it functionally corresponds to SIP servers described earlier. The CSCF can play three roles:

    - *Proxy CSCF (P-CSCF)* is the first point of contact in the visited IMS network. The others only exist in the home network. It forwards the SIP registration messages and session establishment messages to the home network, and also functions as the QoS policy enforcement point (PEP) within the visited network.

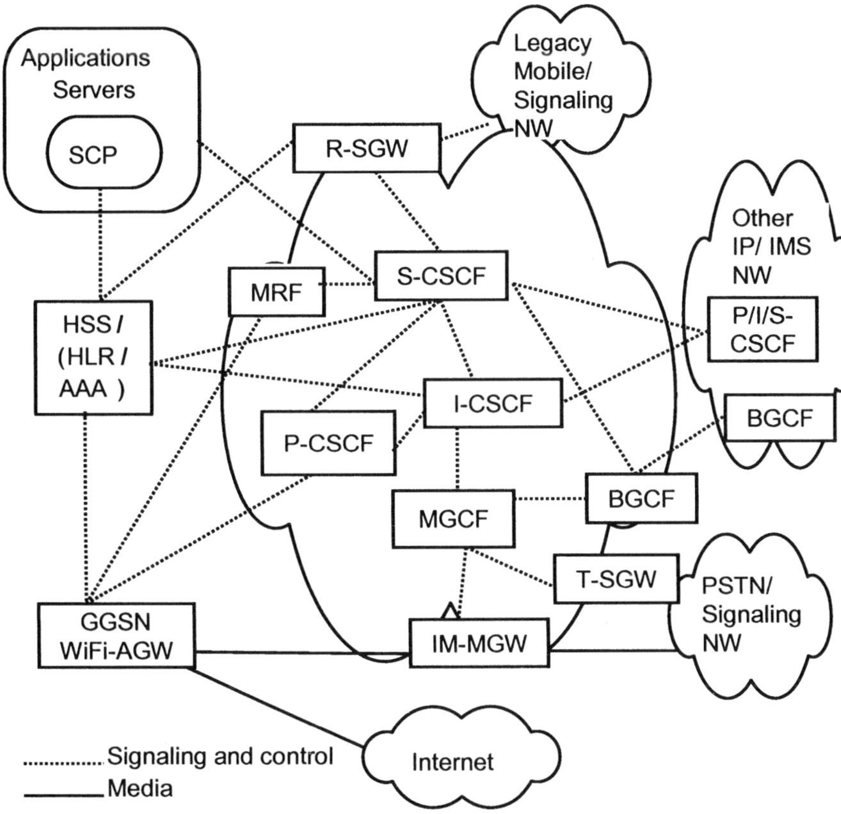

Figure 3.8 3GPP IMS domain architecture. (*After:* [3].)

- *Interrogating CSCF (I-CSCF)* is the first point of contact within the home network from a visited network. Its main function is to query the HSS and find the location of the serving CSCF (S-CSCF). This is an optional node in the IMS architecture, and P-CSCF can perform the I-CSCF function. It performs load balancing between the S-CSCFs with the support of the HSS and also hides the specific configuration of the home network from other network operators by providing the single point of contact into the network. Since it is the gateway into the home network, it supports the firewall function and can collect some form of billing information.
- *S-CSCF* is the node that performs the session management for the IMS domain in the home network; with the help of application servers (AS) and home subscriber server (HSS), it provides the service features.
- *HSS* interfaces with the I-CSCF and S-CSCF to provide information about the subscriber and user profile information. The HSS functionally corresponds to registrar/location server and AAA.

2. *Media gateway (MGW)/media gateway control function (MGCF):* The IMS supports several nodes for interworking with legacy PLMN and PSTN networks. These nodes include MGW, transport signaling gateway (T-SGW), roaming signaling gateway (R-SGW), and MGCF, described later in Section 6.4.5 in the context of routing mobile calls from visiting 3G/WiFi to CS/PSTN. Their functionality corresponds to VoIP Gateway described earlier. Additionally, there is a border gateway control function (BGCF) analogous to the session border controllers (SBC), described in Section 6.9.2, which are used to tie together diverse carriers' VOIP networks, except that within IMS, the BGCF handles only control/signaling messages to other IMS domains.

Media resource function (MRF) functionality corresponds to the media server mentioned earlier, and is divided in two parts, multimedia resource function processor (MRFP) and multimedia resource function controller (MRFC). This provides support for features like support for automatic speech recognition (ASR) and interactive voice response (IVR).

The IMS is important because service provisioning and charging are part of the system rather than service-specific. Data services are offered with managed QoS rather than the best effort; providing a mix of different applications is easy; service roaming is possible; and services are multiaccess, that is, they work across GPRS, UMTS, WiFi, and CDMA2000 rather than being specific to the access network.

The underlying network for the IMS domain is IPv6.

Service Subsystem

The service subsystem includes application servers and service nodes that support value-added services (VAS), intelligent network (IN) services, and roaming.

The VAS is a platform for supporting certain services in GSM, such as short message service center (SMSC) and voice messaging service (VMS). These services use a standard interface towards GSM and may or may not have external interfaces towards other networks.

IN, adopted from fixed network, is a platform for creating and providing more services. It enables service evolution and use of service nodes like service control point (SCP) and changes in the GSM switching elements to integrate the IN functionality. An example of IN service is prepaid subscription.

Roaming is a service that allows a subscriber of one operator to use the services of another operator when inside the latter's coverage area. Roaming over GSM networks has become a key service over the last few years, and it has generated a large percentage of revenue for network operators. Subscribers in a GSM network have an international mobile subscriber identity (IMSI) that identifies them in their carrier's network. For subscribers to roam outside their home network, their carrier must negotiate roaming agreements with other network operators so that the visited network recognizes the subscriber's IMSI and allows roaming to proceed. Besides the roaming agreements, operators need the GSM signaling platform that provides SS7 signal processing and subscriber identification translation to offer roaming services.

Besides GSM roaming, which may be international, interregional, or national, there are other types of roaming:

- *Intertechnology roaming:* Roaming between different technologies (e.g., GSM MAP to CDMA ANSI-41 networks, and 2G or 3G, cellular, and WiFi);
- *Interoperator interservices roaming:* For example, sending SMS or multimedia messaging service (MMS) by a user in one operator network to a receiver in another operator's network.

In this book we will cover intertechnology roaming between WiFi and 3G cellular, but VAS and IN are outside the scope of this book. The underlying networks for services subsystems are SS7 and IP.

3.3.1.3 UMTS Interfaces

The 3GPP has specified several interfaces for UMTS networks. Some of the UMTS interfaces of particular interest in monitoring service assurance are:

- Gb interface between BSS and SGSN;

- lu interfaces between UTRAN and MSC/MSC server/MGW (lu-CS) and between UTRAN and SGSN (lu-PS);
- Gn interface between SGSN and GGSN;
- Gi interface between GGSN and PDN (Internet and MGW, MGCF and CSCF);
- Gp interface between SGSN in one PLMN and GGSN in another PLMN;
- Gr and Gc interfaces between SGSN and HSS and between GGSN and HSS, respectively.

The IMS has introduced addition interfaces. Interfaces of interest for service assurance are:

- Mc interface between MGW and MGCF (also MSC server);
- Mg interface between CSCF and MGCF;
- Mh interface between CSCF and HSS;
- Mm interface between CSCF in one IMS and CSCF in another IMS.

3.3.1.4 Call Flows in GPRS and UMTS Networks

In this section we briefly review the call flows for voice and data circuits in UMTS networks.

Network-Attach Process in GPRS/UMTS Networks

Before a UE can originate or receive a call, it must be attached to a serving network. On powering up, the UE scans for the network and attaches to a serving network. The attach processes for voice and data are different, and there is a joint attach process. For voice service, UE attaches to a serving MSC (MSC-S), and for data service, it attaches to an SGSN. For data services, the UE is always connected to the network but requires additional processes of PDP context setup described later. Here, we consider the attach process for voice and data services. The steps, depicted in Figure 3.9, are as follows:

3. The UE requests that it be attached to the network for the MSC or SGSN.
4. Authentication is made between the UE and the HLR by the BSC and MSC for voice services and the SGSN for data services.
5. Subscriber data from the HLR is inserted into the SGSN and the MSC/VLR.
6. The MSC or SGSN informs the UE that it is attached to the network.

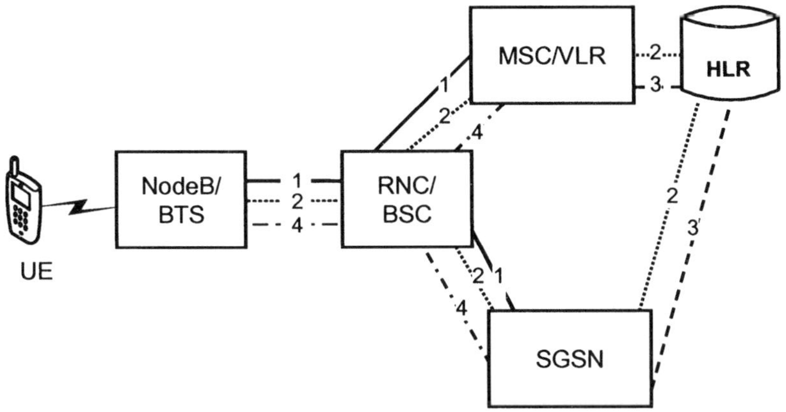

Figure 3.9 Network-attach process in GPRS/UMTS networks.

Voice Call Flows in GSM/UMTS Networks

Figure 3.10 depicts voice call flows for both mobile-terminated (MT) and mobile-originated (MO) calls.

The process for MT calls [Figure 3.10(a)] is as follows:

1. A calling station, in a fixed network, calls a UMTS subscriber.
2. The PSTN forwards the call to the GMSC.
3. Based on the IMSI number of the called party, the GMSC signals call setup to the HLR.
4. The HLR checks for the existence of the called number, and requests a mobile station roaming number (MSRN) from the VLR.
5. The VLR provides the MSRN.
6. The HLR forwards a response identifying the current MSC to the GMSC.
7. The GMSC forwards the call to the current MSC.
8. The MSC queries the VLR for the location range and reachability status of the UE.
9. If the VLR tells the MSC that the UE is marked reachable, then a radio call is enabled.
10. The MSC pages the UE in all location areas (NodeB/BTS) assigned to the VLR.
11. All NodeBs/BTSs page the UE in their cells.
12. A reply is received from the UE in its current radio cell.
13. The RNC/BSC informs the MSC about the response from the UE.
14. The MSC checks security from the VLR.

15. If all necessary security procedures are successful, the VLR indicates to the MSC that the call can be completed.
16. The MSC via the RNC sets up a connection.

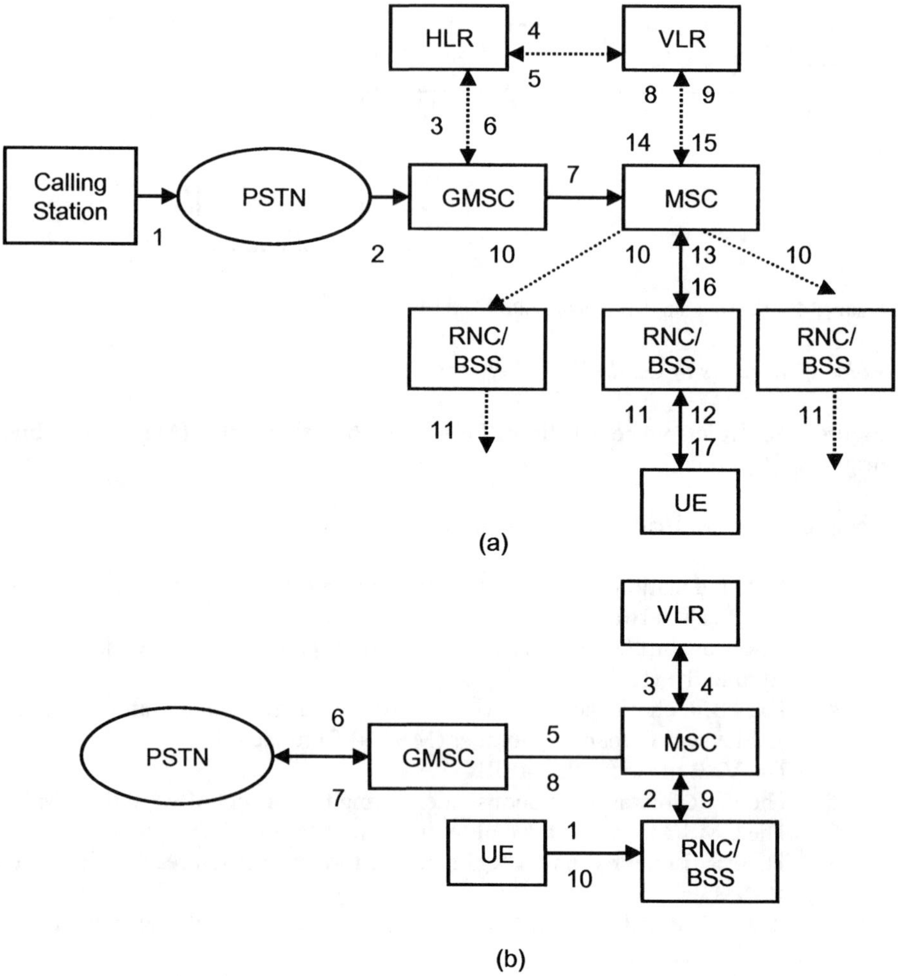

Figure 3.10 Voice call flows in UMTS networks: (a) MT and (b) MO call flows.

The process for MO calls [Figure 3.10(b)] is simpler than for MT calls and is as follows:

- The UE requests a connection from the MSC via the RNC/BSS (1 and 2 in the figure).
- The MSC checks security from the VLR ((3 and 4 in the figure).
- The MSC checks for resources (free circuits) (5 through 8 in the figure).
- The MSC sets up the call (9 and 10 the figure).

Obviously, MO and MT calls will combine steps of both.

Data Call Flows in GPRS/UMTS Networks

Figure 3.11 depicts data call flows in UMTS networks. The basic steps in network attaches and PDP context call flows are as follows:

1. The UE starts the attach process by establishing a logical connection, radio resource control (RRC), between the UE and UTRAN.
2. The UTRAN establishes an lu signaling connection between UTRAN and the SGSN.

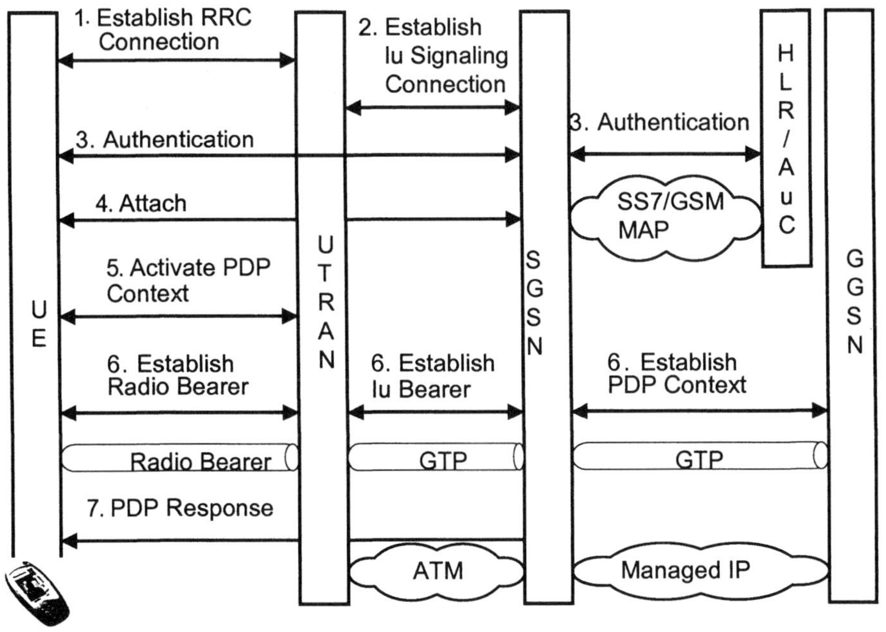

Figure 3.11 UMTS network attach and PDP call flows.

3. The SGSN requests authentication from the HLR/AuC. After receiving authentication from the HLR/AuC, the SGSN confirms authentication with the UE.
4. After UE authentication, the UE attaches to the SGSN; that is, a logical connection between the UE and SGSN is established. This completes the attach process.
5. After the UE is attached to the SGSN, the UE sends an activate PDP context request to UTRAN. This is to request a dynamic IP address for connecting the UE to the Internet.
6. The UE and UTRAN establish a radio bearer channel. The UTRAN and SGSN establish an lu bearer channel by setting up a GPRS tunneling protocol (GTP) tunnel. The SGSN and GGSN set up a GTP tunnel, establish a PDP context, and allocate an IP address.
7. The SGSN sends the PDP response to the UE. The UE can now connect to the Internet and send and receive IP packets.

This data call process can be the source of problems that affect service assurance and quality. Usually the A, lu, Gb, Gn, Gi, Gc, and Gr interfaces are monitored to detect service problems.

3.3.1.5 An Example of Mobile Data Service: WAP/WEB Access over GPRS

This section describes how GPRS networks support data service access to the Internet or enterprise intranets using WAP from a mobile station or using a Web browser from a PC connected over GPRS/UMTS network. First, we describe the WAP architecture, then WAP over GPRS service. As explained later, the Web over GPRS service is a subset of WAP over GPRS service. These services are used as an example to illustrate the basic service modeling approach described later in Chapter 5.

Wireless Access Protocol (WAP)

The WAP architecture allows standard, off-the-shelf Internet servers to provide services to wireless devices and offers many of the same features available in PC-based Web products.

The key features of WAP include a lightweight Web Markup Language (WML) and a script language (WMLScript) designed to create applications for the small displays of handheld devices. WML, an Extensible Markup Language (XML)–family language, has a smaller set of markup tags, making it more appropriate to implement on handheld devices than Hypertext Markup Language (HTML), for example:

- A lightweight protocol stack minimizes the required bandwidth and guarantees that a maximum number of wireless network types can run WAP applications.
- A framework for wireless telephony applications (WTAs) allows access to telephony functionality, such as call control, phone book, and messaging from within WMLScript scripts. This allows operators to develop telephony applications integrated into WML/WMLScript services.
- The WAP is network-independent; it works with almost all wireless air interfaces.
- The WAP is device-independent; applications can be developed using a single standard that will work across various devices. Handset manufacturers can use the same software in all their product lines.
- The WAP is operating system–independent; since WAP is a communications protocol and an application environment, it can be built on any operating system.

From the perspective of content providers and end users, WAP also offers an easy transition to wireless access. With minimal effort, content providers can reach mobile customers with existing applications. It utilizes standard XML and converts it into a form suitable for transmission over wireless networks (WML).

A major limitation of the WAP technology is that the screen is small for displaying complex graphics, and data entry through the numeric keypad can be time-consuming. WAP was designed to allow restricted-function, voice-centric handsets to browse remote information services. Therefore, the specification of WML, in particular, limits the flexibility of WAP-friendly sites with limited capabilities.

WAP Network Basic Architecture

WAP clients access services by sending a request via a URL to the WAP gateway. The URL is used to identify the origin server on which the service is available. The request is sent from the UE using WAP protocols over one of the available bearer networks. The WAP gateway is a WAP to hyper text transport protocol (HTTP) proxy that translates the WAP request into an HTTP request (from binary form to text). The HTTP request is passed on to the server identified by the URL.

The HTTP server may have access to various databases, and other services available in the infrastructure network. Once the request has been serviced, a response is sent back to WAP gateway, which in turn translates it into a WAP response and sends it down to the WAP MS.

As illustrated in Figure 3.12, there are four basic elements to the WAP architecture:

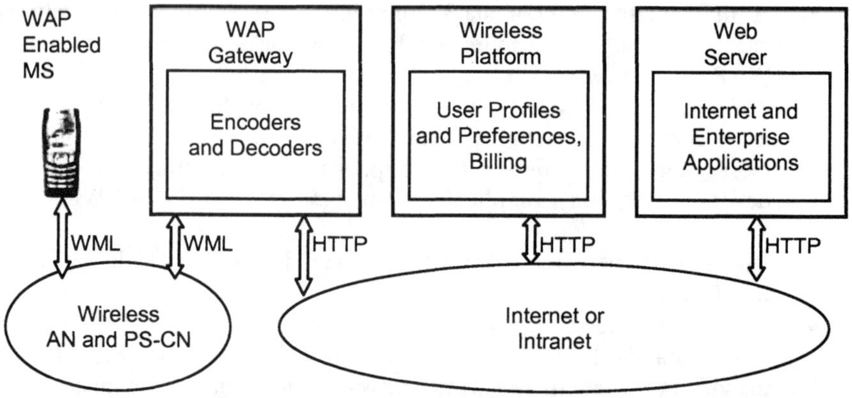

Figure 3.12 The WAP architecture.

1. *The WAP client or microbrowser:* This is a software element that resides in the handheld device and is analogous to a standard Web browser. Minibrowsing with a WAP-compatible browser is somewhat similar to browsing the Web with a desktop browser. The content provided depends on the ISP. Most Websites are simple and only include text. There are no search engines and no downloading capacity. Also, wireless browsers cannot support plug-ins such as Flash, Shockwave, or Real Player. However, more and more WAP compatible Web sites are being launched.
2. *The WAP gateway:* This is the device that communicates with the microbrowser and serves as the translator between its WML code and the standard HTTP requests used by existing Web servers. Much of the browser's processing demand is handled here, offloading this computing task from the handset. The gateway can also be used to track usage and bill for WAP services.
3. *The wireless platform:* This manages user profiles and preferences and other support functions, such as usage or billing. The wireless platforms may be integrated with either the WAP gateway server or with the origin server and may incorporate functionality like content conversion and user profiling to support the integration of wireless and wired access.
4. *The origin Web server:* This includes Internet and enterprise applications running on the same platform accessed by standard PC-based browsers over wireline connections.

WAP/Web Access Service over GPRS Service

Figure 3.13 illustrates WAP/Web access service over GPRS. As mentioned earlier, WAP/Web service is independent of network access. However, for end-to-end service assurance modeling, described later in Section 5.5.2, we need to know how the WAP/Web service is being accessed. Here, we assume that WAP/Web service is using a GPRS transport bearer to access the WAP/Web service. This means that the either the WAP-enabled MS is using a GPRS network to access WAP service or a PC with a GPRS access card is accessing Web service using a GPRS network.

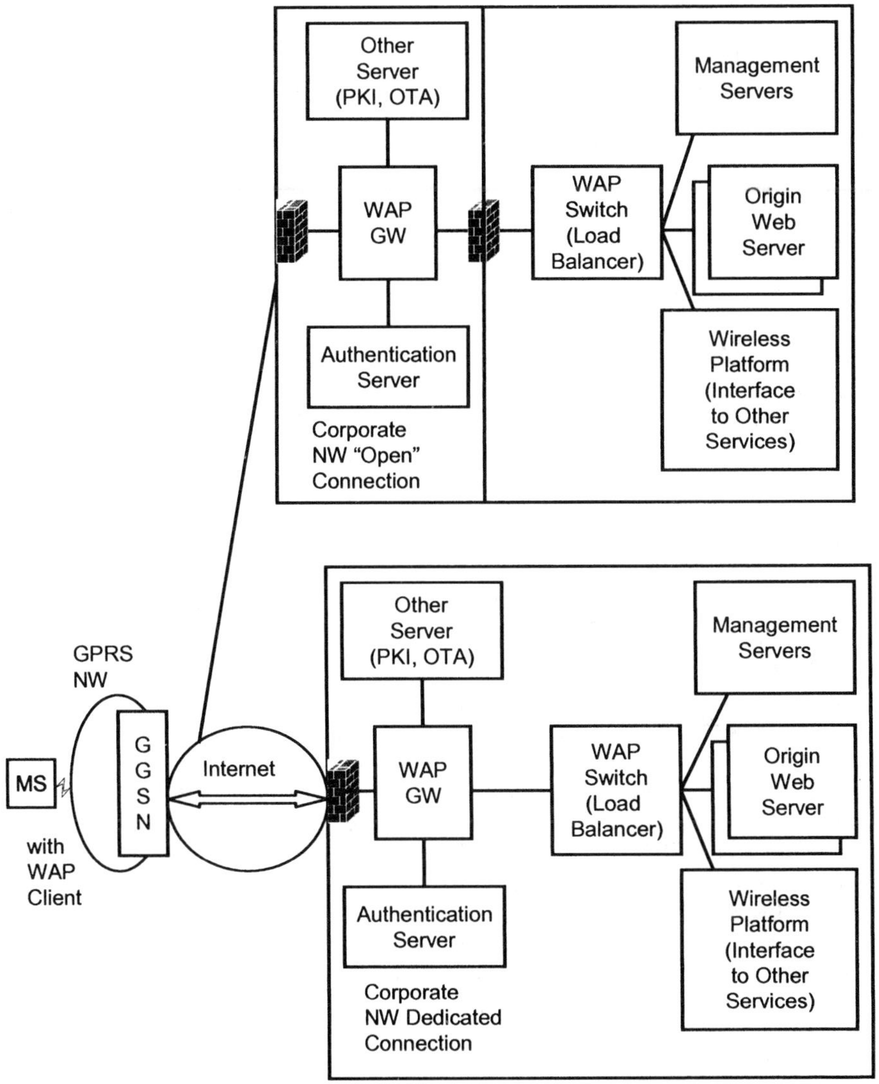

Figure 3.13 The WAP/WEB access service over GPRS architecture.

Web access can be considered a subset of WAP access service. Since for wireless Web access using an HTTP client and PC with a GPRS access card, there is no need to use WML and the WAP gateway, the request for URL access can go directly to the wireless platform and the Web server.

Figure 3.13 depicts two cases of WAP access in an enterprise environment. The first is an "open" connection. This means that any user can access the WAP gateway from anywhere using the public Internet. Initial WAP access will require its own authentication. However accessing other corporate resources or services, such as corporate e-mail, voice mail, or corporate databases, behind another firewall, will require additional authentication and authorization. Usually, to provide redundancy and improved performance, multiple Web servers behind a load balancer (or WAP switch) are used.

The second case is dedicated access from a GGSN. This is a secure connection via a firewall that allows corporate users to access enterprise services using a dedicated connection only from the corporation's home GPRS network.

3.3.1.6 Roaming Between UMTS Networks

In the 3G context, roaming refers to the ability to use voice and data services from a visited network, while users are away from the coverage of their home networks, without any direct dealing with the visited network for billing.

When the GSM standard was introduced, mobile operators used it to promote roaming between European countries and later worldwide. This voice-service roaming was done using SS7 network bridges or gateways and user authentication from visited GSM network via inter-HLR signaling, supported by roaming agreements between mobile operators for revenue sharing. Until GPRS was introduced, mobile operators had only partially installed bridges between their switched networks and data networks.

Now, with the introduction of GPRS and UMTS networks with always-connected capabilities, mobile operators who are able to offer their subscribers Internet access must form relationships with operators from the data world. To carry the roaming data flow, mobile operators either create direct inter-PLMN links using a Gp interface between each other's mobile networks or a solution called GRXs, similar to Internet peering exchange used by the IP backbone operators. The GRX interface is IP based and supports appropriate routing and security protocols to enable a subscriber to access home services from any of the home PLMN's roaming partners.

Before any roaming data session can be set up, the UE in the visited network must perform the functions of network scanning and network attach. These functions are similar to those used in users' home networks however, when the UE wants to attach to a visited network, it must be authenticated from its home network HLR.

After this UE authentication, for data setup, the following steps, depicted in Figure 3.14, are used:

1. The UE sends an activate PDP context request to the visited network's SGSN (V-SGSN).
2. The V-SGSN sends a DNS request (APN name) to the visited networks (V-NW's) network (NW) registrar.
3. The V-NW registrar sends a DNS request to GRX.
4. GRX return a DNS response to the NW registrar.
5. The NW registrar sends a DNS response (APN name and IP address) to the V-SGSN.
6. The V-SGSN sends a create PDP context request to the home network's GGSN (H-GGSN).
7. The H-GGSN checks the user profile from the HLR and creates a PDP context and sends a response back to the V-SGSN.

Figure 3.15 illustrates the roaming architecture for UMTS networks. In this architecture, the border router (BR) is a gateway that ensures secure communication between different UMTS networks, and an inter-PLMN backbone network or GRX network connects different UMTS networks to enable roaming.

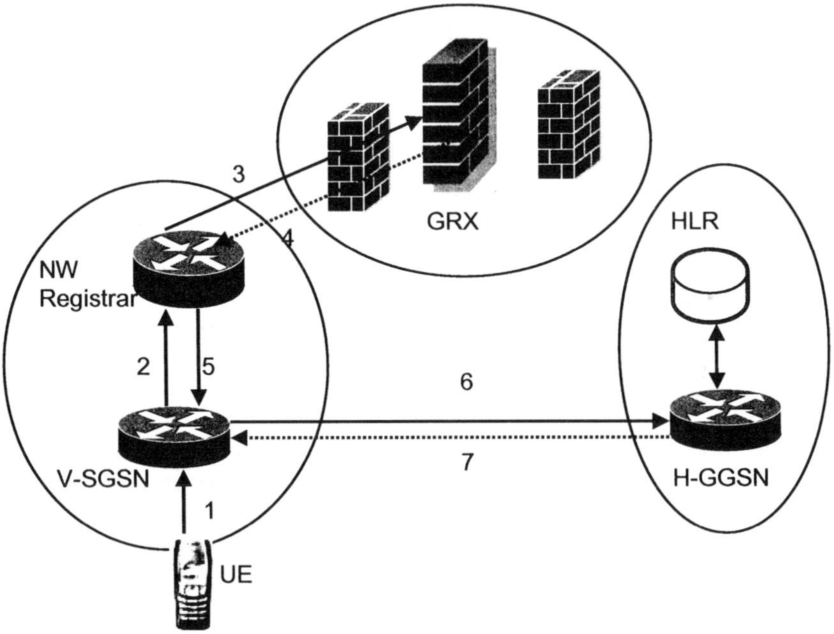

Figure 3.14 Data session setup while roaming in UMTS networks.

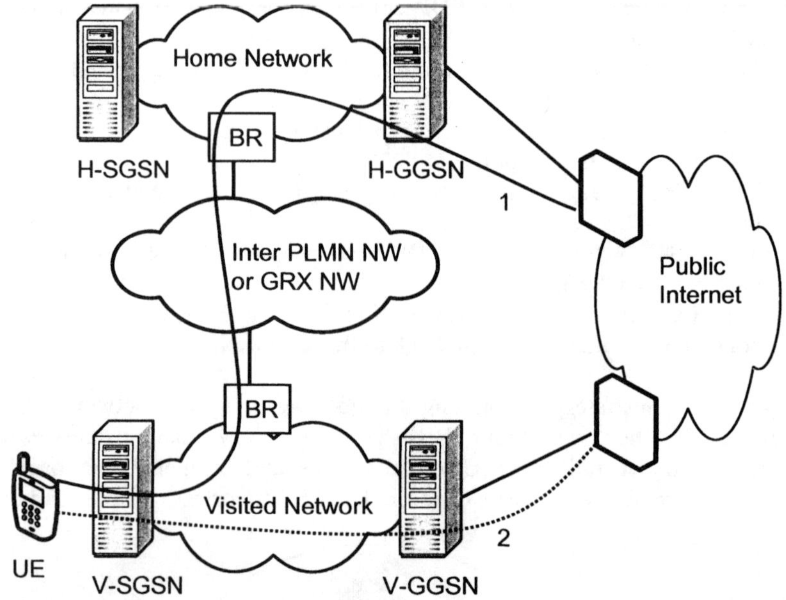

Figure 3.15 Roaming architecture for UMTS networks.

There are two options for accessing the Internet or an intranet. The first is using the UE's home network; this is done by setting up a secure IP link between the V-SGSN via a BR and inter-PLMN backbone or GRX and H-GGSN. This approach has the advantage for the user that in the visited network, the service is from the home GGSN, and all service capabilities and security features are available to the user. The second option is to get Internet access via the V-GGSN. In this option to get secure intranet access may not be available or guaranteed.

3.3.2 3GPP2-Based 3G Networks

The 3GPP2 specified 3G network CDMA2000 as an American standard to meet the International Mobile Telecommunications (IMT)-2000 goals. It evolves from existing cdmaOne services, including speech coders, packet-data services, circuit-data services, fax services, SMS, and Over the Air Activation and Service Provisioning (OTASP).

It has evolved in three phases:

- *Phase 1:* 144-Kbps packet data and voice (1XRTT);
- *Phase 2:* 384-Kbps packet data, voice, and video (3XRTT);
- *Phase 3:* Fixed wireless access at 2 Mbps.

The 1X-RTT is based on IS95A/B technology and has two different implementations, 1X-EVDV and 1X-EVDO.

Although the specifications of 3GPP WCDMA are identical to 3GPP2 CDMA2000 3X, these systems are not compatible. The difference is the chip rate, the frequency at which the transceiver resonates. CDMA2000's chip rate needs to be a multiple of cdmaOne's chip rate, while WCDMA's chip rate has to fit the GSM framing structure. Unlike UMTS's/WCDMA's wide bandwidth of 5 MHz, CDMA2000 is based around a single-frequency 1.25-MHz channel (for reuse of existing spectrum used by IS-95A/B system) band (1X-RTT), or three consolidated 1.25-MHz bands (3X-RTT). Both cdmaOne and CDMA2000 use global positioning system (GPS) signals for network timing.

3.3.2.1 3GPP2: CDMA2000 Architecture

As with 3GPP-based UMTS system architectures, we will not go into details about each of these phases but focus on the CDMA2000 system architecture. Since for WCDMA and CDMA2000 networks, the basic approach for service assurance is similar, and several network components are functionally the same and perform similar functions, here we will describe only the newer elements or differences in function. Table 3.2 summarizes a comparison of mobile network terminology used by GSM/GPRS, UMTS, and CDMA2000.

Figure 3.16 depicts CDMA2000 systems architecture. A brief background of the key components that are different from UMTS is as follows:

- The primary function of the PCF in the RAN is to establish, maintain, and terminate layer 2 connections to the packet data serving node (PDSN).
- The PDSN is equivalent to both the SGSN and GGSN in UMTS and incorporates several functions in one node. The major function of the PDSN is to route packets to IP networks or directly to HA. The PDSN also assigns IP addresses and maintains point-to-point protocol (PPP) sessions to the MS. It also initiates AAA for the MS packet session. Usually, the PDSN incorporates the foreign agent (FA) functions.
- The FA functions support mobile IP roaming and mobility and include functions such as reverse tunneling, registration, and dynamic HA and home address assignments.
- The HA is an important component of mobile IP. It redirects packets to the FA and receives and routes reverse-tunneled packet from the FA. The HA also provides security by authentication of the MS through mobile-IP (MIP) registration and maintains connection with the AAA server to receive subscriber profile data.

Table 3.2
Mobile Network Terminology

Network Component	*GSM/GPRS*	*UMTS*	*CDMA2000*
Base station	BTS	NodeB	BS
Base station controller	BSC	RNC	RNC or BSC/PCF
Circuit switch	MSC	MSC	MSC
Tunnel switch	SGSN	SGSN	—
Network access gateway	GGSN	GGSN	PDSN/FA
User profile/authentication	HLR/AC	HSS/HLR/AC	AAA/HA
Visitor profile/authentication	VLR	VLR	FA
Multimedia support	—	IMS	MMD
Roaming and mobility	SS7/GSM MAP	SS7/GSM MAP/SIP	IS-41 Mobile IP
Back-haul network	FR/Managed IP	ATM/Managed IP	Managed IP
Signaling network	SS7/GSM MAP	SS7/GSM MAP → IP	ANSI-41 → IP

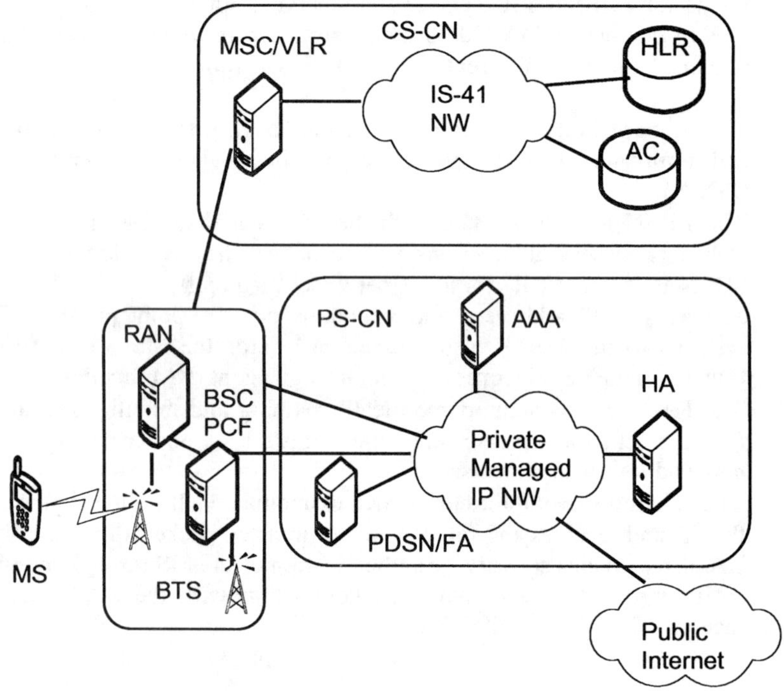

Figure 3.16 3GPP2 CDMA2000 systems architecture.

The AAA server performs at least three types of functions depending on the type of network:

- In the *home network,* the AAA server authenticates and authorizes the MS based on requests from the local AAA server.
- In the *visited network,* the AAA server's function is to pass authentication requests from the PDSN to the home network, and authorize responses from the home network to the PDSN. The AAA server also stores accounting information for the MS and provides user profiles and QoS information to the PDSN.
- In the *intermediary* or *broker network*, the AAA server forwards requests and responses between visited networks and the home network, which do not have bilateral agreements and AAA associations.

There are two other differences between UMTS and CDMA2000 architectures. In UMTS, GSM mobile application part (MAP) supports CS-voice services and roaming, but in CDMA2000 architecture, IS-41 signaling is used. Also, in CDMA2000, roaming and mobility are based on mobile IP protocols. Here we will not go into details about GSM MAP or IS-41 networks, however we will describe mobile-IP-based roaming in Section 3.3.2.5.

3.3.2.2 3GPP2: CDMA2000 Multimedia Domain Architecture

The 3GPP2 multimedia domain (MMD) is a CDMA2000-based wireless network that provides multimedia capabilities based on IP protocols, elements, and principles. The MMD uses SIP as the primary signaling protocol and MIP as a roaming and mobility protocol. Figure 3.17 depicts a simplified view of the MMD systems architecture.

A brief comparison of 3GPP IMS and 3GPP2 MMD follows:

1. Whereas the 3GPP2 MMD architecture has followed 3GPP IMS architecture closely and there are functional similarities between these two architectures, there are some differences as well. The MMD-specific elements introduced are described below, where their similarities and difference are also pointed out.
2. There is a very large overlap in equipment providers to both groups.
3. Both require transport independence to communicate with WiFi hotspots.
4. The primary differences are transport related, not SIP related.
5. The PDSN in MMD corresponds to the SGSN and GGSN in the IMS and access gateway router in WiFi networks.
6. The CSCF in MMD corresponds to the P-CSCF, I-CSCF, S-CSCF functionality in IMS.

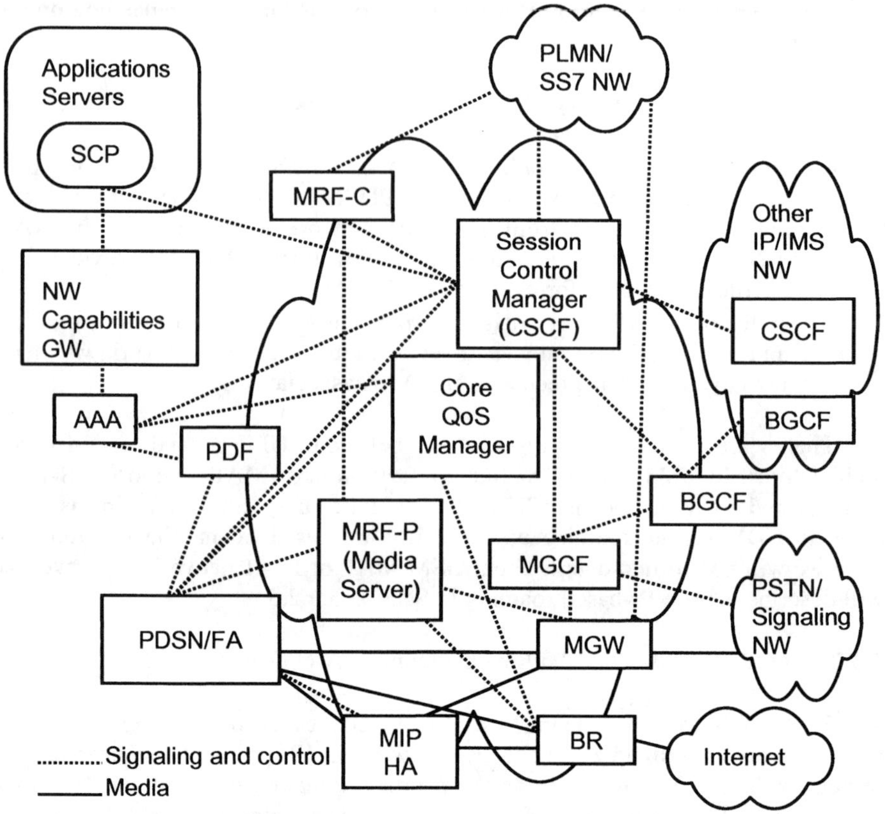

Figure 3.17 3GPP2 CDMA2000 MMD architecture. (*After:* [4].)

7. The MGW and MGCF in MDD correspond to the MGW, T-SGW, and R-SGW functionally in IMS.
8. The AAA and associated databases (not depicted in the figure) correspond to the HSS functionally in 3GPP IMS.
9. The MRF consists of two parts the MRFP and MRFC. It performs multiparty and multimedia conferencing functions. The MRF, together with the GGSN and MGW, controls the bearer and communicates with the CSCF for service validation or for multiparty/multimedia conference sessions.
10. The core QoS manager controls IP QoS for multimedia sessions.
11. In 3GPP, the policy decision function (PDF) is within the P-CSCF for 3GPP2; the PDF is a network element of its own.
12. For 3GPP, the HSS also contains HLR functions; however, for 3GPP2, the AAA function is a stand-alone element.

13. 3GPP2 uses MIP and HA based roaming. The HA tracks the location of MIP subscribers when they move from one network to another. It also receives packets on behalf of the mobile node when the node is attached to a foreign (visited) network and delivers packets to the mobile's current point of attachment.
14. The BR connects the CN with peer networks. The BR performs IP-packet routing, exterior gateway routing protocols, and policing of incoming and outgoing traffic to ensure SLAs between peer networks.

Both 3GPP and 3GPP2 are working on harmonization of the IMS reference model.

3.3.2.3 Some CDMA2000 Interfaces

The 3GPP2 has specified a number of interfaces for CDMA2000 networks. Some of the CDMA2000 interfaces of particular interest in helping service assurance are:

- The A1 interface connects BS/PCF to MSC/VLR.
- The A3/A7 interfaces connect different BSs.
- The A9 interface connects BS to PCF and may be internal to BS if PCF is incorporated within BS.
- The A10/A11 interfaces connect BS/PCF to PDSN/FA.

As with the 3GPP IMS, the MMD has introduced addition interfaces. Interfaces of interest for service assurance are:

- Instead of Mc, it has interface 30 between the MGW and MGCF (also MSC server).
- Instead of Mg, it has interface 17 between the CSCF and MGCF.
- Instead of Mh, it has interface Cx/16 between the CSCF and HSS.
- An equivalent to the Mm interface between the CSCF in one IMS and the CSCF in another IMS has yet to be defined.

3.3.2.4 Data Call Flows in CDMA2000 Networks

In a CDMA2000 network, the MS functions as an MIP client and interacts with the RAN to obtain appropriate radio resources for the exchange of packets. It also keeps track of the status of radio resources (e.g., active, standby, dormant). Upon power-up, the MS automatically registers with the HLR to:

- Authenticate the mobile device for the environment of the accessed network.

- Provide the HLR with the mobile's current location.
- Provide the serving MSC with the mobile's permitted feature set.

After registering with the HLR, the mobile is ready to place voice and data calls. Figure 3.18 depicts a data setup flow. These functional steps are as follows:

1. To register for packet-data services, the mobile sends an origination message over the access channel to the BS. The BS acknowledges the receipt of the origination message, returning a BS acknowledgement (ACK) order to the mobile.
2. The BS constructs a connection management (CM) service request message and sends the message to the MSC. The MSC sends an assignment request message to the BSS, requesting assignment of radio resources. No terrestrial circuit between the MSC and the BS is assigned to the packet-data call.
3. The MSC/VLR checks the authentication of the MS from the HLR/AuC and on authentication, confirms the authentication of the MS to the RAN.
4. The BS and the mobile perform radio resource setup procedures. The PCF recognizes that no A10 connection associated with this mobile is available and selects a PDSN for this data call.

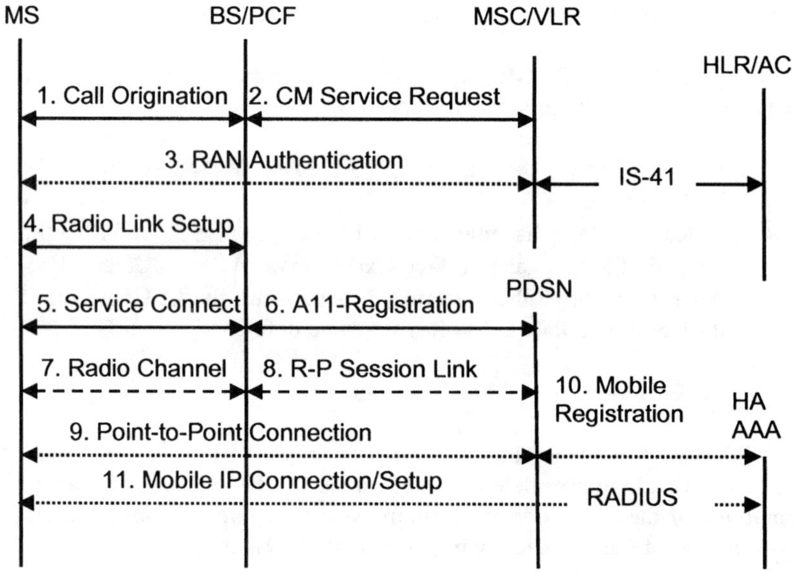

Figure 3.18 CDMA2000 call flows.

5. The MS sends a service connect request to BS/PCF. The PCF sends an A11-registration request message to the selected PDSN.
6. The A11-registration request is validated, and the PDSN accepts the connection by returning an A11-registration reply message.
7. Both the PDSN and the PCF create a binding record for the A10 connection.
8. After the radio link and A10 connections are set up, the BS sends an assignment complete message to the PDSN, and a *radio-node to PDSN* (RP)-session, is established.
9. The mobile and the PDSN establish the link layer (PPP) connection and then perform the MIP registration procedures over the link layer (PPP) connection.
10. After completion of MIP registration, the mobile can send and receive data via generic routing encapsulation (GRE) framing over the A10 connection. The PCF periodically sends an A11-registration request message for refreshing registration for the A10 connection. For a validated A11-registration request, the PDSN returns an A11-registration reply message.
11. Both the PDSN and the PCF update the A10 connection binding record. An MIP connection is now set up.

This data call process can be the source of problems that affect service assurance and quality. Usually the A1 and A10/A11 interfaces are monitored to detect service problems.

3.3.2.5 Roaming Between CDMA2000 Networks

The MS or MN needs to perform four processes:

- System acquisition that attaches (registers) the MS with the wireless network;
- Data connection setup or network bearer-level registration for authentication and authorization to gain access to IP networks;
- User roaming supports (common authentication, user profile access, and billing);
- Terminal mobility support (horizontal and vertical handoffs).

User roaming is based on the user's ability to perform the above four process from a visited network. Thus, users' roaming-related processes are covered as part of these four processes. Terminal mobility is based on the network and the user terminal's ability to maintain user voice sessions, while the user terminal moves from the coverage of one network attach point to another.

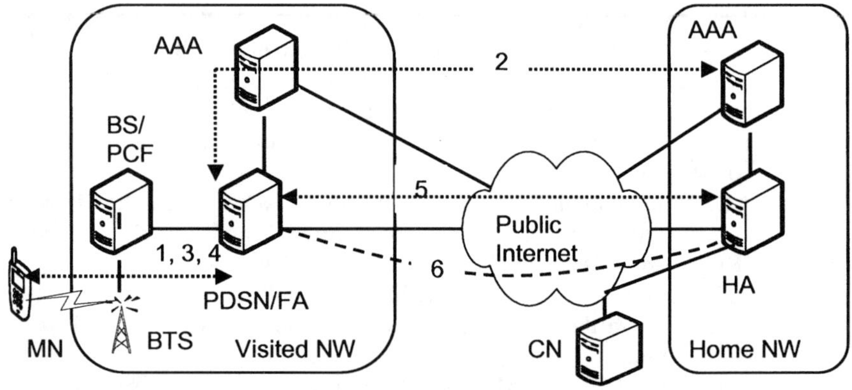

Figure 3.19 MIP-based roaming architecture for CDMA2000 networks.

Here, no mobility-specific processes are involved; it is the ability to maintain the existing processes.

Figure 3.19 depicts the MIP-based CDMA2000 roaming architecture. A high-level view of steps involved in roaming is as follows:

1. The MS, called the MN in MIP terminology, roams into a new (visited) network and sends an origination request to the base station. The PCF determines that there is no data connection associated with the MN and initiates a signal to the PDSN to set up a data connection. The MN negotiates PPP link control protocol (LCP) with the PDSN to establish the data link.
2. The PDP creates a RADIUS access request, assigning challenge-handshake authentication protocol (CHAP) data to the appropriate RADIUS attributes, and sends the attributes to the local AAA server, which proxies the request to the MN's home AAA. The home AAA returns a positive access accept message through the visited AAA to the PDSN. Now the PPP link is established.
3. The MN sends agent solicitation to the PDSN/FA looking for an available FA. The FA responds with a mobility agent advertisement, listing available FAs.
4. The MN creates an MIP registration request, specifying a willing FA as its care-of address (CoA) and forwards the request through the foreign agent to the HA.
5. The FA processes the request and forwards it to the HA. The HA processes the request and responds with a positive registration reply. It also updates its binding tables with the mobile node's new CoA association and its remaining lifetime. The FA processes the reply,

updates its visitor list with a new entry for the MN, and forwards the reply to the MN.

6. A tunnel is established between the FA and the HA. The MN can now send packet data to the host, called the corresponding node (CN) in MIP terminology, using the FA as its default router. The HA can now tunnel traffic sent to the MN's home address to the FA, which de-encapsulates the traffic and transfers it to the MN.

3.4 WIFI-3G NETWORKING INTEGRATION FOR DATA SERVICES

In earlier sections, we described access to the Internet for data services using WiFi networks and roaming between WiFi networks, and using 3G networks and roaming between 3G networks. In this section, we describe integrated use of WiFi and 3G and roaming between these two networks for data services.

3.4.1 Why Use WiFi-3G Data Roaming?

Operators would like to take advantage of growing deployment of WiFi hotspots and their complementary nature with 3G networks. For data services, mobile operators' interest is in roaming between WiFi and 3G networks enabled with dual-band, dual-mode handsets.

WiFi and 3G system integration is important for mobile data service to take off as neither of these can do it alone. The WiFi-3G integration aspects of interest to operators are:

- Access independent services;
- Low-cost infrastructure for high capacity and coverage;
- Better utilization of spectrum resources;
- Standardized interworking functionality;
- Promotion of multimode terminals;
- Flexible integration of new access networks at the IP level.

The complementary nature of the WiFi and 3G technologies in meeting cost-effective business needs is driving their integrated use. While WiFi networks are able to provide inexpensive wireless coverage for high data rates in urban hotspots, 3G is better for providing a very large, nationwide network, where the predominant need is for voice and low-rate data. WiFi will be preferred to 2.5G and 3G because WiFi has much more available bandwidth to the user at a lower cost. Integration of the best of both the WiFi and 3G networks can provide efficient use of radio spectrum, better coverage, integrated voice services, seamless wireless data services, and common billing. It will make wireless multimedia and other high-data-rate services a reality for a large population.

Operators see WiFi as a complementary solution to both 2.5G and 3G. In particular WLAN high-speed access is a good low-mobility complement to high-mobility wide-area cellular. Market needs for hotspots such as high-speed access at airports, conference centers, shopping malls, schools, libraries, and hospitals are now well recognized. By linking WLAN to cellular, mobile operators can provide significant customer value in areas such as:

- Authentication and billing relationships (e.g., a single monthly bill and service support);
- Secure communications environment and trusted relationship;
- Seamless roaming and handoff between WLAN and GPRS/3G networks;
- Access to a wide range of applications and services (e.g., MMS).

3.4.2 WiFi-3G Integration Work in Standards Groups

Both the 3GPP and 3GPP2 standards groups are developing specifications for WiFi and 3G interworking architectures that enable users to access their 2G and 3G data services from WiFi hotspots. The intent is to reuse existing protocols to a large extent, as well as to minimize, impact on WiFi standards.

The GSM Alliance document [5] described base guidelines for GSM/WiFi roaming; however, the current version does not address "tightly coupled" 3GPP architectures. There is not yet a defined standard architecture for 1x-EVDO WiFi interworking via 3GPP2. A loose integration is currently favored in preliminary drafts. The key WiFi-3G interworking requirements are the following:

1. Roaming agreements made between 3G-network operators and WiFi operators give the user the same benefits as if the interworkings were handled within one network.
2. Subscriber billing and accounting are handled between roaming partners.
3. Subscriber identification is done in such a manner that it can be used in both WiFi and 3G environments and in a WiFi/3G integrated environment.
4. The subscriber database could either be shared, or it could be separate for the two networks but share the subscriber's authentication and security information. The subscriber database could be in HLR/HSS (3G terminology) or an AAA server [Internet Engineering Task Force (IETF) terminology].
5. The user should be alerted of any possible degradation of the provided QoS due to change of access network.

There are two other recent developments that are worth mentioning. The first is the development of specifications by the Seamless Converged Communications across Networks (SCCAN) Forum, initiated by Motorola, Avaya, and Proxim in

January 2003. The second is the development of the unlicensed mobile access (UMA) architecture and specifications developed by the UMA consortium formed in September 2004. Both of these are focused more on VoWiFi and CS voice and will be discussed in Section 6.10.2.

3.4.3 WiFi and 3G Integration Scenarios for Roaming

The 3GPP document [6] has defined the following six 3GPP-WiFi interworking scenarios from the data services point of view:

1. Common billing and customer care where a mobile service operator (MSO) is responsible for one bill and customer service;
2. 3GPP system-based access control and charging;
3. Access to 3GPP system packet-switched (PS)–based services;
4. Service continuity;
5. Seamless service provision;
6. Access to 3GPP CS services.

Out of these six, cases 1 to 3 are part of Release 6, and cases 4 and 5 are in Release 7, but no user case has yet been defined for case 6.

Broadly speaking, there are two approaches for WiFi and 3G network integration. One is loosely coupled interworking and another is tightly coupled interworking. In loose coupling through AAA and HLR is used for authentication and authorization. It is for billing purposes only and is available today. Tighter-coupling through GGSN/PDSN, SGSN, or RNC is for seamless roaming and handoff, and operators are looking into deploying it. In the following section, various aspects of these two types of integration are described.

3.4.3.1 WiFi and 3G Loose Coupling

Loose coupling between WiFi and 3G allows for independent deployment and traffic engineering of WiFi and 3G networks. Loose integration makes most sense because it allows enterprise WiFi, public hotspot WiFi, residential WiFi, and operator WiFi access. Figure 3.20 illustrates examples of both loose and tight coupling. For loose coupling, there are two subcases:

1. Open coupling at the CBS-Level (L-1 in Figure 3.20): Here customer care and billing system (CBS) is the common link between two independent WiFi and 3G networks. This is purely administrative integration.
2. Loose coupling at common authentication (option L-2 in Figure 3.20) and the CBS level. This option includes option 1 above, plus common authentication. As depicted in Figure 3.20, common authentication requires a new interface between the AAA-HLR/HSS link and the inter-working Unit (IWU). These interworking functions are of two types:

- Since in 3G networks MAP is used to communicate with HLR, and UEs have UMTS subscriber identity module (USIM) cards for authentication, on the WiFi side, UMTS-IWU is required to support interoperability between WiFi's local AAA (AAA-L in figure) and the 3G CN's HLR or HSS.
- Similarly, since in 3G networks DIAMETER/RADIUS is used to communicate with the home AAA (AAA-H in figure) in the 3G CN, and WiFi UE does not have USIM, on the WiFi side, IETF-IWU is required to support interoperability between WiFi's AAA-L and the 3G network's AAA-H.

It also requires specific NICs in UEs if SIM–card based-authentication is used by WiFi.

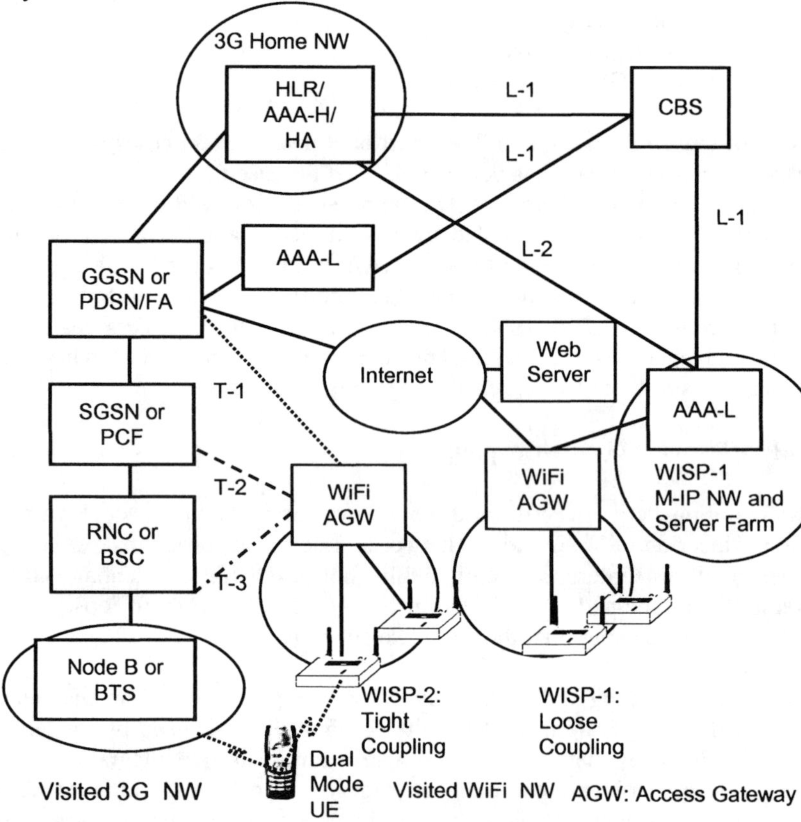

Figure 3.20 WiFi and 3G loose- and tight-coupling architecture.

3.4.3.2 WiFi and 3G Tight Coupling

In tight coupling, the WiFi network acts as another 3G access network. By interjecting the WiFi traffic directly into the 3G core, the setup of the entire network, as well as the configuration and the design of network elements such as SGSNs and GGSNs have to be modified to handle the increased load. Figure 3.20 illustrates the following three options for tight coupling between WiFi and 3G networks:

1. *Tight coupling at the GGSN level:* In this option (option T-1 in Figure 3.20), WiFi, instead of using its own packet-data gateway (PDGW), uses 3G network's GGSN as its gateway to PDN. In this case, the WiFi network is seen as a RAN by the 3G network.
2. *Tight coupling at SGSN routing area management level:* In this case (option T-2 in Figure 3.20), the group of WiFi APs, called ESS, forms a routing area (RA) and is seen by the 3G network as a UTRAN RA. The ESS is connected to an SGSN via a GPRS interworking function (GIF) (within WiFi AR or in a separate unit) using a Gb or Iu-ps interface. The GIF makes the ESS look like a typical RA composed of one cell. UEs scan for a beacon signal to see if the WiFi belongs to the wireless operator. If it does, the UE associates itself to the WiFi and, after that, to the 3G with an RA update. The exchange of packets is via GIF using MAC addresses; and from the 3G CN perspective, hand-over between WiFi and 3G is considered hand-over between two individual cells. All standard 3G protocols operating on the logical link control (LLC) function as usual.
3. *Very tight coupling at the RNC cell management level:* In this case (option T-3 in Figure 3.20), a new lu (RNC-WiFi) interface is introduced such that WiFi is seen as a cell at the RNC level. Again, this requires specific NICs in the UEs.

While such tightly coupled, integrated systems are suitable for the case where both hotspots are operated by the mobile operator, loose coupling with hotspots managed by other operators will still be required.

3.5 CONCLUSION

In this chapter, we first reviewed WiFi technologies, systems architectures in enterprise in hotspot environments, and roaming between WiFi hotspots. Then we described briefly 2.5G and 3G systems architectures for both 3GPP-based UMTS and 3GPP2-based 3G CDMA2000 networks and roaming between these 3G networks. A brief overview of 2.5G GPRS network and WAP over GPRS service was provided. In Chapter 5, WAP over GPRS service is used to illustrate the fundamentals of building service models. We have provided the necessary background, on WiFi and 3G networks and their interworking, network domains and network elements, and roaming aspects, to help explain further details of building service models and VoWiFi/3G services networking and architectures. In Chapter 6, VoWiFi/3G services networking and architectures are built on the WiFi and 3G networking architectures described here. The VoWiFi/3G details in Chapter 6 are then used in developing service models for these services and their use in service assurance in Chapters 7 and 8, respectively.

References

[1] Intel, "Wireless LAN (WLAN) End to End Guidelines for Enterprises and Public Hotspots Service Providers," (Release 1.1), November 2003 http://www.intel.com/business/bss/infrastructure/wireless/deployment/e2e_wlan.pdf.

[2] 3GPP, TS 23.002, "Network Architecture" (Release 5), 2003.

[3] 3GPP, TS 23.228, "IP Multimedia Subsystem (IMS)" (Release 5), 2003.

[4] 3GPP2, 3GPP2.X.S0013-000-0, "All-IP Core Network Multimedia Domain, Overview," 2005.

[5] GSM Association, "WLAN Roaming Guidelines," Document IR.61, Version 3.1.0, August 2004.

[6] GPP, TS 23.234, "3GPP System to WLAN Interworking," Release 6, 2005.

Selected Bibliography

3GPP, TS 23.234 v6.1.0 (2004-06), "3GPP System to Wireless Local Area Network (WLAN) Interworking, System Description," Release 6, 2004.

Alcatel, "Wireless LAN for Mobile Operators, WLAN Beyond the Enterprise," White Paper, http://cnscenter.future.co.kr/resource/rsc-center/vendor-wp/alcatel/T0210-Wireless_LAN-EN.pdf.

Black, U., and U. Black, *Second Generation Mobile and Wireless Networks*, Upper Saddle River, NJ: Prentice Hall, 1998.

Farpoint Group, "Wireless LAN Infrastructure Mesh Networks: Capabilities and Benefits," White Paper Document FPG 2004-185.1, July 2004, http://mithras.itworld.com/download/farpoint/wlan_mesh.pdf.

Garg, V. K., K. Smolik, and J. E. Wilkes, *Applications of CDMA in Wireless/Personal Communications*, Upper Saddle River, NJ: Prentice Hall, 1997.

Lin, Yi-Bing, and I. Chlamtac, *Wireless and Mobile Network Architectures*, New York: John Wiley. 2001.

Minoli, D., *Hotspot Networks: WiFi for Public Access Networks*, New York: McGraw-Hill, 2003.

Chapter 4

OSS Base Platform Functionalities and Technologies

The emergence and growth of numerous new telecommunications companies raises the question of what OSS can do to enhance their competitive position against their counterparts [1]. As most "hard events" (fault events) are well managed by all operators, we propose that "soft events" (performance events) and the associated management mechanisms hold the key to this advantage. The amazing part about this solution is that new values are created through the collation and presentation of data that has always existed within the network. With its efficiency and transparency, this new method may bond a customer to a specific service like electronic glue and become an enabler of new levels of customer satisfaction.

To explain how an integrated, model-driven architecture can fully support ongoing service assurance functionality, we first examine the traditional OSS operations of fulfillment and billing. It is important to paint the entire operational picture before we present a detailed discussion of the service model implementation in Chapter 5. In this chapter, we will examine the OSS processes and flows in the following order:

- OSS definitions, business requirements, and architecture;
- OSS infrastructure, flows, and life cycle considerations;
- The hierarchy of OSS components and their functionalities, including fulfillment, assurance, billing, and revenue support processes;
- The future direction of OSS.

4.1 OPERATIONS SUPPORT SYSTEMS

OSS is a set of software that helps a telecommunications service provider install, monitor, control, analyze, sell, and manage telecommunications-related services. This collection of functionalities might include equipment inventory, service ordering, usage tracking, billing, reporting, and customer relationships.

Fundamentally, OSS provides mechanization for service provider's operations. The OSSs are also called network operations support systems, operations support systems, operation support systems, and operation support software.

From the customer-experience perspective, the task of the OSS is to ensure the specified quality of end-to-end services throughout the service life cycle. This is accomplished by managing the quality of the supporting subservices from a service or third-party providers. Telecommunications networks and the OSS that control them have become much more sophisticated and mature in recent years. The equipment (network elements) have become "smarter" through the use of embedded software controls. This has given rise to a new view of the use of OSS. The robustness of the OSS makes it a competitive tool that can be leveraged by operators as well as customers. This trend is emerging as a new vehicle for customer retention, adding new value to existing customer care for customer-satisfaction improvement, thereby increasing revenue per customer.

The OSS provides methods and procedures to access managed service objects across the entire network infrastructure and manage the business itself. Figure 4.1 depicts the ITU-T's Telecommunications Management Network (TMN) model [2]; demonstrating how an OSS solution is integrated, in building-block fashion from the network element to the business management layers. This figure illustrates how each layer within the structure contributes to the overall success of the business. Expanding on the TMN model, the TMF developed the Telecom Operations Map (TOM) [2], expanding the layer concept to encompass the process implementation and flows (Figure 4.2).

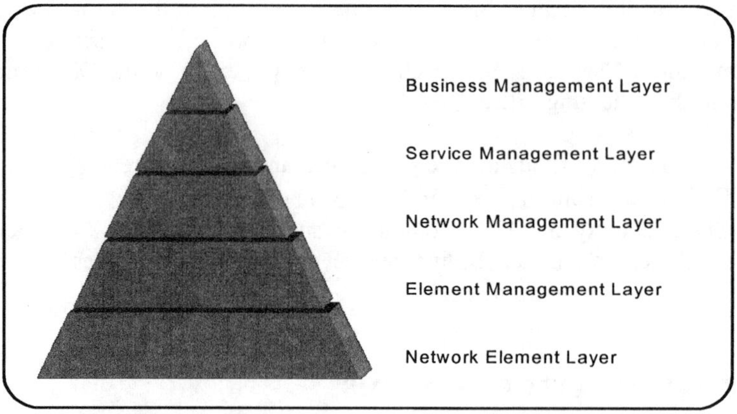

Figure 4.1 ITU-T, Telecommunications Management Network model.

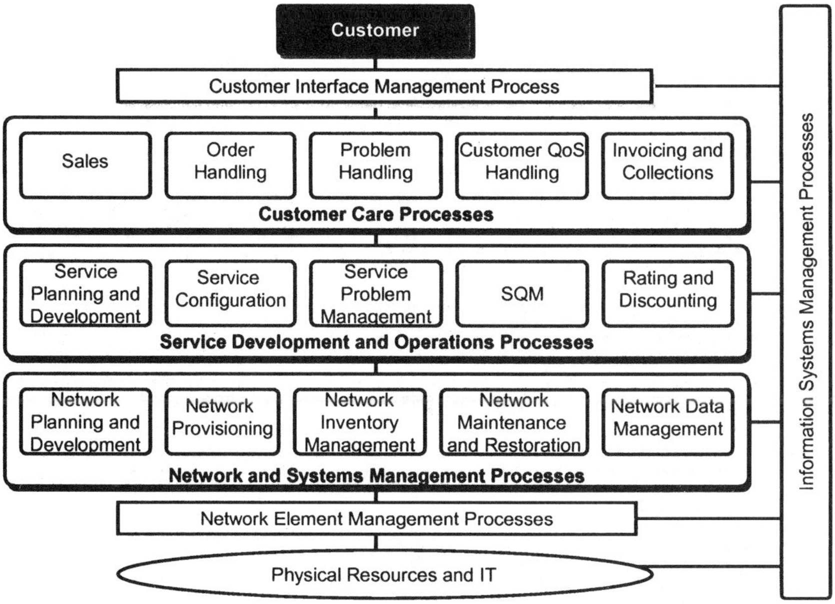

Figure 4.2 TMF TOM. (*From:* [3]. © 2005 TeleManagement Forum. Reprinted with permission.)

The TMF's TOM [3] uses customer interface management at the top layer to drill down to the lower-layer processes. As shown in Figure 4.2, the TOM regroups the TMN model into three basic, end-to-end processes common to any service-oriented business. These three categories are referred to as *fulfillment*, *assurance*, and *billing* (FAB). The fulfillment process deals with timely and accurate provisioning of the customer order. The assurance process deals with maintaining the service through timely responses and resolution of customer- or network-triggered problems. This includes actions to improve the performance of all aspects of a service. And the billing process deals with timely and accurate bill preparation, knowledgeable and responsive billing inquiry support, and timely adjustment-handling and payment operations.

The TOM has been further enhanced in the enhanced Telecommunications Operations Map (eTOM), (see Figure 4.3) [4], which addresses increased complexity in a service provider's business relationships and processes. It specifically adds processes to address the needs of supply chain life cycle management, supplier/partner relationship management, product life cycle management, and infrastructure life-cycle management. The eTOM structure establishes the business process framework and serves as the blueprint for process direction.

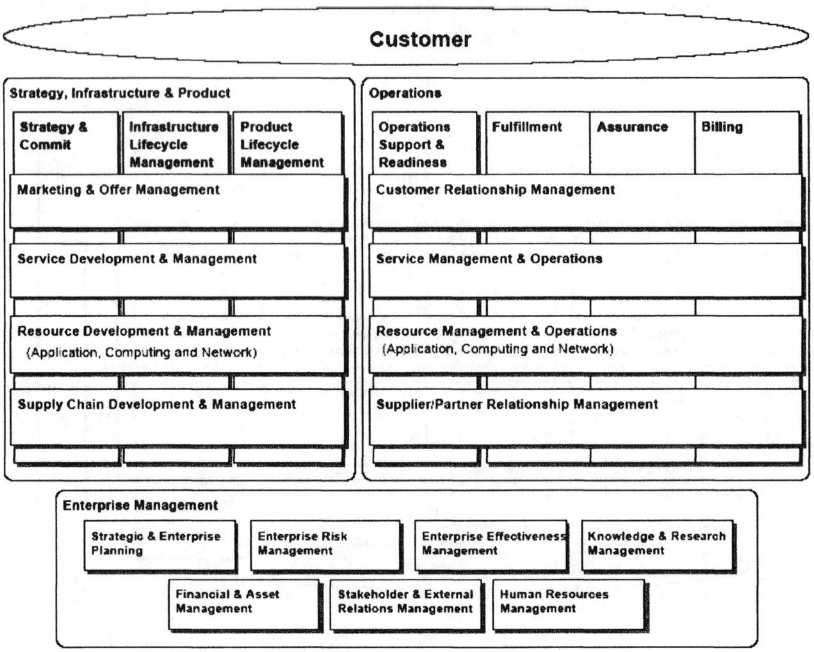

Figure 4.3 TMF eTOM. (*From:* [4]. © 2005 TeleManagement Forum. Reprinted with permission.)

The eTOM represents the starting point for the development and integration of BSS and OSS. Similar to the TOM, the eTOM analyzes all of the business activities of a service provider and categorizes them into different levels of detail, according to their significance for the business. In 2004, The ITU completed the ratification of the TMF's key eTOM as an official ITU standard. The ratification is the last link in the chain of documentation and moves the eTOM to formal adoption status with the publication of the ITU recommendation.

For service providers, it provides a neutral reference point as they consider internal process reengineering needs, partnerships, alliances, and general working agreements with other providers. For suppliers, the eTOM framework outlines potential boundaries of software components and the required functions, inputs, and outputs that must be supported by products. As BSS and service providers' enterprise management are outside the focus of this book, we will use the TOM architecture to illustrate the operational aspect of service assurance.

4.1.1 Alignment to New Business Objectives

Since the network-centric view was, until recently, believed to be the key differentiator, the telecommunications industry still maintains a network-centric culture for quality and reliability of service. Long-term investment and attention to this area have made network quality and reliability a standard expectation for customers. As superior customer care becomes the new differentiator and operators become more customer focused, more attention is being directed toward the changing relationship between customers and services that deliver specific content and the convergence of technologies. The new OSS direction must become an enabler in developing a customer-centric culture among service providers.

In a customer-centric company, the customer care center becomes the conduit to deliver exceptional customer service. New methods must allow the customer care center to participate with any service-affecting process in the entire OSS infrastructure. KPIs and KQIs, whether they are business or network related, will empower the operator's processes to deliver the exceptional customer service expected by the market. The service provider must have the flexibility to segment the market by end customers' expectations or experiences, providing individualized customer care in real time. Real-time or near-real-time KPI/KQI's availability and notification features to both end customers and the operator shift the strategy of customer care from defensive to proactive support.

From the service assurance perspective, most service providers have been focusing on a set of network-facing processes for reporting hard events and resolving network and service issues. With the increased range of services now being offered and the evolution to a competitive market-driven environment, new demands have surfaced in an effort to deliver higher service levels to most valued customers (MVCs). Problem identification, service-impact analysis, and service-affecting issues are dealt with in a prioritized manner from the point of view of the business bottom line and the provider's corporate strategies. Some of the provider's new assurance strategies are expected to perform the following:

- Manage customer revenue with an end-to-end service view.
- Enrich customers' experiences by providing real-time service intelligence.
- Create new market opportunities or capitalize on the existing business with this new differentiator.
- Incorporate tightened service assurance to prevent revenue losses due to customer dissatisfaction.
- Enhance other high-value services and differentiate the provider's brand.

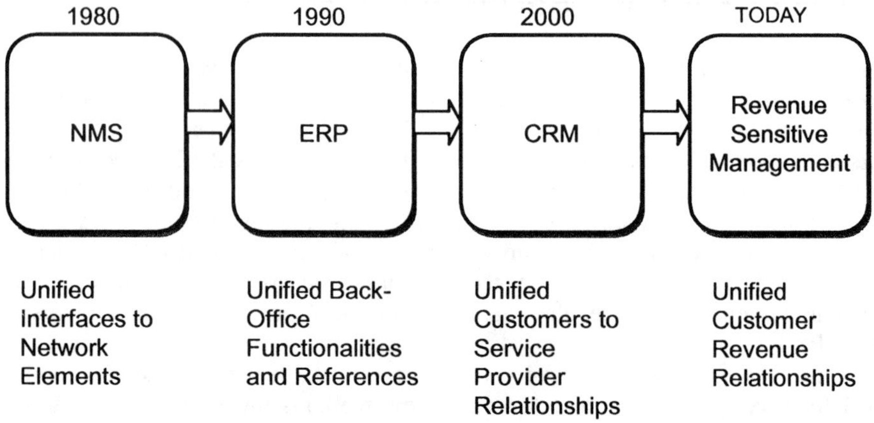

Figure 4.4 The evolution of OSS foci.

Figure 4.4 shows the evolution of OSS focus from pre-1990 to today. OSS emerged from an NMS focus, to enterprise resource planning (ERP), to CRM, and is moving toward *revenue sensitive assurance management* of telecommunications companies. By providing end-to-end service assurance across service offerings, networks, devices, and applications with a revenue sensitive cognizance, revenue sensitive management promises to revolutionize the way service providers do business.

Revenue sensitive assurance management comprises programmable methods that can detect and react to the service quality and provide plug-ins to integrate with existing business-critical systems, such as finance, production, purchasing, personnel, or sales, and suggest actions to protect a service provider's competitive edges. With the new model-driven algorithm engines, service providers finally have the ability to measure many previously unmeasurable service behaviors. This new capability can help operators improve their new market opportunities and quickly launch new service offerings with tangible references. Using the new business mindset, for example, service providers can track technology innovations and react smartly to the service attributes most beneficial to the providers' business practices.

4.1.2 Business Process Considerations

In today's market-driven environment, the service provider has to deliver multiple sophisticated services from their multitechnology, multivendor service (networks) infrastructure, while new demands are being placed on the service-management function to align its activities and procedures to be able to commit to higher levels of service. Below are a number of high-level process requirements where service providers may want to consider design changes.

- *Service-focused operations:* Current NMSs are more network element or technology focused, and existing network performance metrics are different from the customer-perceived service-level metrics. Therefore, the service provider is often unable to evaluate the network from a service or customer perspective. Lack of well-correlated service-level metrics will minimize the visibility of the customers to their end-to-end service experience and limit their ability to express their true troubles. Well-correlated metrics with appropriate performance references can sometimes provide potential-trouble forecast capability that deals with SLA credit in a more proactive manner. As a result, this can enforce probabilistic models of financial impact of outages or troubles. Other than the capability to care for troubles, the service provider can also use service intelligence to identify, for instance, which customers are candidates for service upgrades. How is the health of a service that is newly launched? And what part of the network is causing poor customer perception? These are the areas where service differentiators can be addressed.
- *Connecting back and front offices:* Customer care centers frequently lack timely and accurate information about current service problems. Many times, customers call to complain about a "service outage" before the operator knows about it. Occasionally, both customer specialist and network specialist work in parallel on a disconnected, but related, problem—adding to the cost and time of ultimate problem resolution. Consolidating and integrating service intelligence into a real-time service management view can pinpoint root causes of problems more effectively. Capturing and referencing historical experience to prevent problems repeating the past would be useful for optimizing the problem-resolution process. Depending upon the service infrastructure, integrating third-party service provider network data streams can sometimes increase overall QoS, for instance, the roaming metrics, which would enable the NOC to react quickly to any malfunctions occurring in the field of the services.
- *Bridging the silos of multiple support systems:* There could be multiple support systems in an operator's center, one from each network equipment supplier. Because business focuses from different vendors may vary, it would be difficult to assemble a unified view from the service resources. The user interfaces, the naming of the resources, and the actions all have different meanings in the different systems. A large number of uncorrelated alarms appearing on multiple consoles often compound a problem when it occurs. It is essential that the operator can establish a single, integrated view of each customer; furthermore, all service management applications must share the same customer-related information, associated service instance information, and corresponding

network status details. Integrating service assurance systems with a single service model can enable interoperability with other operations (for instance, trouble ticketing systems) and business support systems to provide the visibility of services for greater operational efficiency.

- *Revenue sensitive management:* During an OSS system's implementation, it is important to keep the focus on how to optimize customer value, maximize profitability, and enhance business agility. Revenue sensitive management should be the driver of the service providers' OSS design. As the power of revenue management is achieved through integration of many different OSS functionalities, it is worthwhile to emphasize here that revenue management should be considered a life-cycle approach in the strategy, infrastructure, and product building blocks discussed in eTOM (Figure 4.3). In other words, when the designers of the service provider are working together and integrated with existing business-critical systems, the elements of revenue management (capture, collection, assurance, and generation; see Figure 4.5) can deliver a powerful new business weapon for creating competitive advantage.

4.1.3 Integrated OSS Architecture

The fact that service representation and customer relationship information reside in the common service model of the OSS enables all operations personnel to share a common view. Operations personnel can use a unified portal to monitor, in real time, the performance of the key processes as well as customer interactions. Customized views for all of a customer's service can be built that would support drill-downs into a specific service or interaction. No matter which operations person a customer speaks with, they should all have the same service view.

The importance of a single view of the customer cannot be understated. The view can be translated into the service provider's enterprise goals or missions to serve their customers. When the provider's service view is in accord with its offering throughout the entire operation, suddenly all employees become touch points with the customer and their messages to the service users concerning the when, what, and how of service delivery and marketing directions will be consistent and factual.

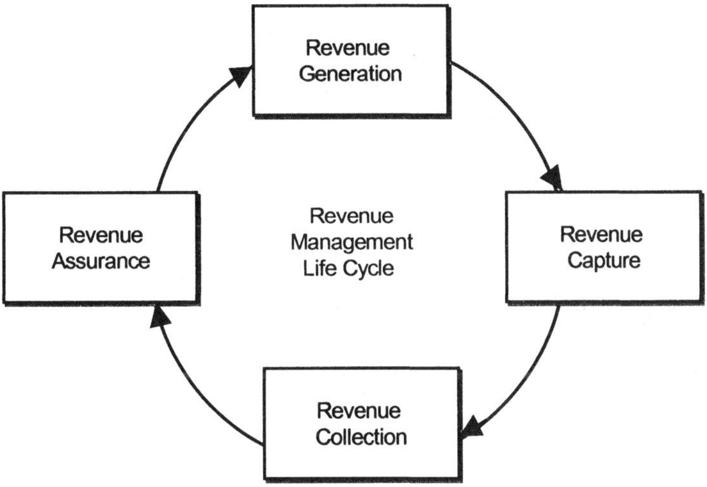

Figure 4.5 The revenue management cycle.

TMF's New Generation Operations Software and Systems (NGOSS) [5, 6] defines an architecture and methodology that can support business, system, and implementation views of OSS and BSS component-based solutions. The goal of NGOSS is to facilitate the rapid development of flexible, low-cost-of-ownership OSS/BSS solutions to meet the business needs of the Internet-enabled economy. NGOSS targets the use of commercial off-the-shelf information technologies, instead of technologies unique to the telecommunications industry, as many legacy management systems have done in the past. This approach significantly reduces costs and improves software reuse and operational flexibility, enabling NGOSS-based systems to support a range of new services and new technology environments more easily. NGOSS emphasizes a service-oriented approach based on the integration of well-defined collaboration contracts.

There are many similarities between the TMF's NGOSS program and the Object Management Group's (OMG) [7, 8] Model Driven Architecture (MDA; Figure 4.6) initiative. Both NGOSS and MDA strive to support cost-effective system specification and component integration through shared information models, and mappings for moving from technology-neutral models to platform-specific implementations. The MDA defines a system specification that separates the system functionality from the actual implementation for a specific technology platform. To accomplish this goal, the MDA defines an architecture for models that provides a set of guidelines for structuring specifications, which are expressed as models.

The idea of using models to drive business in IT is not new. The OMG has been driving such an idea since 1990 with the Object Management Architecture

(OMA). The OMA provided the vision and road map for the problem of integration. Having created the Common Object Request Broker Architecture (CORBA) interoperability standards, the OMG has used them for creating standards in particular application domains. Since 1997, the scope of the organization has broadened significantly. In 1997, the OMG issued several important specifications that are not CORBA based, including the Unified Modeling Language (UML) and the Meta Object Facility (MOF), and later, the XML Metadata Interchange (XMI) and the Common Warehouse Metamodel (CWM).

It is important to note that the MDA is a proposal to integrate all of the work done to date for the OMA, and to point the way to future integration standards.

In the following sections, we will discuss OSS from different viewpoints. In Section 4.2 we will look at the implementation aspects of OSS. System architecture, data and control flows, interoperability, and system design will be addressed. In Section 4.3, we will focus on operational aspect and follow the TMF's TOM model to analyze a typical service provider's process flows and interactions between functional blocks.

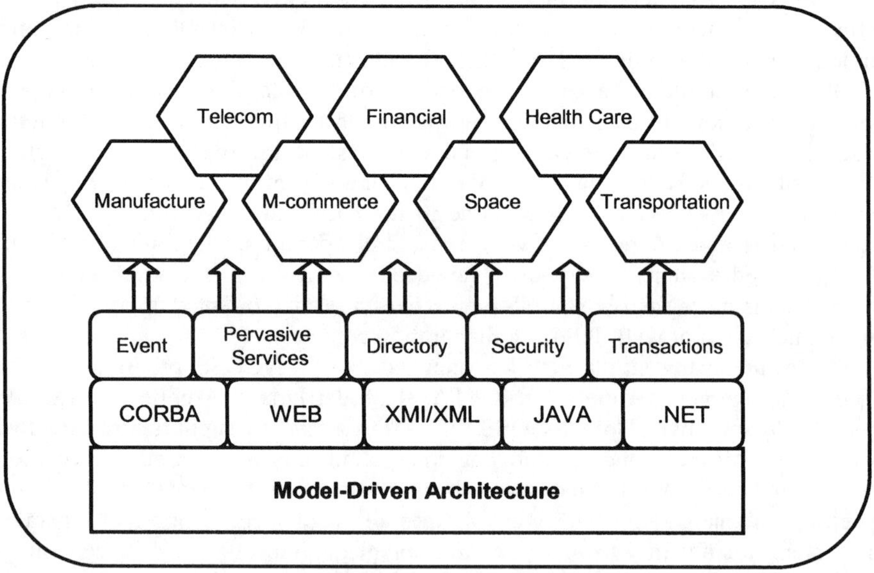

Figure 4.6 OMG's model-driven architecture. (*After:* [7].)

4.2 OSS INFRASTRUCTURE, FLOWS, AND CYCLES

As noted earlier, the model-driven architecture is not an independent task or functional silo of OSS but an integrated feature to ensure the service providers' successful operation on a daily basis. The preceding sections have focused on the business perspectives of model-driven OSS architecture. However, equally important are the technical foundations and resources to support the implementation of the new architecture. The underlying principle behind a successful model-driven OSS architecture is a dynamic and common OSS infrastructure that can augment the end-to-end operations tailored for a new breed of real-time service and interactive applications.

The definition of an OSS infrastructure is the set of common functionality and information supporting the specification, design, implementation, and monitoring of its operational building blocks and its constituent parts. Because service configuration and dependencies are coupled relatively statically across technologies, they can be managed in common. On the other hand, many dynamic interactions, such as traffic policies for IP, rely on a large number of parameters that are technology specific, and a common solution would not be an effective choice. Exceptions may include certain relationships such as hosting whose change must coincide with assignment and activation of circuits; in this case, using a common configuration would be preferable.

As for common functionality for an OSS infrastructure, it is expected to support (1) assignment of planned configurations, (2) activation of services either automatically or manually, (3) reconciliation of reference information regarding service status, (4) visualization of a common service view, (5) reference for other OSS components to exploit common configuration data, (6) consolidation of customer care status, and (7) correlation and isolation of fraud or troubles.

The common information set includes three parts. The first part is *common configuration* information about network logical and physical resources, such as applications, software, equipment, and site facilities. They enable OSS processes to manage reconfiguration, provisioning, fault, and performance information across technologies. The second part is *relationship* (connectivity) information, which is overlaid on the common configuration. It consists of physical or virtual connections set up on a standing basis at the model-creation stage. For instance, a circuit may run between two or more service elements, which may be of different technology types. The combined configuration and relationship picture provides an end-to-end view of the service relationships. The third part is *common status* information about customer care and customer classification. Modern CRM is established on the basis of proactive and interactive care. A common and accurate customer's service-level index and care records can help service providers to provide the right attention to the right customers in real time.

In the following sections, we will discuss the solution approaches for the model-driven OSS architecture, OSS activity life cycles, and OSS control flows.

4.2.1 Solution Approaches

The fundamental strategy for an integrated model-driven OSS is commonality. Common functionality and common information must be coordinated across technologies, while other functions and resources can be managed separately for each technology. It is important to mention that after an integrated model-driven OSS is implemented, all the OSS processes will be so closely related that it would be difficult, for example, to single out fulfillment and billing from the service assurance process. Performance and fault management also use common configuration and service model information as a reference; therefore, the strategies for performance and fault management also influence the choice of solution for configuration management.

Nevertheless, the building-block architecture, such as that used by TOM, would still be valuable to address the relationships between the service providers' business and operations. The additional view point we would like to present in this book is that we would like to consider the entire set of a service provider's operations and business goals so that we may take full advantage of the benefits offered by the new OSS concept.

The possible solution approaches for integrated OSS fall between two extremes, namely bottom-up and top-down solutions. A new solution is suggested to support a combination of the two. Figure 4.7 illustrates the relationships from the service assurance perspective.

As depicted in Figure 4.7, the service assurance process is a cyclic flow. After the service deployment, bottom up, it starts with the ongoing monitoring of network and service elements and usage-data collection. Monitored performance and usage data, using a data-collection infrastructure and data mediation, and network topology and service dependencies allow fault and performance management at the nodal and domain levels.

Further use of the service model helps in event correlation, prioritization, service-impact analysis, and estimation of revenue impacts. Additionally, a bottom-up solution enables a customer-centric workflow with the creation or invoking of automated actions within workforce and customer-management systems. After this identification of service and revenue impacts, top-down solutions include performing SLA management and drilling down for troubleshooting to find the root cause of problems, and restoring service. A combination of bottom-up and top-down solutions has the potential to yield the most useful and accurate customer correlation and improvements in efficiency and productivity.

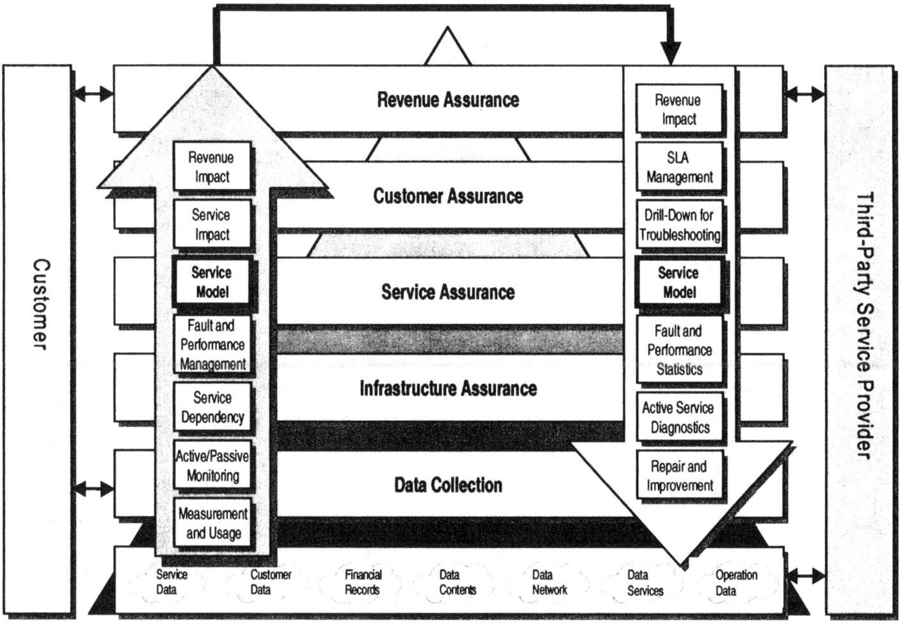

Figure 4.7 Bottom-up and top-down assurance flows.

4.2.1.1 Approaches to Technology Management

There are three approaches to managing telecommunications technology, and the appropriate solution might be a combination of these approaches.

- *Single common solution:* At one extreme is a single solution to manage common information across all technologies and aspects. The OSS infrastructure would provide a single point of interface with a common configuration that would activate and detect circuits, equipment, and possibly software. It would also provide service-related information such as alarm analyses, performance analyses, common user views, and workflows. Optionally, this information could also be made part of an integrated solution. The complexity of this implementation comes from (1) a lack of flexibility to handle arbitrary equipment details and connection types, (2) dynamic validation rules to manage resources, and (3) a closed or nonconfigurable interface for system interactions. To manage a customer-centric service that includes an end-to-end view may substantially increase the difficulty and cost of solutions.

- *Separate management solution:* At the other extreme is a separate tailored management solution or suite to manage individual technologies in isolation. This means each solution would maintain its own records of managed service resources. Because each technology has a dedicated suite, the interface for the logical design model (element configuration management) and assurance functions (performance and faults) for each technology can be isolated and is very straightforward. The implementation of common views or references, in particular the service configuration and service model for cross-technology solutions, would need separate interfaces with each sub-service or system. This option might be appropriate if the service provider decides to manage different technology subsets independently at the service level. This is particularly true for small or specialized third-party service providers. Generally speaking, this option presents potential complexity and very high development costs when multiple interfaces with other technology managers or integrated solutions are needed.
- *Service-based common solutions:* Between the above two extremes are intermediate solution approaches. Services can be divided by technology (or related technology groups) and common operations can be handled by different sets of vertically integrated solutions. Here the service provider manages different technologies with dedicated systems, which would allow data regarding performance, fault, and configuration to be synchronized with the general solution. In reality, some technology-oriented configuration functions still need to be synchronized between the separate solutions. Workflows would either be separate for each type of sub-service or would need to interface with each configuration system. This solution coordinates across a collection of element and network managers and provides more specific functionality and information than the single common solution. Any overlaps of information between subsets can be coordinated by the interfaces.

Depending upon service providers' corporate strategy, some primary service providers can use their market size as power to enforce their solution providers (vendors) to develop common solutions following open interface standards or even influence the specification of interface standards. For other service providers, option three would be more preferable because of its flexibility and potential dynamic to move to a true common architecture.

4.2.1.2 Integration with the Operational Processes

As discussed in the previous section, *infrastructure references management* helps the service provider to manage configuration integrity across service lines, and provides a portal for all OSS processes to access unified configuration information. Infrastructure references can be divided into the following two categories:

- *Service resources:* Common configuration is defined bottom up, starting from the network elements and ports utilized to the service-level features that are explicitly configured. All managed objects defined as structured records in the common service resources database should be displayed graphically. The database can be automatically reimported from service and network configuration managers either automatically or triggered by a workflow process. Service resources managers should extract common configuration information from element-level configuration systems. The element-level configuration systems bridge the gap between the logical model configuration and vendor-specific executions. An autodiscovery feature can assist the service resource manager to keep up-to-date on all configuration changes. It should be able to cross-check integrity with the records entered by the workforce or workflow managements and to reconcile discrepancies either by predefined rules or by manual investigations. Where major rollouts or upgrades may impact operational elements or services, the relevant configuration sets would be executed via change control either by automatic workflow management or manual coordination.
- *Service model:* The purpose of the service model is to consolidate common information from multiple source components and provide a top-down reference for services. The service model could be implemented as a virtual or physical model, or as a combination of the two. In principle, separate logical models could be used for each technology or specific customer group. A service model should allow the service manager to drill down to the lowest level of subservice where quality needs to be explicitly managed. A complete service model would include common configuration, service quality monitoring, and workflows for troubleshooting, changes, and possibly upgrades. From a functional perspective, it would provide a reference of status for service resources made available by various OSS. As part of a revenue sensitive solution, it should include an indication of revenue impacts. The impacts can be based on a direct classification of usage or an indirect inference by weighted metric calculations. The dependencies of a service model should be qualified (e.g., availability) or quantified (e.g., percentage of degradation) by the behaviors of the service being monitored. Change control workflow, fault management, or performance management can trigger the states of dependency. In order to manage a large number of metrics, a comprehensive service model should provide the ability to define the baseline thresholds automatically based on historical data and

support what-if scenarios for case study, planning, and troubleshooting purposes.

4.2.2 OSS Information Flows

An SLA can be used to specify business relationships for internal, external, or third-party players. For the service provider to perform effective service management, it is important to clarify the responsibilities of internal or third-party sub-services that participate in end-to-end services for ensuring the specified end-to-end QoS levels. There are three types of OSS information flows, and depending upon the service providers' corporate strategy and business focuses, each flow may have a number of categories. Figure 4.8 summarizes the major flows.

The configuration information flow reflects the life cycle of a configuration change. This flow may be triggered by (1) introduction of a new service, equipment, or technology, (2) changes to the existing service capacity, (3) upgrades to the existing QoS levels, or (4) fixes to faults or technical problems. Much of this information enters the OSS in the form of service orders. The configuration status and activities are recorded in the workforce management system or trouble ticketing system.

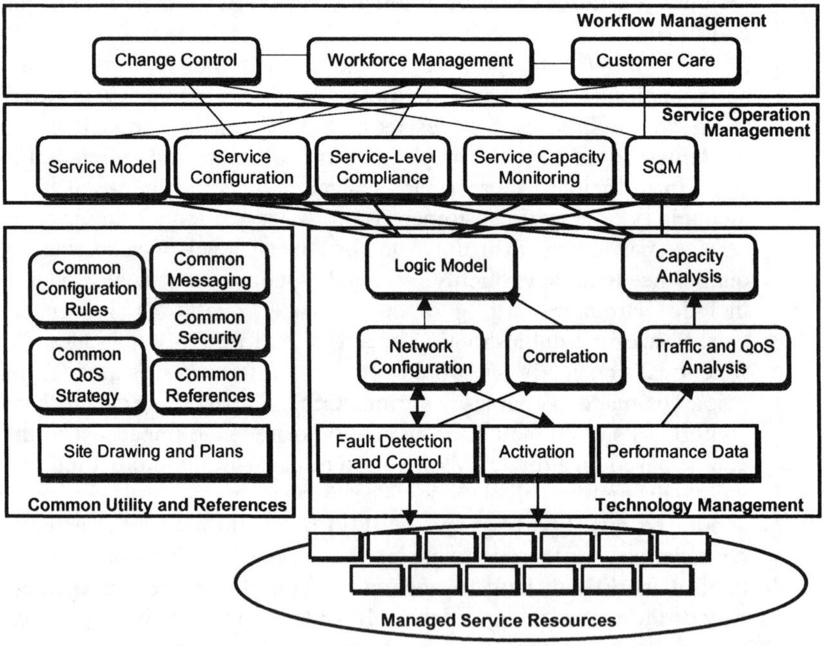

Figure 4.8 OSS information flows.

The result is a closure of a service order or trouble ticket. Customer surveys that may be performed after the completion of this service should be kept in the CRM system.

The faults information flow determines symptoms and isolates hard events such as technical problems that impact services. The fault information flow may be triggered by (1) configuration errors, (2) capacity bottlenecks, or (3) equipment or connection failures. The event data collector gathers the service events associated with particular parameters of components. The fault agent processes events locally. If the agent decides these events represent problem conditions, they can be forwarded to the fault manager. The fault manager can apply root-cause correlation or deduplication rules to identify the problem cause and decide what actions to take. Recovery functions are called when a response to a fault occurrence is required. A notification is then sent to an administrator for investigation or to other OSS components for further actions.

The performance information flow is similar to the fault information flow, but the processes of making the collected information sensible for appropriate performance actions are much more complex. The performance manager must first determine that service degradation has occurred. To accomplish this it must correlate information from many sources and apply appropriate algorithms. A separate determination must them be made regarding an appropriate course of action. Fundamentally, the performance manager would collect the following information: (1) bandwidth/speed, (2) utilization, (3) delay, (4) availability, (5) test results, and (6) outages/errors. These measurements are compared with predefined QoS threshold specifications. Rules and algorithms are then applied to create improvement recommendations. Trending of historical information is also useful to predict future performance and provide proactive maintenance recommendations. These various analyses and recommendations are presented in the form of trouble tickets to support the optimization of configuration and to upgrade and plan for capacity. In other applications, the performance manager monitors operational performance such as MTBF or MTTR for SLA compliance. A more intelligent performance manger can proactively request additional performance data from network data collection or network test management for further analysis.

4.2.2.1 Design Considerations

OSS information flows at the service level are largely driven by the overall strategy for managing the network; hence, they form the overall approach to an integrated OSS. The key strategy decision is whether to manage services by an end-to-end or domain specific approach.

An *end-to-end service instance* is envisioned as a round-trip interaction or session, experienced by a specific user, and the result of such interaction generates one or more usage records. Regardless of how many different service resources or technologies support the service instance, customers only experience the end-point service quality. Any bottleneck that occurs in any section of the end-to-end information flow will impact the entire usage experience. In order to assess a fair user experience against an end-to-end service instance, the service provider typically monitors their MVCs or most valued classes of customers to evaluate the service availability and quality against specific service levels. The goal is to produce a unified customer-centric view of information flow for configuration, fault, or performance. Collaboratively gathering measurements from different sources and performing vendor-neutral and technology-independent analysis for an end-to-end service can be a challenge, especially for new technology integration. Service dependencies, performance, and faults would all be managed from an end-to-end perspective and integrated with the detailed management of all the supporting technologies. This strategy assumes that the dependencies between different types of subservice (e.g., interconnect to third-party content) are as important as the internal dependencies.

The end-to-end service instances can also be managed in combination as *domain services.* Domain services are of different types. They can be distinguished by access domain, transport domain, and application (content or information) domain, for instance. Each domain may further be broken down to lower-level subdomains, particularly if the managed resources have commonality in service natures. It may also involve managing the same technology separately under different types of subservice. The purpose of this method is to manage the quality of the lower-level subservices that are specific to the domain attributes. The domain-service approach is commonly used in new technology as a transition stage to a more integrated solution. This strategy assumes that cross-dependencies between services can be managed manually. If the procedures and measurements from a domain so uniquely differ from other domain services, the service provider may decide to keep the entire OSS flow independent in order to save the effort of integration. The disadvantage of keeping a domain service as a long-term operational procedure is its inflexibility regarding integration, plus the number of reports required to construct a customer-centric service view as they are measured independently.

The end-to-end service instance view is a requirement for most primary service providers as they are responsible for end-customer SLA compliance management and customer care (interaction). The domain-service view is more suitable for third-party service providers or providers targeting customers with special interests. It can also be a staging management means toward a more mature integrated solution.

4.2.3 OSS Information Flow Life Cycles

OSS life cycles [2] address the business processes used in the management of streamlined tasks. These include manual or automatic processes such as billing, CRM, or workforce management.

4.2.3.1 The Customer Care Life Cycle

The customer care life cycle is triggered by a sales action or inquiry, shown in Figure 4.9 (1). The order takes into account the provisions for a specific instance of customer service. The ordering process is responsible for taking care of many customers and many different services, with many orders to add, delete, or change to reflect the service provider's offering (2). This implies the need to support high transaction rates in the service providers' customer care systems for order automation processes. Modern customer care systems allow the service provider to activate a customer's service remotely. The content may be fed directly from a common configuration or may be entered manually. The problem handling and performance reporting processes are central to the CRM (3). Real-time customer assessments on QoS and trouble management create inputs to customer retention operations (4). A successful customer care life cycle is essentially driven by satisfactory QoS levels, accurate invoicing, and a proactive response to customer concerns (5) that circle back for returning purchases for additional or new services (1).

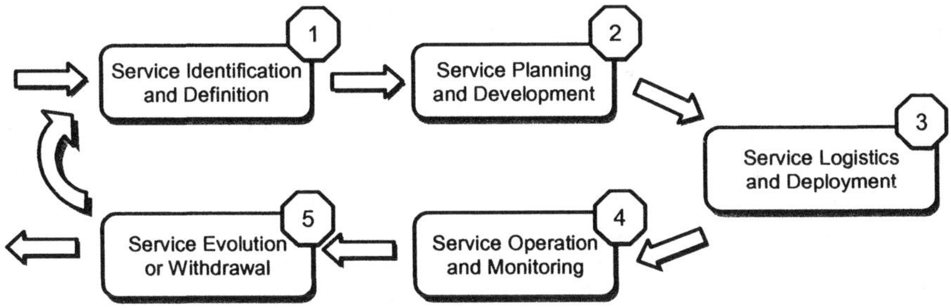

Figure 4.9 The customer care life cycle. (*From:* [2]. © 1999 TeleManagement Forum. Reprinted with permission.)

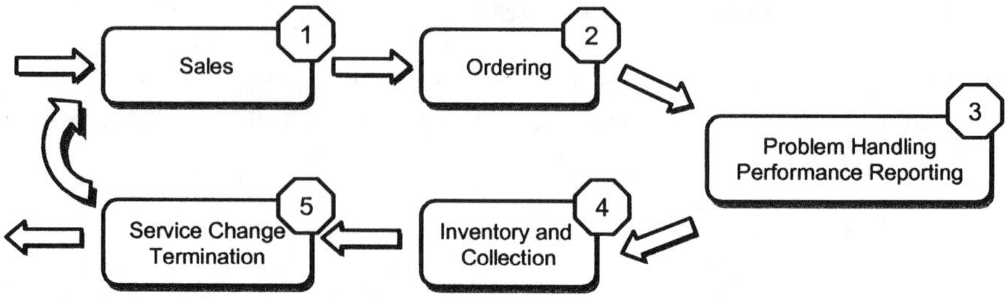

Figure 4.10 The service management life cycle. (*From:* [2]. © 1999 TeleManagement Forum. Reprinted with permission.)

4.2.3.2 The Service Management Life Cycle

The service management cycle located in the middle row of the TOM model forms a longer-period life cycle driven by the introduction, modification, and withdrawal of service offerings. This life cycle (see Figure 4.10) is triggered by the service provider's decision to offer new service or service classes, either as the result of market research or customers' needs (1). It involves creating the specific policies, rules, processes, and data templates used to configure and select service products. Deployment (2) of the service typically involves multiple tasks over a period of time and might involve such things as the RAN supply orders for third-party equipment, and so on. The progress of deployment must be synchronized with the corresponding customer-order workflow and corresponding workforce management in order to improve overall efficiency (3). Service operation and monitoring are responsible for dispatching and tracking relevant rollout or upgrade workflows.

Service quality management (SQM) manages customer expectations against monitored service performance in accordance with the SLAs (4). Over time the service reaches the end of its life because of competition or changing technology and the service provider must either upgrade the service or retire it (5).

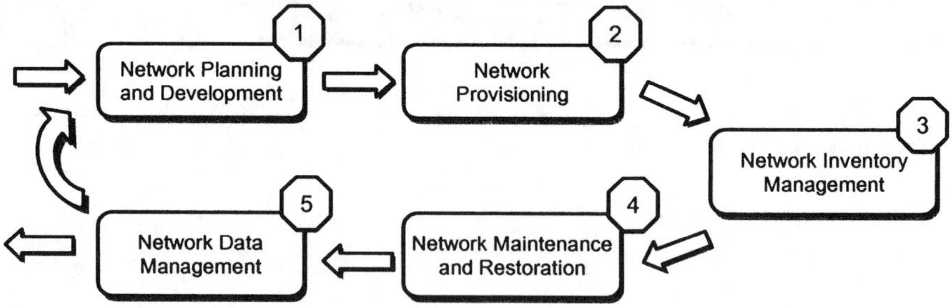

Figure 4.11 The network management life cycle. (*From:* [2]. © 1999 TeleManagement Forum. Reprinted with permission.)

4.2.3.3 The Network Management Life Cycle

The network management processes form the lower layer of the TOM model and support both the customer care process life cycle and the service management life cycle. The main challenge of a single common solution lies with this cycle, because NMSs have to support business processes as well as technologies and equipment from multiple vendors. These are often heterogeneous in nature and difficult to integrate. Different elements or sub-NMSs make assumptions about who owns specific data and who may access or change it. Different operators choose to manage data in their own environments in different ways, and many of their own systems make similar assumptions about location and ownership. Customizations to satisfy interoperability would make industry-level collaboration virtually impossible and result in life-cycle support issues for network management and service management. This concern applies to the entire life cycle of the network management process [see Figure 4.11, from network planning and development (1) to data management (5)].

4.2.3.4 Life-Cycle Interaction

The interactions of the service management life cycle, customer care life cycle, and network management life cycle are depicted in Figure 4.12. The service management life cycle includes the operations of the other two life cycles, and this encompasses a wide scope of activity. For the purpose of explanation, we will use a configuration change to illustrate the interactions between processes.

The service development process uses the appropriate logical design model (service configuration) to prepare high-level designs and to predict potential traffic scenarios (performance reporting) of a new technology. The traffic scenarios are based on trends from capacity history analysis (network data management), plus forecasts from marketing. Using engineering tools, the designers can play what-if scenarios for a different design (of service configuration and SQM). This design, in turns, is translated into detailed configuration definitions, traffic policy (network development and planning), and associated actions (network provisioning, maintenance, and restoration). The workflows will maintain the status of the configuration change to control its progress. This may include the design, plan, and completion status (problem resolution). Figure 4.13 shows an example product life-cycle management (PLM) process interaction from eTOM [9] that involves more functional processes than in Figure 4.12. However, depicting a complete set of end-to-end process interactions for PLM may sometimes require a significantly complex diagram and flows.

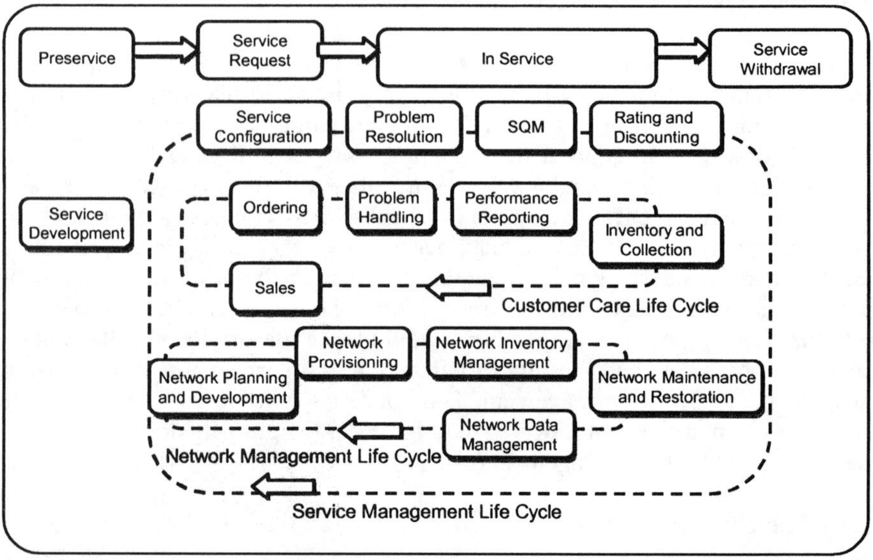

Figure 4.12 TOM life-cycle relationships. (*From:* [2]. © 1999 TeleManagement Forum. Reprinted with permission.)

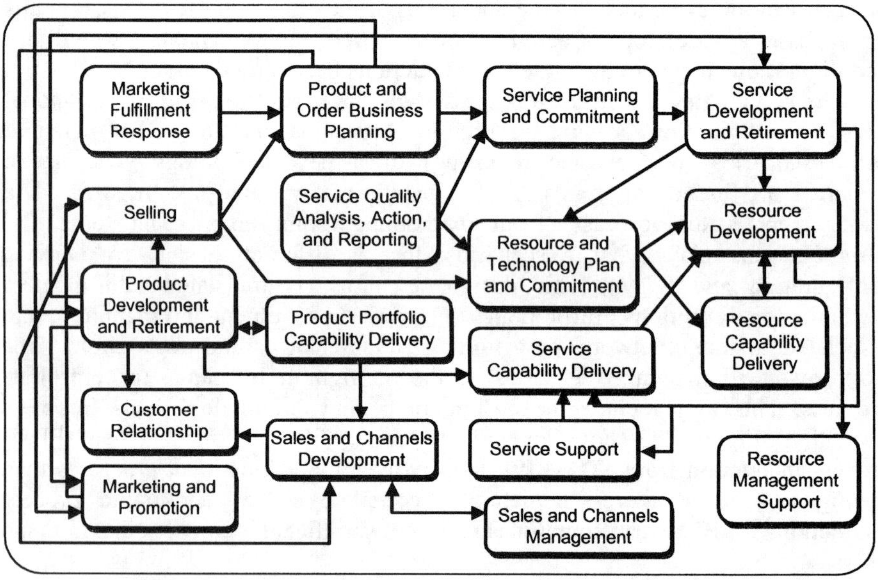

Figure 4.13 Example of PLM process interactions. (*From:* [9]. © 2005 TeleManagement Forum. Reprinted with permission.)

4.2.3.5 Design Considerations

In order to achieve business workflow integration, the service provider should consider the life-cycle interoperability aspect because of their cross-dependencies. The IT industry has gone a long way toward system interoperability and introduced a variety of solutions. Solutions can be categorized as loosely coupled integration or tightly coupled integration. Loosely coupled business workflows can be defined as extensively integrated but loosely connected workflows.

Tightly coupled business workflows can be defined as hard-coded and customized business processes workflows. Based on the above definitions, this section further elaborates on the concepts of loosely coupled and tightly coupled processes.

A loosely coupled solution has a modular architecture in which the interfaces of each system or process incorporate a limited number of publicly defined standards and protocols to support interactions. The business rules and workflow, as well as the technical product implementation, are based on abstract business object modeling, using a language such as XML. Loosely coupled solution architecture has the benefit of flexibility and can minimize the requirements for shared understanding and interoperability. The solution can be implemented with either the bus or the portal concept.

An example of bus solution is products based on the enterprise service bus (ESB) from the Java Messaging System (JMS). There are two kinds of messaging: synchronous and asynchronous. Synchronous messaging involves a client that waits for the server to respond to a message. Asynchronous messaging does not wait for a message from the server. With publish and subscribe message passing, the sending application or client establishes a named topic in the JMS broker or server and publishes messages to this queue. The receiving clients register (specifically, subscribe) via the broker to messages by topic; every subscriber to a topic receives each message published to that topic. The relationship between the publishing client and the subscribing clients can be either point to point or one to many.

An example of a portal implementation is service-oriented architectures (SOAs) based on Web services technology. The Web services architecture requires three fundamental operations: publish, find, and bind. The service providers publish services (messages) to a service broker. The service requesters find required services using a service broker and bind to them. Web services components and message flow are shown in Figure 4.14.

Because of the separation of concerns between description, implementation, and binding, the architecture style defining an SOA provides unprecedented flexibility in responsiveness to new business integration requirements. OSS process flows can now be supported by an open infrastructure of these exposed services in composite applications.

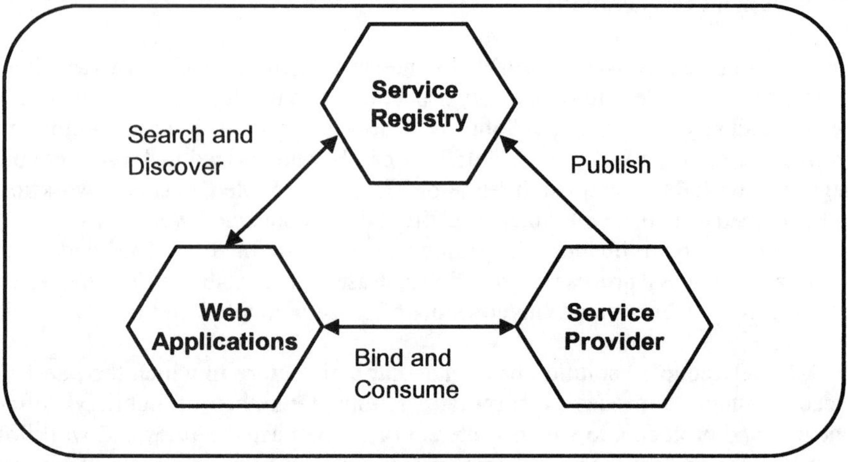

Figure 4.14 Web services components.

For a loosely coupled solution to succeed [10, 11], many technical challenges must be resolved. As loosely coupled architectures are built based on peer-to-peer relationships, any system can link to any other system. In a complex deployment environment, chained (or networked) applications may present issues such as security, manageability, accountability, reliability, and discovery.

Table 4.1

Differential Dimensions

	Loosely Coupled	*Tightly Coupled*
Data integration	Less dependent	Closely dependent
Process integration	Medium	High
Interdependency	Medium	High
Number of partners connected	Many	Few and important
Roles of partners	Core or secondary	Core
Typical technologies	XML/Web services, JMS	CORBA, DCOM, RMI, and so forth
Relationships	Leverage	Bottleneck/strategic

Recently, a major service provider's early mass deployment of Web services technology revealed the problems of performance and measurability. Additionally, traditional transactions using a two-phase commit approach (all of the participating resources are gathered and locked until the entire transaction can take place, at which point, the resources are finally released) may not work well in an open environment where transactions can span hours or even days. Even though such architectures look rather error prone, problems can be avoided by good communication between organizations during the design phase.

On the other hand, a tightly coupled solution comprises rigidly defined communication protocols and instructions. It normally requires a complex implementation or language-specific efforts if the systems have incompatible interface definitions, or worse, no interface definitions (i.e., only low-level programming capability). The advantages of this solution are its high speed and high security in that the risk of transmission errors is very low. If business flows require the associated systems to have a clear tier relationship and rigid interactions, the tightly coupled architecture will be a preferred option, even if it requires a high investment in information infrastructure. Vendors or system integrators employ standards in an effort to minimize the level of customization when dealing with a tightly coupled architecture. Some well-known interfaces include CORBA/Internet Inter-ORB Protocol (IIOP), Distributed Component Object Model (DCOM), Java Remote Method Invocation (RMI), and Remote Procedure Call/External Data Representation (RPC/XDR).

In conclusion, loosely coupled systems have a few well-known dependencies, and tightly coupled systems have many unknown dependencies. The choice is thus more subject to what the service provider is trying to accomplish. Table 4.1 lists the comparisons of the two architectures in different operational dimensions.

4.3 OSS INTEGRATION

The 15 subprocesses of the TOM (Figure 4.2) provide templates for telecommunications service providers to design service or network operational processes. Using the building-block structure, the system designers can concentrate on a few consecutive processes instead of designing the entire business definition. For instance, the system designers can define a new business line by starting from the sales process and continuing horizontally through the customer care process.

By moving horizontally in the customer care layer from the sales process to the invoicing and collection processes, the designers complete the customer care life cycle. Once the horizontal process flows are complete, the designers review the processes against the vertical flow-through processes of FAB. The vertical processes represent the flow-through of how the customer connects to, and interacts with the network. An example of a service assurance process might start with problem handling, then flow through customer QoS management, service problem restoration, SQM, network inventory management, ending with network maintenance and restoration.

Note that in joint service arrangements, service providers may not implement the entire OSS processes. However, it is crucial that the service provider at least have a complete end-to-end knowledge of the service flow and have appropriate procedures. Because many activities in the traditional fulfillment and billing processes influence service assurance, it is appropriate to review the key elements of these processes.

4.3.1 Fulfillment

This section illustrates a typical flow of a fulfillment process [9, 12] from receiving a customer inquiry, to service order creation, to the configuration of the service and its installation. The service fulfillment process starts with presale activity and ends with on-time and correct installation of a customer's requested service (Figure 4.15). Support information is required to manage any SLAs, to provide for problem or trouble management, and to produce an accurate bill. Depending upon the individual service provider's process, orders can be placed through the sales process or directly through the order management process. For bulk sales or resellers, the service provider would use a dedicated sales function to manage customers. The different flows are shown as steps 1, 2, and 14 in Figure 4.15. Interfaces to the customer are shown, as well as the output interfaces required to support service assurance (e.g., trouble or problem management, SLA management) and billing processes. Interfaces are required with other service providers or network operators when the service offered to a customer involves joint service arrangements.

It is not necessary to complete the entire fulfillment flow for services that have preassigned service arrangements. For configured and tested facilities that are preprovisioned, the network provisioning flow can be skipped, for instance.

In other cases, more process activities may be required for reasons such as SLA requirements or a complex bundled offering. Which flows are necessary is determined by the provider's operational processes and policies.

Additionally, security and test management are critical and can apply at every interface. Both functionalities can be applied as a subprocess across the entire OSS or can be managed as a dedicated process. In the following sections, we will discuss the associated functional blocks individually.

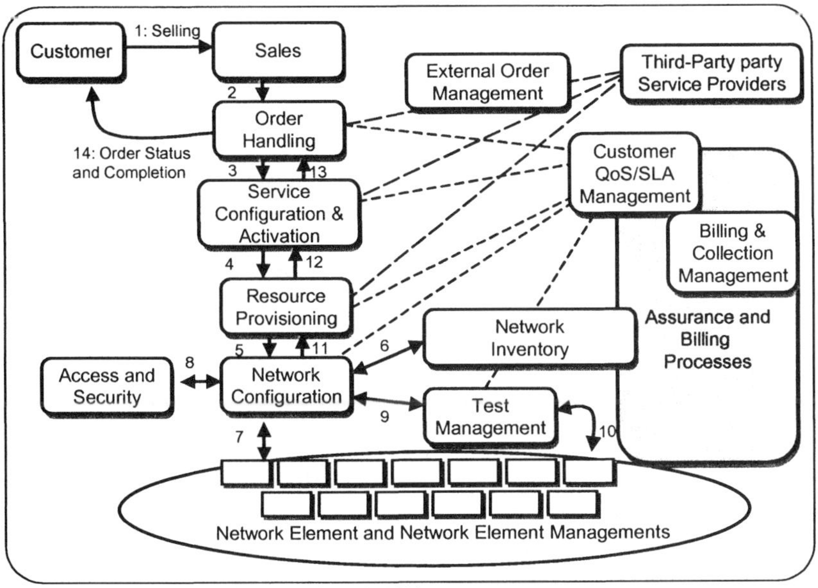

Figure 4.15 Fulfillment flows. (*After:* [9, 12].)

4.3.1.1 Element Management and Network Mediation

An EMS is responsible for managing a particular vendor's network equipment. Many equipment vendors supply an EMS along with their network elements (NEs). In the case where an EMS is provided by the NE vendor, the service provider may choose to use the stand-alone EMS to manage the network and the services at the initial deployment stage.

Alternatively, service providers may use their own NMS to administer the entire network domain. In either case, full interoperability between the NMS and vendor's EMS is required.

The purpose of having a network mediation layer is to encapsulate the knowledge of network devices into one system layer so that the upper-layer OSS can function with generic network or service object models and need not be exposed to vendor device specifics. The network mediation layer can either communicate with the NE directly or communicate with the EMS. Thus, an upstream OSS need only model a particular network device or domain in a generic way. A cohesive mediation implementation will minimize the complexity and number of systems required to interface directly with network devices. The information colleted by the mediation layer should cover measurements of performance, fault, and usage data. These are used to support OSS functions such as alarm monitoring, performance monitoring, capacity management, traffic engineering, and fraud management.

4.3.1.2 Sales

The functional goals of the sales process are to respond to market needs quickly, sell the correct service to suit the customer's need, and set appropriate expectations with the customer via SLAs, if necessary. To accomplish this, the processes should be able to assess the needs of each customer and be capable of educating the customer about the service products that match those needs. A complete sales process should include the activities of selling and field support that can create a match between the customer's expectations and the service provider's ability to deliver. Some service providers implement the sales process as pure sales; some include various levels of technical sales or back-office support. As the sales intervals are sharply decreased, technical sales support is engaged early in the cycle (this may include preorder work and interfaces) to reduce costly order errors, especially for customized solutions.

Sales functions can be organizationally aligned by geographical area, industry, or account size. The sales process typically starts with identifying a potential customer or customer need and ends with the closure of a sale. Support activities may include refinement of customer needs, planning of service details, and billing arrangements. Recent e-commerce activity has created a new channel for the service provider to conduct routine selling activities on the Internet. This method makes a customer-enabled interface possible and also opens up less costly customer care contact points.

4.3.1.3 Order Handling

The order handling process (Figure 4.16) starts by accepting a customer's order for service, whether from the customer directly, the customer's agent, or a third-party service provider. It tracks the progress of the order, keeps the customer informed, and supports changes when necessary.

As orders can include new, change, cancel, and disconnect instructions for all or part of a customer's service, this process must include a completion acceptance from the customer. The process should ideally include a follow-up or notification to ensure that the service is working properly. A completed order should include sufficient information to build or update a customer account record in trouble or problem handling, performance reporting, and billing processes and systems.

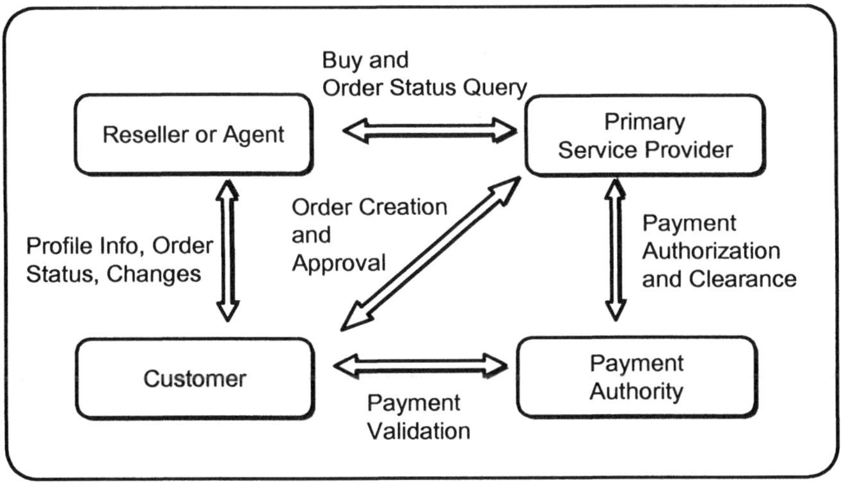

Figure 4.16 Sample flow of order handling.

This process is also accountable for initiating and receiving credit information for new customers according to the service provider's business rules. When customization is available for complex bundled offerings, this process should also support preliminary feasibility inquiries and optional pricing estimates. Sometimes this requires the development of an order plan. Modern CRMs typically include the entire order handling process and tracking capabilities for preorder activities and can perform channel management and market research.

4.3.1.4 Network Provisioning

The network provisioning process [2] starts with a configuration (or reconfiguration) or installation request from the network management layer processes or the service configuration process (see Figure 4.17). This process includes the configuration of the network to ensure that network capacity is available and is ready for provisioning, maintenance, and tests to ensure operational readiness. This process essentially administers the logical network and interfaces with the network inventory manager for physical installation or implementation in the network or associated services. The physical installation or implementation may include network and service additions, changes, deletions, and configuration changes. For a new service, the result of the process is the network and associated operations' being logically configured.

If a network resource has been preconfigured, the provisioning process can be managed directly through service provisioning or customer care processes, as the physical resource and should be ready for service. In order to keep the inventory identifiable throughout the OSS process, the network provisioning process is responsible for assigning and administering identifiers for provisioned resources and making them available to other processes.

The Figure 4.17 demonstrates the sequence of operations, starting with a network-provisioning request (1) from the service configuration process within the service management layer and finishing with the configuration result (10). It then shows the start of monitoring (11) with messages being sent to the service configuration and network data management processes, respectively.

4.3.1.5 Service Configuration System

There are three triggers for the service configuration process. The first is a service infrastructure need to maintain performance or to add service-specific capacity; thus it is not customer specific. The second trigger is customers' requests for service installations or service configurations. The third trigger is a response to a reconfiguration request due either to a customer demand or a problem resolution after the initial service installation. The service configuration process interacts with the network provisioning process and the network inventory process to perform physical implementation or installation work in the network or an associated service resource.

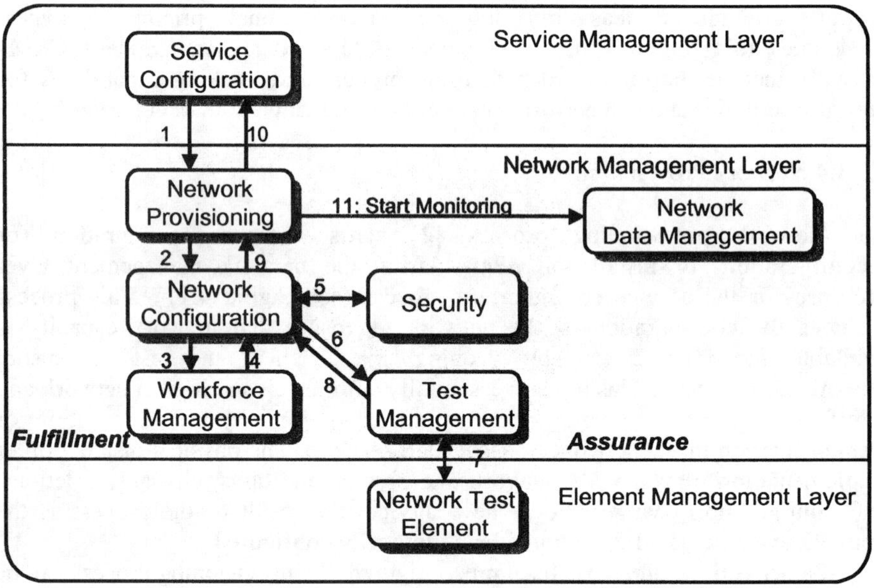

Figure 4.17 Detailed flow of network provisioning. (*After:* [9].)

4.3.1.6 Network Configuration and Routing

The network configuration and routing process is responsible for installing the initial network configuration and subsequent network reconfigurations. If physical actions are required for any configuration-related work, this process will issue work orders to the workforce management process to coordinate such tasks.

When an order from network provisioning indicates that a reconfiguration (e.g., rerouting) is needed in the operational network, this process will need to coordinate with high-level processes to apply business rules for the utilization of the network in order to design the service appropriately. To ensure data integrity, the network configuration process has to maintain the configuration information stored in its routing and connectivity tables and keep it synchronized with the network configuration information stored in network management and administrative systems.

4.3.1.7 Network Inventory

Inventory functionality includes the ability to add, change, decommission, assess the status of, and reconcile inventory related data. The network inventory process is responsible for installation and administration (acceptance) of the equipment in accordance with the physical implementation of the network. This process also manages spare parts, software and hardware upgrades, and repair process records that may impact the network provisioning process. The trigger for this process is typically a work order request for installation often initiated by a network problem. It can also be triggered by the repair, fault, or spare part subprocesses. Some service providers outsource portions (e.g., site work, repair, software upgrades, or equipment or software inventory) of this process to suppliers.

4.3.1.8 Testing

The test process is responsible for verifying the operational status of a service and determining the cause of any faults. This process maintains a collection of hardware and software test suites and a database for test scenarios, which should identify the component under test and the expected test results. The test scenarios should cover the goals of the test, test criteria, test approach, testing process, test schedule, test routines, result collection, and result expectations. Depending upon path and equipment characteristics, the test results should align with the service metrics specified in any associated SLAs. It should also coordinate schedules with the trouble ticket management process to provide traceability and audit capability against all activities, such as MTTR.

4.3.2 Service Assurance

The service assurance [9, 12] cycle may begin with changes to an SLA that may require updates to the offering. In this case, the service assurance process will identify and update the service rules or metrics (e.g., per new configuration) according to the SLA. Even though the flow may be similar to the change process in the fulfillment phase, it differs in that now the order status may need to be monitored per the SLA metric (such as MTTR).

Alternatively, service assurance triggers may come from internal trouble or problem management systems or from a customer complaint. Regardless of where the triggers are from, it is the service assurance process's responsibility to perform isolation, restoration, and repair and to provide information to billing for credits, if required.

Figure 4.18 shows a possible sequence of activities in response to a service-detected problem.

As shown, there are two ways that a potential service-affecting problem can be identified: the *hard events* (1) through the alarm notifications or *soft events* (2) by synthesis of service performance measurements, such as a degradation indicator through network data management. Network data management collects and processes all performance- and behavior-sensitive measurements, such as traffic, delay, and usage.

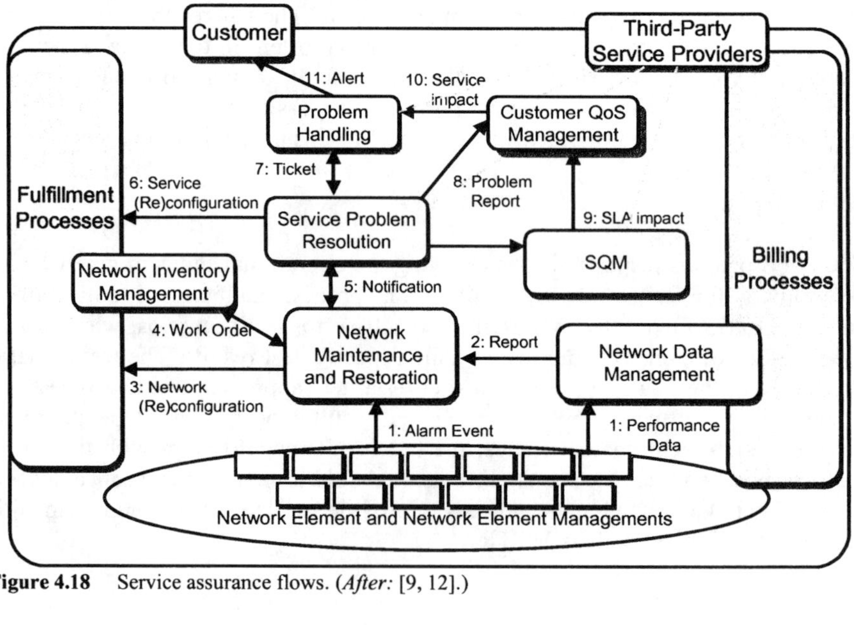

Figure 4.18 Service assurance flows. (*After:* [9, 12].)

There are two main trends in the area of service assurance. First, most service providers are trying to drive their service assurance processes to become increasingly proactive through being triggered by automation rather than by customer input.

This is important for improving service quality and customer perception and for lowering costs. The maturity level of OSS automation depends upon how effectively the service provider can collect the needed service intelligence and the tools to react to complex service behaviors.

Second, customers are demanding more control and more self-controlled service support. This trend is causing a major shift to interactive support through automation, including the ability for the customers to see and act on their "own" service performance. The value for the service provider and customers is mutual. Allowing customers to view information over this new interface not only opens the door for customers to customize their needs, but it also opens a window for the service provider to understand the customers' changing service preferences. If a service is provided as a joint effort with third-party providers, the main service provider can use this new interface to allow partners to monitor and support the service provided. This driver, again, needs more mature controllability and intelligence, as described above.

4.3.2.1 Network Data Management

Network data management [2] is responsible for the collection of usage data and events primarily for the network maintenance and restoration process, billing process, and other service-management-level applications. Figure 4.19 shows the network data management flow after a service is provisioned (1) and the trigger to network data management is activated.

There are two main uses for the performance and network behavior information in the network maintenance and restoration process. The first is to conduct network performance and traffic analysis and optimization (5). For more intelligent OSS, changes in traffic conditions or equipment failures (3, 4) may trigger changes to the network via network provisioning for the purpose of traffic control. Reduced levels of network capacity or performance can result in multiple requests, such as to network planning for more resources, to network provisioning for a reconfiguration, or to EMSs for specific actions in the elements themselves (8). Second, the collected data with appropriate analysis will be useful to support proactive customer care such as supporting proactive trouble ticket management via the network maintenance and restoration process (8).

The network data management also provides usage information to billing processes for rating and discounting (10, 11).

Other than the above two paths, this process also provide service intelligence to verify compliance with SLAs and QoS levels (9). Because SLAs are not known at the network management level, QoS specifications must be translated into indicators and corresponding thresholds. After translation, this process must ensure that the network performance goals are tracked and that notification is provided when they are not met (e.g., a threshold is exceeded or performance degrades).

4.3.2.2 Network Maintenance and Restoration

The network maintenance and restoration process is responsible for maintaining the operational integrity of the network. The process may start with an infrastructure identified problem, a customer identified problem (provided via the service management layer), or an analysis of monitored infrastructure information. It has four major functionalities:

- Maintains historic data of network problems and performance.
- Determines the options available to restore service in response to faults by taking into account the availability of network capacity, equipment, and workforce. Real-time detection mechanisms are usually implemented within the network protocols and devices. Fault localization is typically achieved through algorithms that compute a possible set of faults, while fault identification is done by testing the hypothetical faulty components.

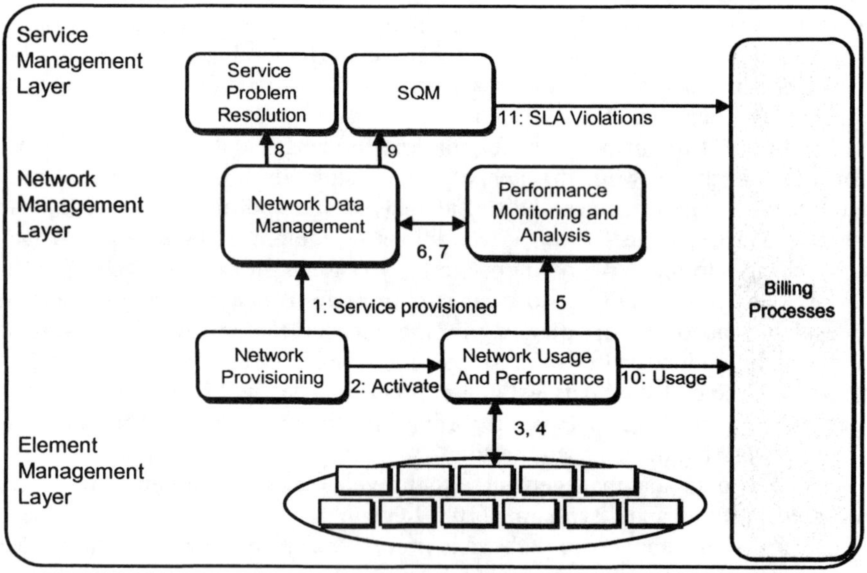

Figure 4.19 Detailed flow of network data management. (*After:* [9, 13].)

- Restores or repairs customer affecting troubles quickly by initiating tests, analyzing the root cause of problems, and notifying the SQM or network provisioning of the need for corrective action. The analysis can also propose the solutions to traffic- or usage-related problems that are reported by customers or by a fault in the network.
- Identifies problems in the network prior to their becoming customer affecting in accordance with network performance requirements.

4.3.2.3 Service Problem Management

The service problem management process is triggered by a service problem identified by (1) the service infrastructure, (2) a customer-specific problem or service issue from the problem handling process, or (3) the analysis of service and network trouble data. If the identified problem affects multiple customers, resolution may include immediate reconfiguration and corrective action or longer-term service design changes.

If the problem is referred by the problem handling process, this process is then accountable for providing expertise and support to the problem handling process to resolve a customer-specific problem. If the problem is from the service infrastructure or the analysis of service and network trouble data, it may need alarm correlation to detect symptoms, isolate the problem, and repair malfunctions in the network. The network problem correlation function provides powerful tools for the identification of faults and traffic problems in the network. It can use problem reports from the network maintenance and restoration and network data management processes. Based on this information, this function can then request actions in accordance with predetermined service policies. The network root-cause location function identifies the location of the fault and determines whether it is service affecting (Figure 4.20). It must implement immediate fixes, if required, or identify quality improvement efforts. A trouble report will then be created containing all relevant details of the fault for the problem handling process.

4.3.2.4 Problem Handling

The problem handling process is responsible for receiving service complaints, resolving them to the originator's satisfaction, and providing meaningful status on repair or restoration. The goal for this process is to identify proactively the majority of problems and to resolve these problems before impacting the service users.

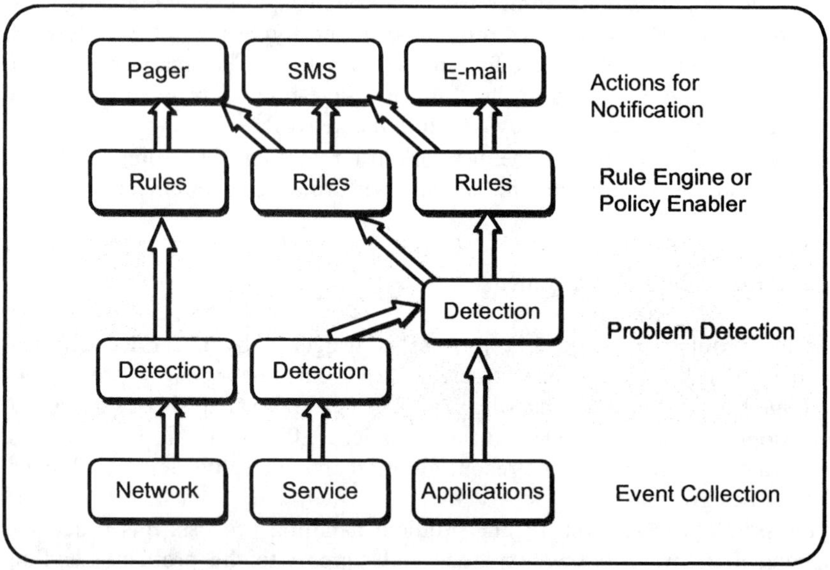

Figure 4.20 Detailed flow of fault management.

Problem reports can be originated from the network or customers. Several reported problems may be related to a single fault, or a service resource failure may be caused by a failure from elsewhere. Proactive problem handling begins with a service-resource-generated problem and the creation of a trouble ticket.

The ticket can optionally be used to notify the service users in the event of a pending service-affecting disruption. Proactive management also includes working with customers on planned maintenance outages.

The problem handling process (Figure 4.21) uses the trouble ticket to trigger the root-cause location process to localize the fault, work with the service problem resolution process to resolve the problem, and provide status on repair or restoration activity. A complete trouble ticket, at this stage, should describe activities, dependencies between activities, plans, and time frames for repair tasks. Correlation functions can apply to the repair records to inhibit duplicate activities from subsequent related complaints. This process should be concluded with a completion acceptance by the customer and an internal record to support both SLA reporting and outage credits, if applicable. A follow-up contact to the customer to ensure repair quality should also be part of the process.

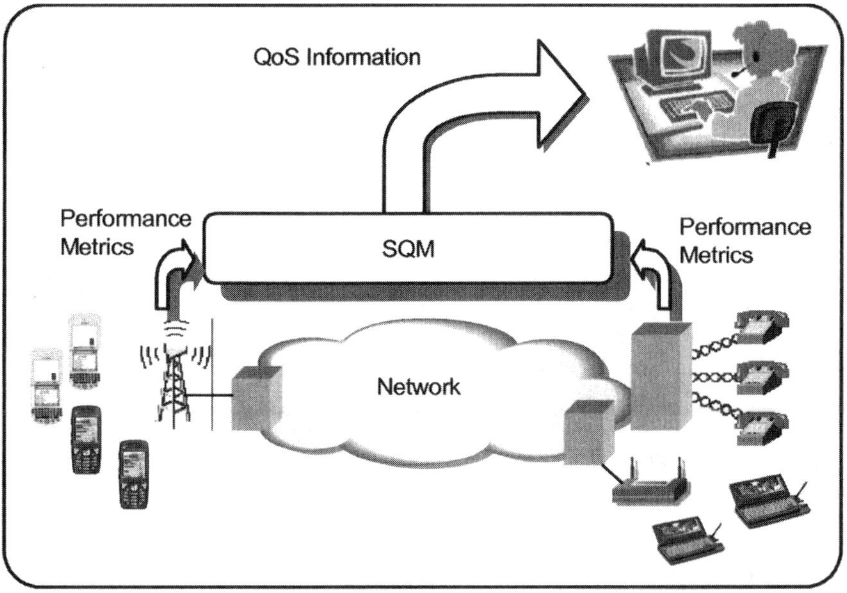

Figure 4.21 Problem handling.

4.3.2.5 Service Quality Management

SQM manages the service life cycle from service introduction to retirement. Depending upon the service provider's OSS architecture, the service-level manager (SLM) function may be implemented as part of an SQM subsystem. In either event, service models play a key role here to relate the customer's view of the service to the actual service or network configurations.

Not to be confused with the SLA management discussed in Chapter 2, the description of SLM in this section addresses a subsystem of SQM operations.

Quality measures and costs for a specific SLA agreement are defined within and tracked by the SLM. It is important that SQM have a real-time capability to track the service levels, problems, improvements, and costs for management purposes. Because of the nature of new technologies, pure connectivity service is no longer the only option for customers.

Often time, bundled services can enable the service provider to create more value-added offerings. Source quality statistics may need to be aggregated into service indicators, based on service dependencies. Therefore, SQM monitoring should incorporate quality statistics from multiple sources to infer the quality of services as delivered, hence to identify or prioritize problems. This information about whether the service results meet or exceed the committed operations objectives can be referenced by internal management or by customers through the customer QoS process.

In additional to the ongoing reporting capability, SQM is also responsible for managing service performance to quality and cost targets. If improvements are required for the service resource to maintain service levels per service class, SQM will provide recommendations and track the progress of the improvements or alert the sales process to slow sales.

In the case where SLM is part of the SQM process, the business objective of SLM is to allow the service provider to manage its customer SLAs. In Chapter 2, we discussed the value of SLAs and their applications to the service provider's business operations. Similar to SQM, the SLM system is not an independent process; it coordinates the SLA negotiation, determines appropriate QoS levels to take care of customer performance requirements, provides customizable QoS reports, and tracks the rebates and reconciliation due to service violations. Additionally, SLM provides network usage analysis by using the usage data collected by the data-collection function. It performs arithmetic and statistical calculations to evaluate traffic handling. SLM performs trend analysis on the traffic data and compares historical and predicted future levels against established threshold levels. The trend analysis enables early identification or predictions of problems before they become service affecting. This can result in a request to trouble ticket management or notifications to service managers.

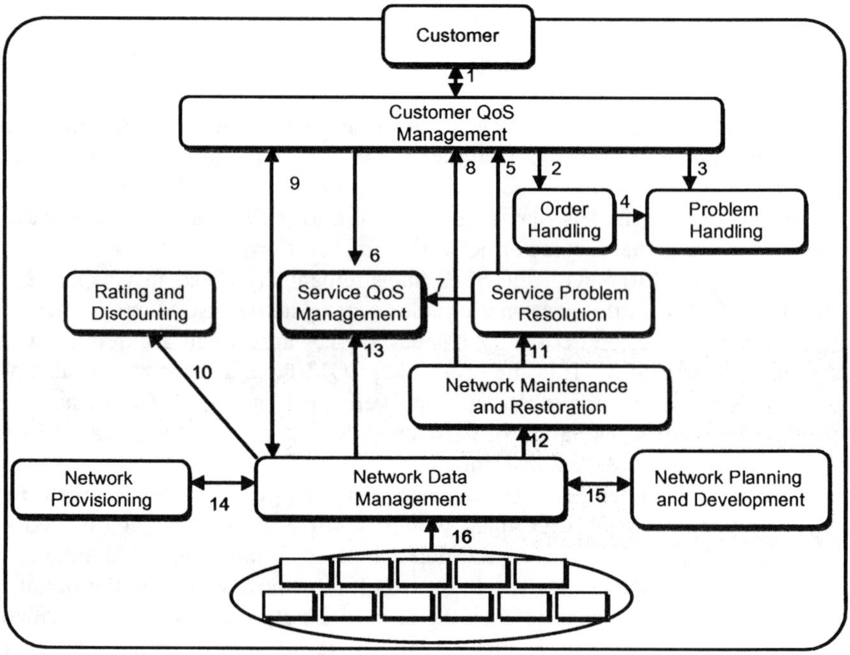

Figure 4.22 SQM flows and interfaces. (*After:* [9, 13].)

Figure 4.22 depicts the data flows and interfaces of performance analysis, management, and reporting for SQM and SLM [13].

4.3.2.6 Customer QoS Management

The customer QoS management process can be triggered by performance reports from the service problem resolution process, the SQM, or third-party service providers. QoS includes network performance as well as performance across all of a service's parameters [e.g., orders completed on time (OCOT) and MTTR]. The goal of this process is to provide effective monitoring and reporting on action planned and taken to assure service levels that meet specific SLA commitments or standard commitments to the customers. The reports of service quality information to both the customer and the service provider's management should cover the complete parameters of the services provided, including any developing capacity problems and customer usage patterns. As a system, this process should provide standard and predefined, as well as exception, reports, including overviews and performance of the service against any SLAs. Figure 4.23 shows a sample of a good overview report.

This process must respond to performance inquiries from the customer and proactively generate performance and quality reports for internal management. For SLA and QoS violations, the customer QoS management process should support notification to problem handling routines regarding service impacts and pass information to billing routines for credits.

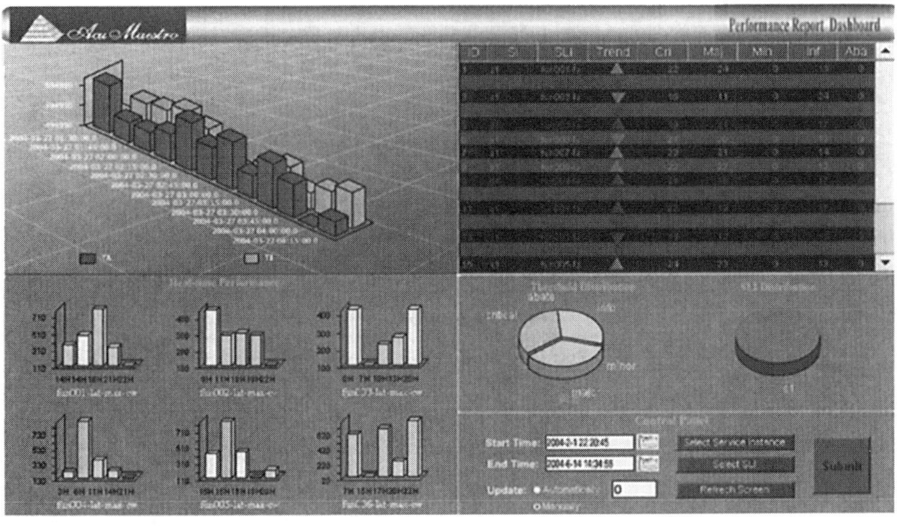

Figure 4.23 Sample SQM dashboard (AcuMaestro BASE problem summary).

4.3.3 Billing and Revenue Support Process

The billing and revenue support process [9, 12] can be triggered by a new or updated customer account or the registration of an SLA with a customer. It issues (4) and corrects (3) bills, including application of credits for SLA violations, supports customer account or billing inquiries, and manages accounts receivables, including payment collections. Figure 4.24 shows a typical sequence of activities to generate a bill with flat-rate elements, usage charges (1, 2), and possible SLA adjustments. For service promotion or marketing reasons, the service providers may apply discounts or rebates to a specific customer's bill with or without service outages or SLA breaches. The decision can be made in accordance with different service types, promotion plans, customer relationships, and company policies or customer contracts.

When a service is provided by a combination of different service providers, usage and other billing data may be aggregated by the primary service provider. In this case, the primary service provider will present one bill to the customer.

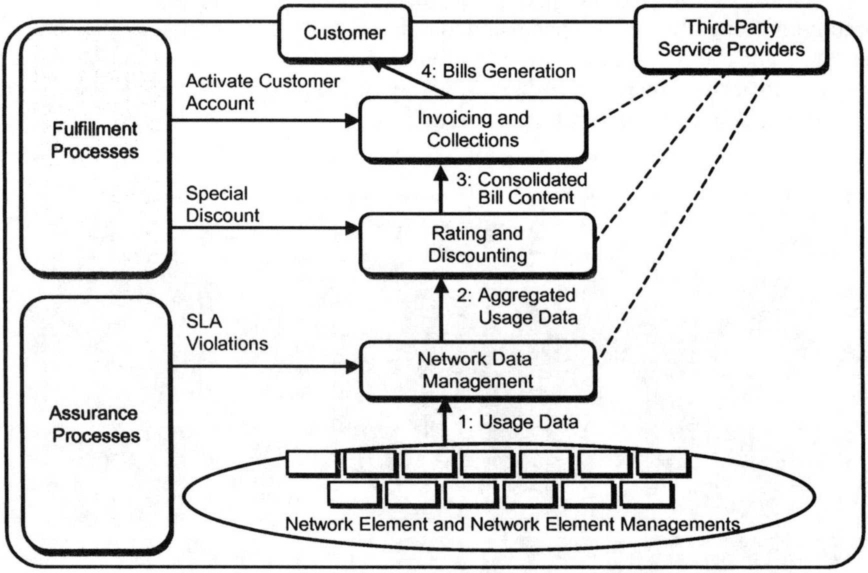

Figure 4.24 Billing and revenue support flows. (*After:* [9, 12].)

4.3.3.1 Rating and Discounting Process

The primary function of the rating and discounting process is to match usage to a customer charging record so that the invoicing and collection process can generate correct billing invoices. This process is triggered by registering a specific customer's identifier for determining usage and appropriate discounts, charges, or credits. To accomplish this, the process needs to apply the following:

- Correct rating rules to usage data on a customer-by-customer basis;
- Discounts defined by the ordering process;
- Promotional discounts and charges;
- Rebates or charges from SLA compliances;
- Resolutions to unidentified and zero-billed usage-cases.

As with other processes, service providers may provide rating and discounting functions for other providers as a service. For joint service arrangements, billing, invoicing, settlements, and reconciliation between service providers may be involved.

4.3.3.2 Invoicing and Collection Process

The invoicing and collection process is responsible for providing a correct bill, sending invoices to customers, processing their payments, performing payment collections, and resolving any billing-related problems.

This process begins with input from the order handling process to set up a customer account (including flat-rate and nonrecurring charges); it accepts rated usage from the rating and discounting process to render a total bill. In addition, this process handles customer inquiries about bills, provides billing inquiry status and is responsible for resolving billing problems to the customer's satisfaction.

For joint service arrangements, service providers may provide invoicing and collection functions for other providers as a service. In such cases, billing, invoicing, settlements, and reconciliation between service providers may be involved.

4.3.4 Fraud Management Process

Fraud management [14] becomes a key OSS function, especially for the mobile service providers, because a considerable amount of revenue can be lost due to service fraud. This is because technically the mobile service provider does not know the location of the *end of the wire*, which would lead to the home of a fraudulent customer. In the roaming case, a roaming visitor is not the service provider's customer; therefore, the service provider does not have complete information to assess fraud. As a result, fraud prevention is largely out of the control of the customer's home service provider, and end-to-end fraud protection has to rely on the service provider who is handling the call service.

Furthermore, fraud can happen in any OSS process flow. This characteristic also makes it difficult to represent fraud management easily. Typically, fraud management in mobile networks includes at least the following functions:

- Classify fraud risk level on a per customer basis using demographic and credit information from the problem handling and rating and discounting processes.
- Update the fraud risk level to the rating and discounting processes based on usage payment behaviors.
- Detect fraud patterns by accessing the current records in the rating and discounting, problem handling, customer QoS management, and invoicing and collection processes.
- Suspend the service of customers with a high fraud-risk level by reconfiguring the service and updating problem handling policies.
- Consult the home provider to assess fraud to determine if the visiting customers (i.e., roamers) should be suspended.

Figure 4.25 shows the occurrence of fraud detection and prevention flows in the OSS processes.

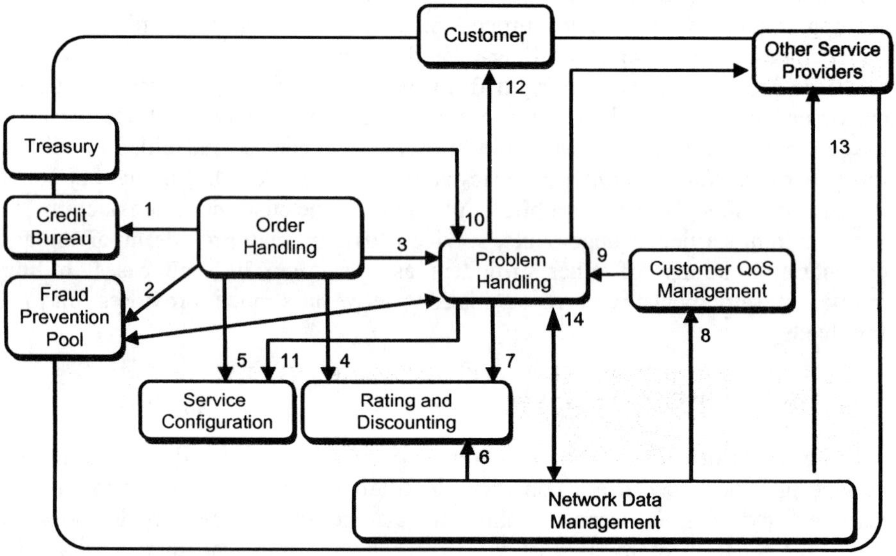

Figure 4.25 Fraud management flows. (*After:* [9, 14].)

4.4 CONCLUSION

What does the future of OSS development hold, and how will these systems bond customers with the service providers? As OSS matures to become more of an automatic end-to-end control process, customers will actually become an integral part of the solution. They will gain the ability to pull and push customized information tailored to the needs of their businesses. The key from the customer's perspective will be seamless, real-time management capability using the same service intelligence used by the service providers.

We have seen a tremendous evolution in OSS within the last few years. Through equipment vendors' contributions and solution providers' efforts, the latest OSS developments have successfully enabled service providers to provide automatic configuration and provisioning of their technologies.

New methods such as Web services have created a new service infrastructure, allowing service providers' system information to be openly communicated. The efficiency of intersystem communications and new channels, as the result of the new business infrastructure, has made such customer services possible. Customers can directly interact with customer service representatives or technicians to answer questions, locate information, purchase capacity, self-provision, and reconfigure existing service parameters online.

Web access is not about cost reduction of customer care operations but rather about giving the customer a choice to be an interactive part of the service-specification processes. So what is next after automatic service deployment? We surmise that it lies within the service fulfillment process—the automation of service assurance. With the new emphasis, OSS will be the vehicle that will carry interactive participation to new usefulness.

For service providers to support the future features as described, they should internally create an intelligent operation environment. We think that the *service model* will become the common reference for services and subservices. It will hold the information about services that is needed across multiple OSS components (e.g., quality monitoring, alarm analysis, and trouble and change) and for non-OSS components (e.g., customer SLA, customer order, and problem management). This information will include the value or importance of services (e.g., revenue, specific customers); their dependencies on subservices, hence on the underlying common configuration; and their planned or unplanned *status* (e.g., preoperational, normal, degraded, or unavailable). A set of programmable and actionable scripts that can drive meaningful activities based on the service intelligence available in the built-in service model framework will create actions both proactively and reactively. Upon successful deployment of such an intelligent operations environment, the OSS will have many forms of service delivery, from the traditional customer care center to the customer's direct control over portions of the service infrastructure. Service providers' operations will have open access, capable of integrating various vendors supporting software and hardware, including other providers' OSS. Closely integrated different technologies, such as land-line or wireless applications, will enable on-demand customer care from anywhere on the globe. The ability to access all service instances and their relationships with customers (through multiple channels) in a single location has provided market segmentation down to the individual level. This individualization creates new values for the customer via a fully integrated, comprehensive service infrastructure. The service providers' benefit is created by customer loyalty and retention achieved through a new level of service and increased satisfaction.

References

[1] "SMART TMN Technology Integration Map—GB 909," TeleManagement Forum, October 1998.

[2] "Network Management Detailed Operations Map—GB908 Version 1.0," TeleManagement Forum, March 1999.

[3] "TOM Application Note: Process Reengineering, Development and Management—Simple Methodology Steps—GB910A," TeleManagement Forum, September 2000.

[4] "Enhanced Telecom Operation Enhanced Operations Map (eTOM)—The Business Process Framework—For the Information and Communications Services Industry—GB 921," Release 4.0, TeleManagement Forum, March 2004.

[5] "The NGOSS Technology-Neutral Architecture—TeleManagement Forum 053, Version 3.0," TeleManagement Forum, April 2003.

[6] Strassner, J., et al., "TeleManagement Forum White Paper on NGOSS and MDA Version 1.0," 2003.

[7] "OMG Model Driven Architecture," http://www.omg.org/mda.

[8] "Model Driven Architecture (MDA) Document Number ormsc/2001-07-01, Architecture Board ORMSC," http://www.omg.org/docs/ormsc/01-07-01.pdf.

[9] "Enhanced Telecom Operations MAP (eTOM)—The Business Process Framework for the Information and Communications Services Industry, Addendum F: Process Flow Examples, Release 4.0," GB921F, March 2004.

[10] Wong, C. Y., H. Hvolby, and J. Johansen, "Why Use Loosely Coupled Supply Chains?" Aalborg University, Denmark, http://www.cip.dk/fileadmin/template/publicweb/Docs/on-line_publications/publications/Why_use_loosely_coupled_supply_chains_01.pdf.

[11] "J2EE and Message Bus Based System Architecture," Unixpros, LCC, http://www.eclipsenetworks.com/papers/JMBUSwp.pdf.

[12] "Telecom Operations Map— GB910," TeleManagement Forum, March 2000.

[13] "Service Quality Management Business Agreement—TMF506," TeleManagement Forum, February 2001.

[14] "TOM Application Note. Mobile Services: Performance Management and Mobile Network Fraud and Roaming Agreement Management—GB910B," TeleManagement Forum, September 2000.

Chapter 5

Service Model Fundamentals

Wireless operators are constantly facing the challenge of improving the quality of wireless services. As many different services emerge, quantifying and assuring service quality becomes an increasingly complicated task for most operators. It is an ad hoc task with little coordination among operations managers. In a nutshell, today's service management is made up of isolated NMSs plus an IT management environment. Network management tasks consist of collecting a lot of performance data, generating weekly or monthly reports, and logging large numbers of events or alarms. Data is mostly generated by a number of disjointed EMSs, or in some cases, by individual NEs. In the service and application areas, traditional IT management platforms such as Hewlett-Packard's HP OpenView, Computer Associates' Unicenter, or IBM's Tivoli are popular platforms for monitoring and logging server and corporate LAN status or alarm events. However, there is usually little correlation between the IT management platform and other EMSs. For each isolated domain, true service management lies in the hands of the personnel taking care of their particular domain (application, core network, radio network). Different domains are normally handled by different organizations, which are operated independently and, most likely, with little interaction. There are no integrated and correlated views of end-to-end service quality, and there are hardly any consistent efforts towards assuring service quality.

Most recently, driven by the desire to migrate to a service-centric management paradigm, many wireless operators are actively moving towards a model-based approach to service management. In this chapter, we focus on the fundamentals for the design and construction of such a service model. We first survey various service-oriented management models in the literature. We then present a targeted service model based on existing work but with the adaptation necessary for solving the service assurance operations problems in a carrier-grade mobile operator's network. The service model concept will be viewed as the cornerstone of future service management, and is also the foundation of the following chapters, when VoWiFi and 3G integration operation is described.

5.1 DRIVING FORCE: WHY IS IT NECESSARY TO HAVE A SERVICE MODEL?

Mobile operators are faced with the following challenges:

- Managing services in such a way that they have direct impact on revenue assurance;
- Being able to introduce new services quickly that have a shorter life cycle than traditional services;
- Moving from managing networks to managing services that are delivered by the underlying networks;
- Moving from islands of management systems to integrated management;
- Managing suppliers or third-party vendors, who provide portions of the services or networks.

These challenges have the following implications for service assurance:

- Quality from the user perspective is more important, and thus, end-to-end QoS monitoring and assurance are needed.
- Moving from network-centric performance management to service-centric management means the service needs to be represented and modeled in a more systematic fashion. The model uses information from network management but also requires information from the service and business layers. Instead of management of the "health of the network," the focus is to manage the "health of the service" from the user perspective.
- QoS management outsourced to third-party vendors who provide and manage portions of the network needs to be dealt with as part of the overall business and service management. Correspondingly, both external and internal QoS metrics are required.
- Different organizations within a provider need to coordinate and have the same end goal on service management and assurance. The goal needs to be clearly defined and conveyed to various organizations so that a consistent plan is agreed upon.
- Fault, network performance, and user QoS perspectives are closely related. It is necessary to have a common model and integrated process to handle these entities in a consistent manner.
- It is more important to learn how to manage "soft faults" rather than "hard faults." Hard faults are straightforward to define: working or failed. However, soft faults or performance problems are much harder to detect and characterize. When a network delay exceeds the threshold of 100 ms, how do we actually measure that, and how should the threshold be

defined? Does one declare a performance violation if the threshold is crossed once in 15 minutes?

- When the performance degrades, its impact from a revenue perspective needs to be tracked, preferably before customers notice it.
- For service planning purposes, it is necessary to predict where the bottleneck is in order to prioritize and schedule upgrades in an effective manner. However, resource management now has to take account of different classes of service quality due to the mixing of video, data, and voice.

Evolving to a new service management infrastructure will take time, conscious effort, and commitment from service providers. Starting with the seeds of a service model infrastructure will be a crucial step towards the management-paradigm shift, which will have far-reaching benefits.

5.2 SERVICE MODEL IN A NUTSHELL

In Chapter 2, we described the process flows related to service management and, in particular, service assurance. The description there is generic and is applicable to any service. However, a number of service assurance aspects in the processes and flows have a lot of commonality with respect to different services. For example, in order to manage a service in such a way that the customer benefits, one has to identify a set of attributes that end users care about. It is desirable that these attributes be defined in such a way that they could be reused for different services and new attributes can be built on top of the defined ones. Such modular and extensible design will save a lot of engineering effort and time in the long run.

The set of desirable end-user attributes also forms the basis of many assurance functions. Service management will then be focused on optimizing this controlled set of customer-focused attributes. We will see later in this chapter that these are the KQIs that will be crucial for service quality assurance.

A key aspect of a service model is to facilitate the automation process and the tedious job of collecting the necessary data for analysis. It also provides an intelligent structure for organizing the collected data and distributing it to support different OSS applications. A desirable service model also contains the right tools to help the operator to look at the relevant information and make an intelligent decision as to what the priority is. As an example, an alarm has been detected in one of the switches as an "overloading" problem. Since the switch is in a remote site, attending to the problem requires a truck roll. Suppose this situation happens on a Friday afternoon at 4 p.m. The questions at the NOC are: Is the problem severe enough that such a dispatch is essential? Can it wait until Monday, when the whole repair crew is available? An experienced operations engineer who is familiar with the switch environment may be able to pull together all the relevant information and arrive at the right decision in a short time. However, such an experienced person may not be there to make such a decision. On the other hand, with the assistance of an intelligent service model, a less experienced person may be able to prioritize the impact of such a switch alarm properly and recommend the correct decision in a short period of time.

From an operations perspective, a service model is capable of capturing much of the experience of the experts in service assurance. The challenge of designing a useful service model is designing it to support not only today's known applications but also future new services. Due to the volatility of services, this future proof capability is a necessary requirement.

The idea of using a formal service model to support various operations functions is a relatively new concept and is undoubtedly evolving. However, the need to move to a customer- and service-centric operations paradigm is well recognized by many leading providers, and the incorporation of service-model-based management has been initiated by a number of advanced operators with others following the trend.

5.3 SERVICE MODEL DESIGN CONSIDERATIONS

To deal in a systematic, scalable, and reusable fashion with the service management issues described above, it is desirable to define a service management framework. In this section, we describe such a framework based on the concept of a service model, which is designed with the following criteria.

5.3.1 Managing Complex Services

One of the major reasons for developing a service model is to use it to manage the increasing complexity of mobile services. Mobile service is becoming more complex because it is basically a superset of Internet services. Compared to circuit switched voice, mobile services have a much shorter history. Consequently, many characteristics and user requirements are not stable or understood. This is compounded by the fact that the life cycle of mobile data services is much shorter than providers are accustomed to. For mobile services, further complexity arises as a result of the mobility nature of the service. Issues related to handoff, roaming, and the capacity limitation of the RAN all add complexity to the management problem.

5.3.2 Supporting Reusability

Due to the proliferation of mobile services and many different possible quality grades of services, it is impractical to reengineer a service management model each time a new service is created and offered. Providers must adopt a framework that can be reused with minor modifications when new services are offered. As we will describe in the following sections, reusability is possible since most services can be decomposed into a few fundamental components. The decomposition aspect is a key characteristic of service models.

5.3.3 Adapting to Different Business Needs

The framework must be flexible so that it can be adapted to satisfy the needs and requirements of various customers and business models, including various requirements demanded by different administrations and stakeholders in the value chain. That is, we should not be required to design a totally different model for each stakeholder in the value chain as described in Chapter 2.

5.3.4 Bridging Services and Networks

The service model forms a significant element in bridging services management and traditional network management. It directs and provides proper guidance in the allocation of resources and attention to relevant and important aspects of network management. It is therefore imperative that the service model encompass both service and network components in an integrated fashion rather than dealing with them separately. As such, the service model must be designed to be used by different organizations responsible for only one technology domain, yet that will need access to information beyond its management domain to derive an end-to-end view.

5.3.5 Supporting Future OSS Extensions

The framework has to be able to support various operations and management functions. It should initially support traditional OSS functions and has the capability to extend to support additional OSS functions when they become feasible in the future. For example, the initial model may not support many security functions; however, when monitoring and control functions for security are added later, the framework should be flexible to support them without changing previously defined components. Another example is that new radio access technology (e.g., 802.16 technologies) will continue to evolve, and the service model should be configurable to deal with new technology without significant change.

5.4 RESEARCH IN SERVICE MODEL METHODOLOGIES

There are many varieties of service models described in the literature. Each approach focuses on certain aspects of service management. We will first survey these approaches and discuss how they can be applied to solve service management problems. As this book focuses on assurance principles, our description of a service model is naturally biased toward solving the assurance problem. In this context, our survey of the literature is also related to service quality assurance.

5.4.1 Measurement Navigation Graph

The measurement navigation graph (MNG) [1] model, first proposed by Hellerstein et al., is based on a directed, acyclic graph whose nodes are measurement variables. The goal of the MNG is to provide a framework for identifying performance problems. In the MNG, network measurements are represented by nodes, and the relationships between the measurements are indicated by directed arcs. An example of an MNG is shown in Figure 5.1.

In Figure 5.1, the service total response time (top node) depends on three measurements (second-level nodes): CPU response time, input/output (IO) response time, and paging response time. Each of these measurements further depends on lower-layer measurements (third-level nodes). Note that the service time that affects two parent measurements of IO and paging is a result of sharing resources between the user's IO and paging system IO.

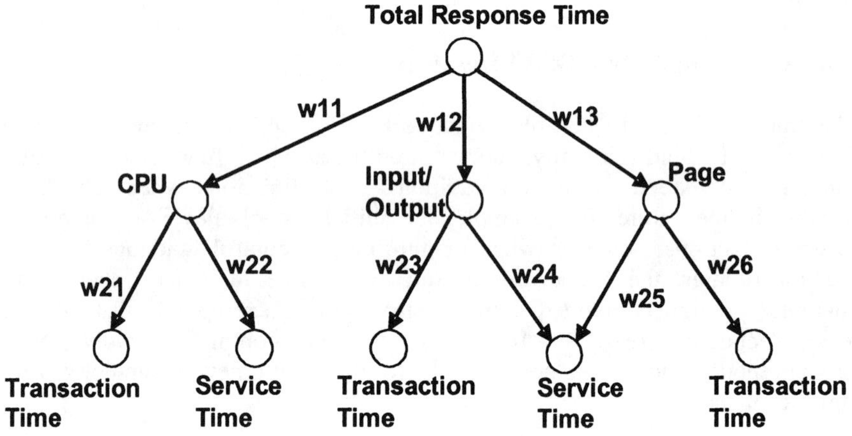

Figure 5.1 Measurement navigation graph model. (*After:* [2].)

To identify performance problems, the approach used in [2] is called quantitative performance diagnosis (QPD). QPD assigns weights (w's in the figure) to the arcs in the MNG. These weights represent the proportion of the "responsibility" of each child measurement with respect to the performance degradation of the parent measurement. The weights are determined based on a linear model of the relationship between parent and child measurements, as well as some of the statistical parameters of the measurements such as mean and standard deviations.

The MNG and the QPD provide a useful analytical model for performance problem diagnosis. However, this formulation alone does not provide a complete solution to the service assurance problem since it only focuses on measurements without discussing how these measurements are related to the service or service components as a whole. Moreover, if the relationship between measurements is not available in a detailed analytical manner, the QPD cannot be used directly.

5.4.2 The Internet Service Model

The Internet service model proposed by Smith et al. [3] takes a different approach from that of the MNG. In this approach, a service model targeting the Internet was proposed. The Internet Service Model is based on a directed graph linking the computer service components.

The model also defines the measurements needed for monitoring the health of the service. An example of the Internet service model as described in [3] is shown in Figure 5.2. Figure 5.2 shows a service model that addresses many OSS functions that are relevant to service assurance. Compared to the MNG approach, this model focuses on end-user services and less on network performance analysis. However, the model does not address how the measurements of various components are related to each other, and it also does not suggest how data are to be processed to support assurance functions. In addition, this model does not deal with the topology of a network. Topology information is very often required to deal with root-cause analysis of end-to-end service problem and capacity and traffic planning.

5.4.3 Service Model with Focus on Root-Cause Analysis

Another class of service models focuses on solving the problem of root-cause analysis. The HP OpenView event correlation system (ECS), the SMARTS InCharge model, BMC Software's PATROL Diagnose solution, and SeaGate's Nerve Center are commercial products addressing the root-cause analysis problem. These products are conceptually similar, and most of them depend on causal and temporal relationships between events. But they are very different in how they correlate events. We will briefly describe two approaches, one approach from SMARTS [4], and another from Lucent [5].

5.4.3.1 SMARTS InCharge Model

InCharge is an ECS developed by SMARTS [4], supporting an object-oriented network modeling language called Managed Object Definition Language (MODEL) and a correlation engine based on an approach using a codebook. Event relationships and network configuration information are encoded in MODEL, an extension of CORBA IDL (Common Object Reference Broker Architecture Interface Definition Language), with additions of new syntactic constructs to specify semantics that cannot be specified in CORBA IDL, such as relationships, events, problems, and causal propagation. Given a MODEL specification, InCharge then extracts a causality graph from it, converts the graph into a codebook, and uses the codebook to perform correlation. InCharge supports two types of events: problems and symptoms. The basic philosophy of InCharge is that every problem in a networked system has a unique signature. The signature consists of a cause and multiple symptoms. The symptoms can be detected by alarms or events. They include those that are inside the faulty component and those that are in related components. Symptoms of different problems may overlap, but the unique combination of symptoms (i.e., the signature) uniquely identifies the problem.

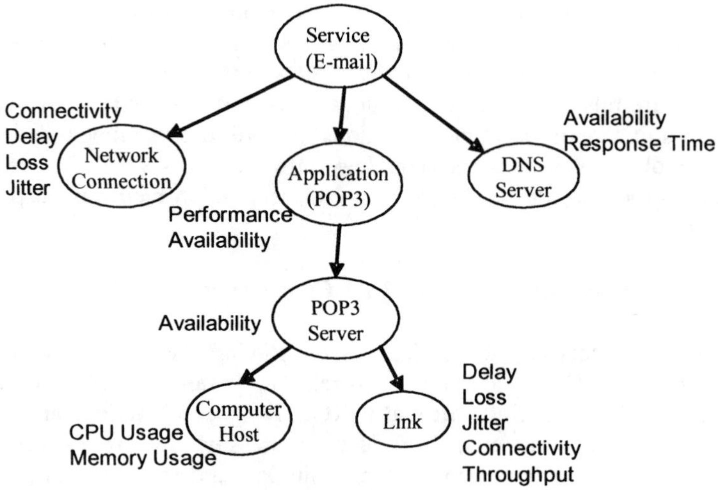

Figure 5.2 Internet service model. (*After:* [3].)

InCharge captures all the unique combinations of symptoms and their correlations to the problem in a large table called codebook in which the axes are the problems and the symptoms, respectively. By matching the signature of the problem, the root cause is uniquely identified.

One problem of the codebook approach is that the size of the codebook becomes increasingly large for practical networks and services. InCharge deals with this scalability problem by using only a subset of the symptoms to identify problems. InCharge claims that not all symptoms are required to identify a problem uniquely. This is a reasonable assumption if the right number of symptoms of a root cause is discarded.

While the InCharge approach provides a practical way to correlate cause and effect (problem and symptom), it does not explicitly use a very important piece of information (i.e., the temporal correlation), in the codebook designs. In some cases, in which temporal effects play an important role, such as statistical fluctuations of performance parameters, the temporal correlation is expected to produce better results. This aspect of InCharge is also pointed out in [5]. In general, the InCharge model is expected to work better for hard faults than soft faults.

5.4.3.2 Temporal Processing

The work published by Hasan et al. in [5] extended the event correlation idea to include temporal correlation. In this approach, events are characterized by their occurrence in time. Instead of describing an event e as e = [on or off], it is now characterized as $e_1(t) = [on(t_0 - t_1), off(t_1 - t_2), on(t_2 - t_3), \ldots]$, where on and off represents occurrence and nonoccurrence of an event, and $t_1 - t_2$ is the duration from time t_1 to t_2. As such, the temporal characterization of events provides a history of the occurrences of the faults or performance related events. Hasan et al. [5] suggested the use of a temporal event specification language that supports temporal operators such as e_1 followed by e_2 ; e_1 follows e_2 within 2 minutes, and e_1 not within a certain time interval. Correlation of temporal events follows a similar approach as causal correlation, except now the events are described with a time component. For example, one can have the following event description: if $e_1(t)$ is true and $e_2(t)$ is true, then $e_1(t)$ causes $e_2(t)$.

Temporal characterization and correlation provide more powerful tools for dealing with problems that occur transiently, which are more common in performance assurance. However, the work described in [5] did not directly address how practical performance events should be defined or how to apply temporal processing to solve service-level problems.

5.4.4 Service Model Literature Summary

The above description of the research areas related to service model spans the last decade. The early work in MNG and QDP laid the foundation of analytical models. This approach had its root in queuing theory. The Internet service model addresses customer- and service-driven needs. It is a large step in the right direction. However, the Internet service model lacks analytical capabilities and does not incorporate network topology for more detailed analysis. Subsequent root-cause models based on event correlation have provided the foundation for analytical analysis. However, the commercial versions of these models normally focus on fault event analysis and pay less attention to performance analysis. Introduction of the time domain event correlation is a needed addition, but so far, there are few practical systems deploying such time domain tools in a systematic manner. Very often, it is up to the practicing engineer or planner to create ad hoc tools to solve these problems. Our goal in this chapter is to take all these models and extract the right pieces suitable for the design of an integrated service model. The subsequent service model aims at solving the more general service assurance problem, especially in the context of the emerging mobility services. Our vision is that ultimately, carrier-grade providers will evolve their OSS assurance environment to a service-driven management platform based on the integrated service model concept.

5.5 SERVICE MODELING DETAILS

In this section, we focus on defining a service model that provides a framework for solving the service assurance problem. Our presentation will follow the work of [6], which is influenced by some of the earlier work of the Internet service model, the MNG, and various root-cause analysis approaches.

Our approach is analytical in nature and starts with a graph-structure representation of a service. The graph structure is hierarchical and captures the essence of important relationships between service components and customer needs. In addition, the graph structure also facilitates the implementation of many assurance algorithms. With the graph structure defined, the next step is to identify the service components in a systematic manner. We suggest performing an analytical decomposition of a service to identify the critical subservices. Subservices are further broken down into other lower-layer subservices, including network and physical components. To create a service model, all the components need to be instantiated, which means that the service model is logically associated with the real network and relevant assurance data can be fed into the model. The real value of a service model will best be illustrated with a set of assurance OSS functions. Thus, we will next describe a set of assurance-related tools in conjunction with the service model and discuss how these tools can be used to solve various service assurance problems. The rationale for the assurance tools will be described in the context of some theoretical algorithms that rely on some fundamental statistical analysis tools.

5.5.1 Graph Structure of the Service Model

At the conceptual level, the service model provides a systematic way to organize service and network data. The essence of this data organization is that certain critical aspects of a service and their mutual relationship are explicitly represented and stored. These critical aspects are modeled as service components. In general, service components can be logical or physical. A digital session and e-mail applications are examples of logical components, whereas a host computer and a router are examples of physical components. Service components are directly or indirectly related to each other. It is important to capture these relationships and make them available to OSS applications in a systematic manner.

More formally, a service model can be represented by a graph structure. As shown in Figure 5.3, the nodes of the graph denote service components, and a directed arc indicates a relationship between two components. A solid arrow indicates a dependence relationship or impact relationship, and a dotted arrow indicates a containment relationship. The two fundamental relationships are dependence and containment. When we say component *A* depends on component *B*, we mean that the operational state or performance of *B* influences the operational state or performance of *A*. In notation, we write $A \to B$, where A is the parent node and *B* is the child node. The dependence relationship is useful for many assurance applications and makes up the fundamental structure of the service model. The reverse dependency relationship is the impact relationship, such that if *A* depends on *B* ($A \to B$), implies that *B* impacts *A* ($B \to A$). The impact relationship is denoted by a solid arrow, except that it points from a child node to a parent node.

The second relationship that we model is the containment relationship. Containment has its original root in configuration and inventory management. It applies to the description of the relationship between two physical components or between a physical and a logical component, but not between two logical components. We say that a physical component, *X*, contains another component, *Y* (physical or logical), if *Y* is contained inside the physical enclosure of *X*. We use a dotted arrow ($\cdots >$) to represent the containment relationship. An example of the containment relationship is a host computer that contains (has) an Ethernet interface card or a router containing a Dynamic Host Configuration Protocol (DHCP) function (logical). A use of the containment relationship is in root-cause analysis. For example, when a Web access service is not performing properly, the root cause may be a malfunctioning interface card. With the containment relationship, the source of the problem can be clearly identified and located with respect to the service model.

A few characteristics of the service model are worth mentioning. First, we choose a graph representation rather than a tree structure (which is more common in representing computer file structures) because a graph allows a many-to-many (and many-to-one) relationship, whereas a tree only allows a one-to-many relationship. A graph structure is thus more versatile since it can represent a shared resource. For example, suppose both an e-mail application and a DNS naming service are hosted by a single server, thereby sharing the same computer resource; that will be best represented in a many-to-one relationship of a graph structure. Second, the directed graph should be acyclic. This means that if *A* depends on *B* and *B* depends on *C*, then *C* cannot depend on *A*. Otherwise, the relationship will become inconsistent. A similar acyclic restriction also applies to the containment relationship. Third, the service model graph does not have any connotation of temporal significance. Otherwise, the structure will become too complicated. This does not restrict the service model from being applied to deal with temporal analysis, as we will discuss the role of temporal analysis in the context of root-cause analysis in Section 5.7.5. Finally, we do not place quantitative constraints, such as weights, on the dependence relationship, although the model does not preclude the assignment of weights to attributes inside the component. Without this constraint, it is easier to combine different service components with each other.

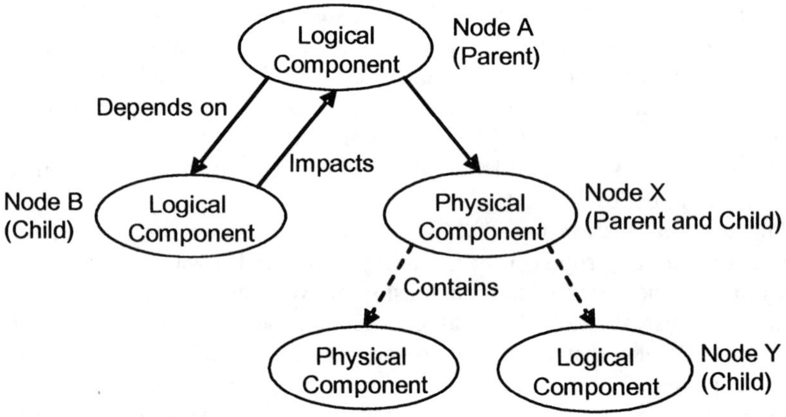

Figure 5.3 Service model as a directed graph.

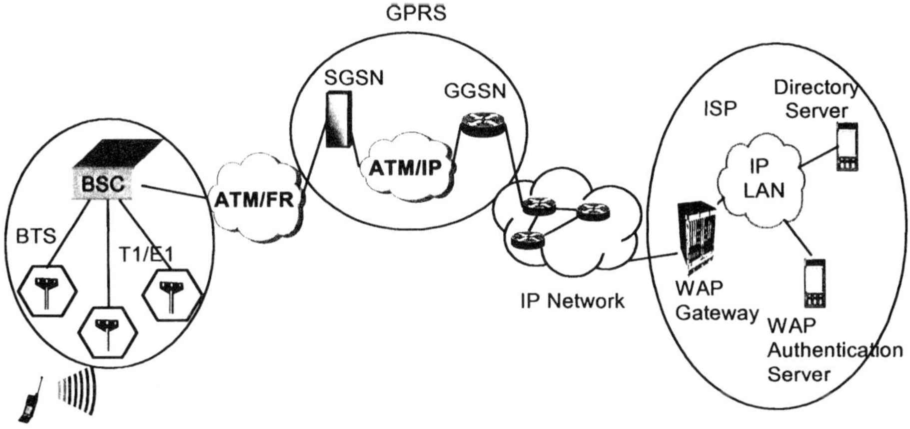

Figure 5.4 End-to-end view of WAP over GPRS.

5.5.2 Service Modeling

There are multiple ways to model a service. In this section we look at some fundamental building blocks and general guidelines for building a service model. These building blocks are useful for supporting automated, scalable, and effective modeling. The best way to describe the service model is to refer to a real network and service when possible. In the following sections, we describe the service model concept with respect to a referenced WAP service over a GPRS [7] network. A high-level end-to-end view of WAP over GPRS is shown in Figure 5.4. Refer to Section 3.3.1.5 for more detailed description of the WAP architecture. In the following, we will build up the service model for WAP over GPRS step by step, based on the kind of graph structure and dependency relationship that has been discussed. In addition, we will introduce the concept of service decomposition, an essential modeling technique used in breaking a complex service down into simpler subcomponents.

5.5.2.1 Modeling Subservices

We start with the idea that a service is usually made up of many subservices, which in turn are made up of lower-level subservices. For example, a DNS address lookup is a subservice, and the authentication of a mobile user is also a subservice. The first step towards building a service model is thus a simple decomposition of a service into a number of subservices, as illustrated in Figure 5.5.

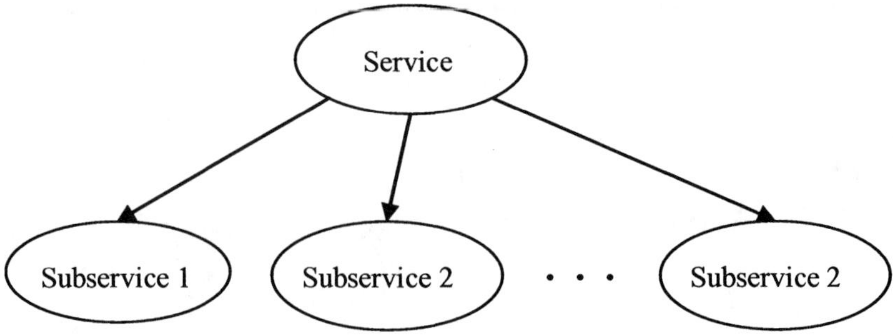

Figure 5.5 Decomposing service into subservices.

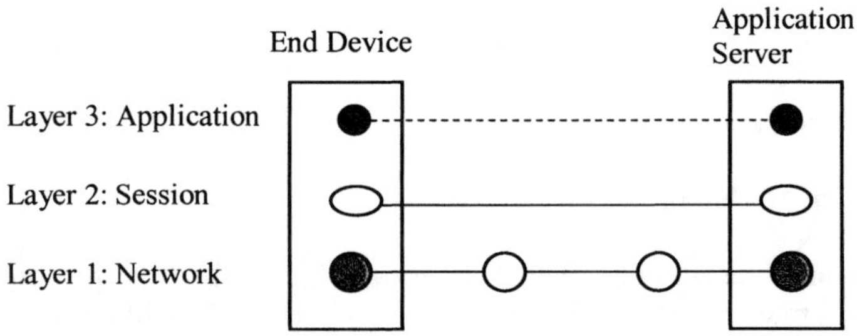

Figure 5.6 Simplified protocol stack.

The question is: How does one determine what the subservices are? There are two effective ways to identify subservices: (1) by examining *the protocol stack,* and (2) by studying *the execution phases* of a service. We start with identifying a simplified, but general, protocol stack (we do not need the details of the protocol for service modeling). We use a three-layered representation of the stack as shown in Figure 5.6. For each layer corresponding to the protocol stack, we consider the following subservices:

- *Applications:* Authentication and authorization;
- *Digital session:* Setup and teardown of a session;
- *Network:* Data transfer.

Depending on the specific service, some of these subservices may not be present. By examining the service phases, it will become clear what subservices are needed in the service model.

As shown in Figure 5.7, the network layer has two subservices, one related to transmission of GPRS data and the other related to the data transfer of the WAP application. They are grouped under a subservice called data transfer. The session layer has two subservices, MS authentication, and admission control, grouped under a subservice called PDPContext setup. The application layer also has two subservices, WAP authentication and WAP session association, grouped under a subservice called service log-on. This decomposition is illustrated in Figure 5.8.

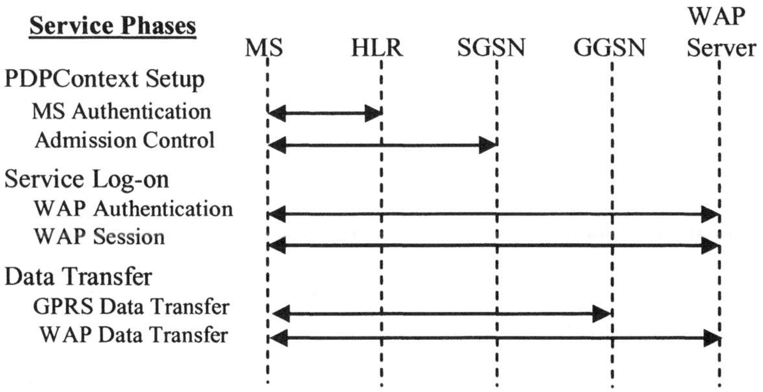

Figure 5.7 Service phases.

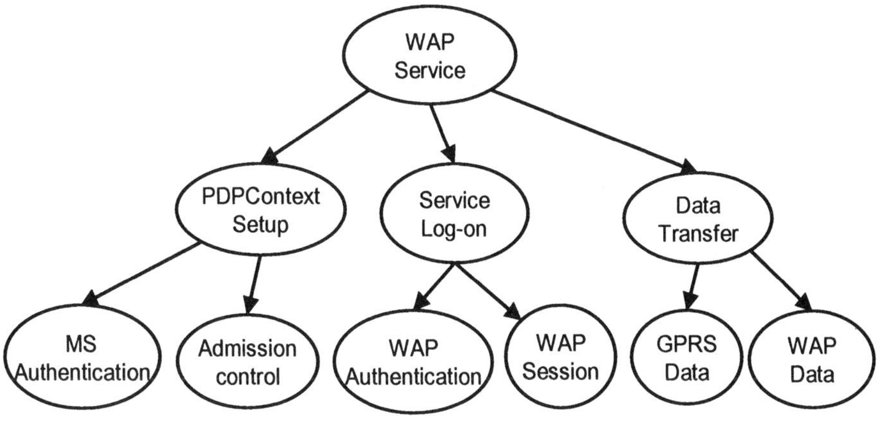

Figure 5.8 Decomposition of WAP service.

5.5.2.2 Physical Component Modeling

The service model described so far is logical, with no physical resource explicitly involved. The next step in creating the service model is to decompose further the logical subservices into components implemented by physical devices and network resources. A general guideline of this decomposition is shown in Figure 5.9. This decomposition follows the observation that, in general, a subservice depends on a server (of that subservice), a network, and an end device. As an example, the WAP authentication subservice can be decomposed into a WAP authentication server, a network (IP LAN) component carrying the authentication messages, and the mobile station. Another example is the MS authentication subservice, which is decomposed into an HLR, an SS7 component, and the MS.

Based on the general guidelines for the subservice decomposition, we can now complete the service modeling of a service as shown in Figure 5.10. In Figure 5.10 we have made a simplification by omitting the MS subcomponent. Note that the service model can be further decomposed into lower-layer components, depending on the OSS application and the resources available.

5.5.2.3 Modeling of Server, Server Cluster, and Application Software

Modeling the application and server part of the service is significant to the overall service model. This is the part that is usually managed by the IT organization. Defining a clear and practical model is important as many of the service and performance issues are related to the application servers. The service model needs to be able to address the issues of distributed server application, load balancing, and database access, as described next.

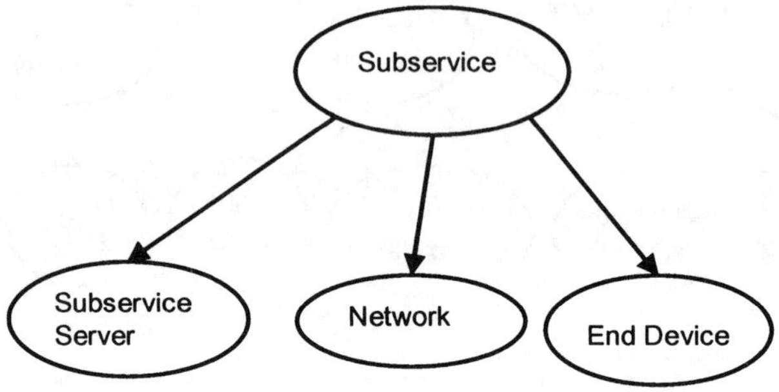

Figure 5.9 Decomposing a subservice into physical entities.

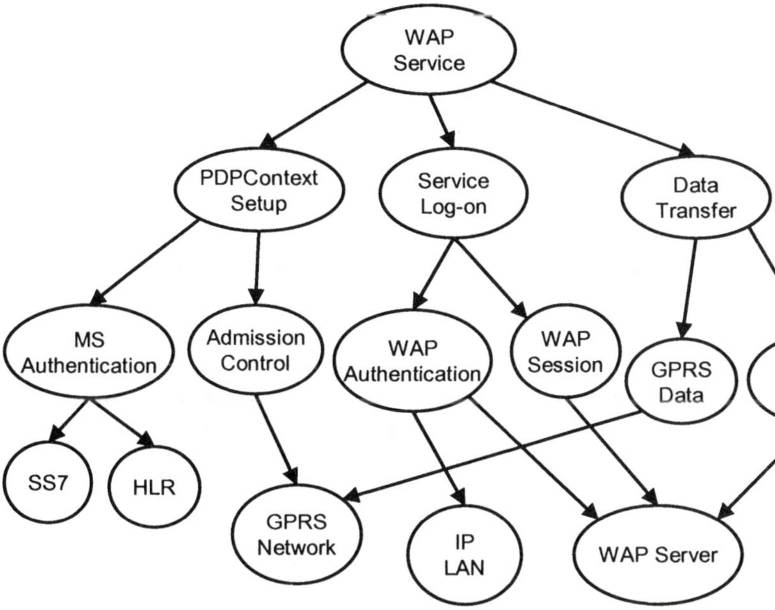

Figure 5.10 Decomposing a WAP service.

Our model is based on a server cluster service component, which represents a single point of entry from the mobile client perspective, where client requests may be handled by either a single server or by multiple servers in a load-balanced server cluster. A cluster can be implemented by Linux Network Information System (NIS), which manages a network of computers. A practical example of a server cluster is a Simple Mail Transfer Protocol (SMTP) server cluster, which uses the DNS round-robin mechanism to balance incoming SMTP messages among a number of SMTP servers. The cluster can consist of a single host with no load-balancing software or of multiple hosts with load-balancing software. The term load balancing is used in a high-level context, which refers to a system that balances the traffic load among multiple servers. For example, it does not necessarily imply the use of a multiprocessor computer host where the host operating system balances central processing unit (CPU) load between the multiple processors. An example of a server cluster with load-balancing capability is shown in Figure 5.11.

A server cluster can have performance alerts, load-related performance alerts, availability alerts, and load-misbalance alerts. Performance and load alerts are triggered by poor performance or high load in the software subcomponents. A misbalance alert is triggered when one or more of the child server software components experiences significantly different load levels from the others.

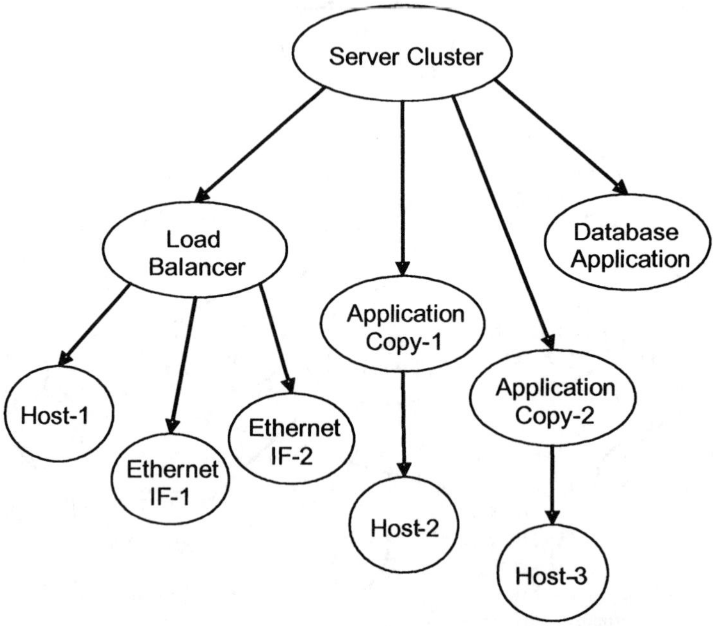

Figure 5.11 Server cluster service model.

Distribution of software applications is modeled as multiple application components underneath the server cluster, as shown in Figure 5.11. The copies of the application may be hosted by different computers that can be geographically apart. Information related to management of the distributed applications is handled by the cluster component.

In many applications, the server needs to access a remote database, which is implemented by a separate database file management system such as Linux Network File System (NFS). Many performance issues, such as server crashes, queue overflows, and access denial, can also be modeled as attributes of the server cluster component.

5.5.2.4 Modeling of Networks

A service model should be as detailed as necessary. So far, we have described a crude modeling of a network. In practice, since many assurance issues are network related, it is desirable to decompose the network component into more fine-grained network-related subcomponents.

From the modeling viewpoint, it is convenient to identify three common topologies of a network: (1) a star structure, (2) a mesh network, and (3) a shared-medium structure. Examples of a star structure include access networks of the traditional telephone network, the asymmetric digital subscriber line (ADSL) network, and the fiber-to-the-home (FTTH) network. Mesh networks include IP and Frame Relay networks. Examples of shared-medium structures include the Ethernet, radio access, and the cable network. From a service modeling viewpoint, a combination of these structures can be used to model most of the known network topologies. From the GPRS (also see Section 3.3.1.1) routing area example shown in Figure 5.12, we see that all three types of network topologies are deployed. In Figure 5.12(a), each GPRS consists of one SGSN, which is connected to multiple BSCs via a frame relay network (mesh).

Referring to Figure 5.12(a), each BSC is connected to multiple BTSs via T1 or E1 facilities (star). The BSC, T1/E1 transport facility, and the BTS groups together form a grouping called the BSC network. Finally, the handsets are connected to the BTS via radio access (shared). The service model of the GPRS routing area is shown in Figure 5.12(b), which shows a hierarchical structure consisting of a fundamental building unit as shown in Figure 5.13. This building block begins with decomposing a network module (say, the *n*th level) into three subcomponents: a network node, a transport network, and a network group. The network node aggregates the sessions and prepares them to be transported to the next module. It corresponds to the center point of a star topology. Usually, the network node also has intelligence for setting up sessions and has multiple physical interfaces with the transport network.

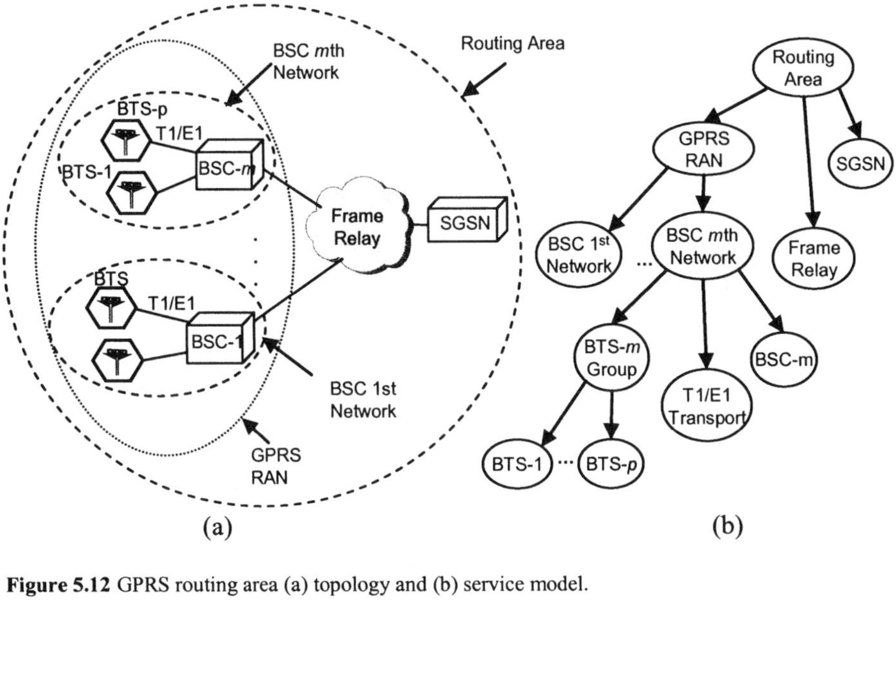

Figure 5.12 GPRS routing area (a) topology and (b) service model.

The transport network subcomponent provides connection and transport between the network node and lower level network modules. Examples of transport networks include frame relay, ATM, and IP networks.

The third subcomponent, the network group, is an abstraction of all the lower-level networks. This group can be further decomposed into the next hierarchy of network modules. As we can observe in Figure 5.13, both the star and the shared medium structure are used to create the fundamental network model building block. Based on this building block, a service model for a hierarchical network structure can be generalized as shown in Figure 5.14.

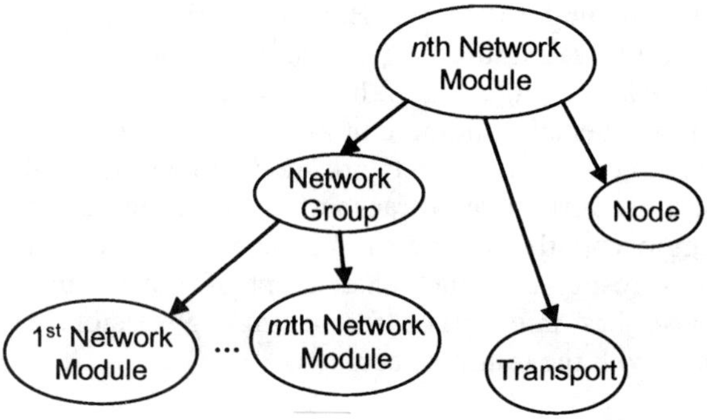

Figure 5.13 Basic building block of a hierarchical network.

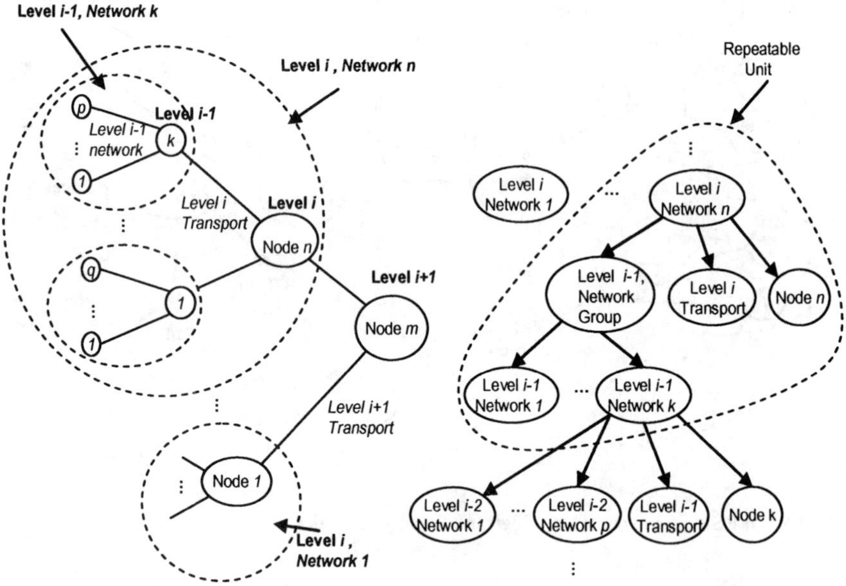

Figure 5.14 Hierarchical network and service model.

5.5.2.5 Containment Relationship

So far we have been describing the dependence relationship, which targets supporting the assurance OSS function. Very often, a performance problem may be due to misconfigured equipment or attributes. A service model will greatly reduce the complexity of debugging if it allows easy access to configuration information. Configuration can be conveniently supported via the containment relationship, which is defined between a managed object associated with a physical device and another managed object that can be a physical or logical entity. For example, an application running on a computer is contained inside the host computer. In this case, the containment relationship is defined between a physical object (host computer) and a logical entity (a software application). The containment relationship allows:

- Configuration of managed objects;
- Grouping of resources to support various OSS functions;
- Inventory of replaceable objects (including both hardware and software objects).

It should be noted that a managed object can be either:

- A separate component (e.g., a software application);
- An attribute of a component (e.g., CPU).

An example of the containment relationship is shown in Figure 5.15, where a host computer contains two applications (logical managed objects). An example of an application is a SIP proxy server, where SIP messages are routed or redirected. A second application residing on the same host can be a SIP registration server storing the location of the SIP user.

The host computer (container object) stores the inventory information about the application software (contained object), including the type of software, version number, backup location, software update and download URL, date of installation, and configuration parameters.

Container objects are usually physical objects, but contained objects can be logical or physical. An example of a physical contained object is an Ethernet interfaces, as shown in Figure 5.16. Here the Ethernet interface is modeled as both a component and a contained object. Compared to the model in Figure 5.15, the two Ethernet interfaces are brought outside the host computer component because they have different parents; Ethernet 1 is the child of application 1 and Ethernet 2 is the child of Ethernet 1.

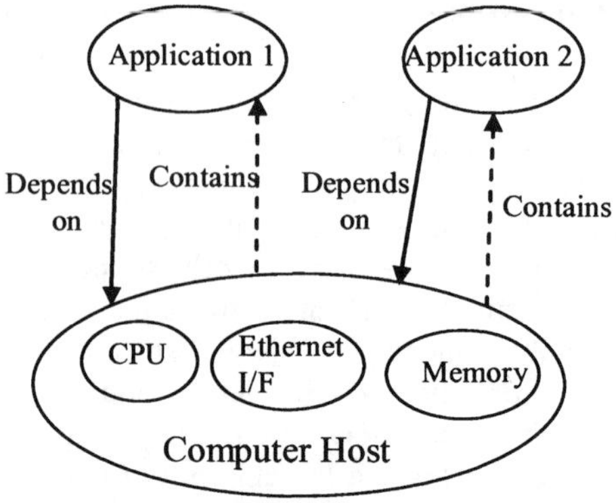

Figure 5.15 Containment relationship.

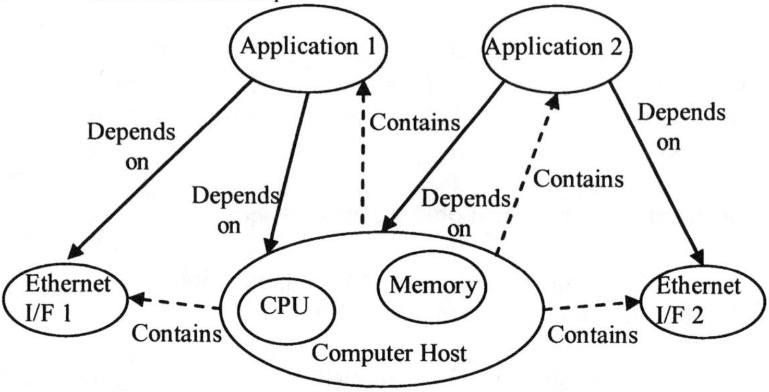

Figure 5.16 Ethernet interface as a contained managed object.

5.5.3 Anatomy of a Component

Having laid out the structure of the service model, we are now ready to describe the key functions supported inside a service component, as well as the interfaces between components. In designing these functions, we consider issues such as modularity, user programmability, scalability, and ease of use.

5.5.3.1 KPIs Inside a Component

From the assurance viewpoint, we characterize a component with respect to three generic categories of KPIs: availability, performance, and usage. The grouping is useful and necessary for operations of the service model for supporting different OSS applications. A KPI group also provides a summary of the overall condition of a component, making it more scalable and more suitable for reuse. Variations in the details and the semantics of the KPIs are hidden within the component.

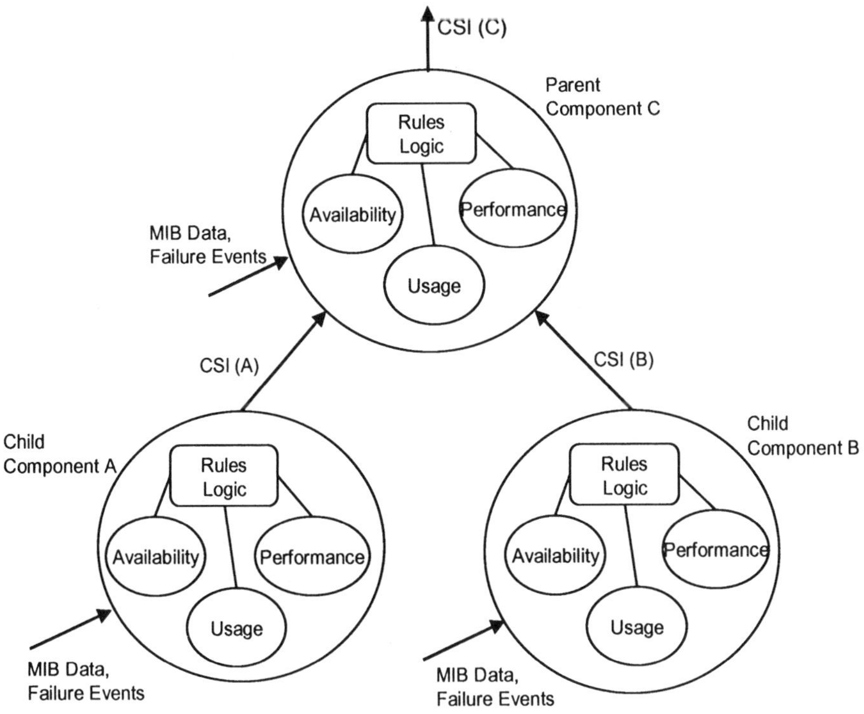

Figure 5.17 Anatomy of a component. (*From:* [6]. © 2004 Telcordia Technologies. Reprinted with permission.)

As illustrated in Figure 5.17, the three groups of KPIs are:

1. *Availability:* An indication of the level of availability of the component. Three levels are defined: First, the *component is totally down* (3), an example is a hardware failure condition. Second, *all key performance indicators of the component are poor* (2), meaning the component is still up, but very poor performance is expected in all performance measures. Finally, the *component is available* (1).
2. *Performance:* A three-level performance indicating the overall performance of the component. Levels are: *severely degraded* (3), *degraded* (2), *and normal* (1). The difference between performance degradation and degradation as a result of an availability problem is that "performance" problems are usually related to traffic load and network resource balance and are not *fault* situations. Moreover, a performance condition usually goes away when the load situation is relaxed. On the

other hand, an availability problem usually signifies that there is some abnormal condition, which does not go away in time.

3. *Usage:* A general term covering "usage of the service" [e.g., measured by call detail record (CDR)], or usage of the network resource (e.g., total number of PDPContexts within a measured duration), or usage of computer resources such as percentage utilization of CPU resources. *Usage* does not directly indicate service quality, but it provides a good indicator to the root cause of a performance problem and is also very important as an impact indicator as it tells how many users may be affected. Three general levels can be used to quantify usage: *high usage and performance affecting* (3), *high usage but not performance affecting* (2), and *normal usage range* (1).

5.5.3.2 Interfaces Between Service Components

A service component can be viewed as an object with input, output, and state. The state of a component is represented as its component status indicator (CSI), which provides a summary of the status of the three KPIs described above. When a component is in a healthy and normal condition, the CSI is kept internal to the component. When the component is in some abnormal condition, the component informs its parent component(s) via a CSI alert. The abnormal conditions include:

- Performance data exceeding thresholds;
- Faults and exceptions;
- Receiving CSI alerts from child components.

The criteria for the generation of a CSI alert is decided by a set of rules (see Figure 5.17) that is configured inside the component. Rules are usually in the form of "IF condition THEN action" statements. More details of the rules will be discussed in Section 5.7.4.

Structure of a CSI Alert

A CSI alert needs to carry information so that higher-level assurance functions can uniquely identify where and how the alert was generated. An example of a CSI alert is as follows:

MMSC_CSI_Alert:

[CSI_ID=MMSC_Cluster: Performance_Level = 3; Handle = 000110, "% of messages successfully delivered < 98%"; Time_stamp = 4.31.2005.14:23:34]

In this example, CSI_ID identifies the specific component, which is a multimedia message service center (MMSC). When a number of conditions that trigger a CSI alarm are identified, one or more handles is generated. A handle identifies a specific QoS alarm. This is not to be confused with the CSI alert, which is generated after the rules are evaluated. An optional field of text describing the alarm corresponding to the handle can also be included. Also, a time stamp specifies the date and time when the CSI_ID was triggered. A plot of the CSI trigger as a function of time will give a trail of the severity (frequency of occurrence); it also provides time series data for time domain correlation with other CSI alerts for trouble analysis. A further description of the use of CSI to deal with impact analysis will be given in Section 5.7.4.

5.5.4 Basic Components Template

So far we have described the structure of a service model in a conceptual manner. We have described how a service can be decomposed into logical components and how these logical components can be further decomposed into lower-level physical components, such as computer hosts or networks. We have also described how the dependence and containment relationships between components form the basic building blocks of a service model. Next, we will describe the functional design of a basic component of a service model. We regard this as a component template design. The defined component template has built-in functions so that a service model can be "created" by binding together the necessary components. A basic component should satisfy the following design criteria:

- Component identification;
- Service model structure identification;
- Support of QoS monitoring;
- Support of assurance functions;
- Life-cycle management;
- Notification;
- Communication between components.

In the following section, we describe why these functions are needed and how these design criteria are satisfied.

5.5.4.1 Component Identification

This consists of an object identifier (OID) that uniquely identifies this component and the component type (service, logical, physical).

5.5.4.2 Structure Identification

This consists of a list of dependency parent component OIDs and a list of dependency child component OIDs. In addition, this also gives a list of components that are contained by the current component and the component that contains the current component.

5.5.4.3 Monitoring Support

This consists of a description and configuration data for a data-collecting agent, which can be an EMS, a probe, or another monitoring device. For example, the data collector may be an SNMP server, or it can be a database storing the performance data.

5.5.4.4 Software Life Cycle

This identifies the version of the software, time and date of installation, and the status of software installation (pending with date, active, inactive, scheduled to be serviced).

5.5.4.5 KQIs, KPIs, and Management Information Bases

This identifies a set of management information bases (MIBs), KPIs, or KQIs that will be collected with respect to the current component. MIBs are raw data collected from external monitoring devices. KPIs are usually selected indicators defined specially for the current component. They are usually computed from a selected set of MIB variables. KQIs are similar to KPIs, except KQIs are usually more closely related to direct user perception of the service. Usually, KPIs from multiple components are processed and computed to form a KQI.

5.5.4.6 Component Status Indicator

This is an indicator that summarizes the overall performance status of a component. As described before, the CSI has three attributes: availability, performance, and usage. More details on the structure and usage of the CSI will be discussed in Section 5.7.4 when impact analysis is described.

5.5.4.7 Rules Engine

Within each component, there is an analysis engine that analyzes the KPI or KQI, fault information, and CSI from other components and decides on the state of the CSI. A user can program assurance and QoS policy rules in the form of scripts.

The rule engine executes these scripts and makes a component behave according to the policy. This provides flexibility and programmability of the service model. More about scripting and rule execution will be described in Section 5.7.4.

5.5.4.8 Notification

When a CSI is in an alarmed state and the CSI has a high priority, a notification can be generated. The notification can be in the form of an e-mail, page, or short message sent to the maintenance personnel, or it can be a visual or audible alarm signal. When a notification is triggered, human attention is usually required.

5.5.4.9 Communication Between Components

Sending KPI information between parents and children is necessary since the status of a parent component depends on all the states of its children. However, due to the scalability problem, it is desirable to reduce parent-child data transfer to cases when it is essential. The service model is designed in such a way that KPI information is passed only when one or more of the child components is in an alert status.

5.6 COMPUTATIONAL TOOLS FOR STATISTICAL ANALYSIS

A service model provides a structure that links the service-level attributes with those of the components in the lower layer. Once the relevant data is collected and organized with respect to the structure of the service model, additional processing and data mining are required to extract useful information from various parts of the service model to satisfy assurance applications. In this section, we will provide a fundamental mathematical framework for the service model and establish useful building blocks suitable for the analytical study of assurance problems. We will see that by representing KPIs as random variables, we are able to apply many powerful tools borrowed from the disciplines of statistical analysis and signal processing to solve difficult problems.

We start with some fundamental definitions of random variables, random processes, and their statistical properties. Then, we show how to model KPIs and KQIs as random variables and random processes. This introduction to the statistical modeling of KPIs and KQIs is important since a wealth of knowledge in statistical processing and estimation theory can potentially be applied to various OSS functions, including the definition of SLAs, detection of SLA violations, identification of the root cause of performance problems, identification of network or server-capacity bottlenecks, and prediction of trends. As an example, an SLA may include a condition that the measured IP packet delay should be no more than 30 ms for 99% of the packets. Compared to an SLA of 25-ms delay for 95% of packets, which one is harder to meet? Another example involves the issue of how to set the threshold of a KPI in the context of other related KPIs corresponding to related components of a service model. A random or ad hoc setting may cause the generation of alarms too frequently or risk the chance of missing severe problematic events. The statistical framework discussed in this section will provide answers to many of these questions. Many of the statistical processing capabilities have long been effectively applied to other disciplines, such as economic studies, image and speech processing, and communications applications. Here we shall show how these statistical tools can also be effectively applied to solve OSS problems within the service model framework. Readers who are familiar with this topic may skip to Section 5.7.

5.6.1 Statistical Tools and Properties

A random variable x generally takes on values from $-\infty$ to $+\infty$ and is characterized by its density function $p(x)$, which can be used to compute the mean and the variance as

$$\text{mean} = E[x] = \int_{-\infty}^{\infty} xp(x)dx \tag{5.1a}$$

and

$$\text{variance} = \sigma_x^{\ 2} = E[(x-m_x)^2] = \int_{-\infty}^{\infty} (x-m_x)^2\, p(x)dx \tag{5.1b}$$

where σ is the standard deviation, $E[x]$ is the expected value of x, and $p(x)$ is the density function. The density function provides a straightforward way to compute the probability of x, assuming a value in the interval between a and b, which is given by

$$\text{Probability } [a < x < b] = \int_{a}^{b} p(x)dx \tag{5.2}$$

Although the density function $p(x)$ can be used to generate all the statistical properties of the random variable, it is usually not available. Two special cases are the normal (or Gaussian) distribution, where the probability density function is given by

$$p_{normal}(x) = \frac{1}{\sqrt{2\pi}} \exp\{-(x - m_x)^2 / 2\sigma_x^{\,2}\} \qquad (5.3)$$

and the uniform distribution, where the probability density function is given by

$$p_{uniform}(x) = \begin{cases} 1/w, & \text{for} -w/2 \le x \le w/2 \\ 0, & \text{otherwise} \end{cases} \qquad (5.4)$$

The normal distribution has been used extensively to model physical and other engineering phenomena. In many situations, it is frequently used as an assumption, due to its mathematical tractability and reasonably good approximation to many practical situations. The uniform distribution is generally used when bounds on the variable are known a priori but no other information is available.

5.6.2 Relationship Between Two Random Variables

In many applications, it is useful to measure the relationship between two random variables. We will see that the ability to measure quantitatively the relationship between two random variables will also be useful for assurance functions with respect to the service model.

Given two random variables x and y, with (x_1, y_1), (x_2, y_2),…, (x_i, y_i) being the outcome or the observation, an effective way to display the relationship between x and y is by using a scatter plot, as shown in Figure 5.18, where the first (left) plot shows a positive correlation (i.e., when x is large, y is large) between random variables x and y and the second (right) shows a negative correlation (i.e., when x is large, y is small).

By inspecting the scatter plots, we can infer that there is a roughly linear relationship between x and y. The degree of this roughly linear relationship is a measure of how tightly x and y are correlated to each other. Note, however, that a scatter plot does not provide any information about the causal relationship between the pairs of realization (x_i, y_i).

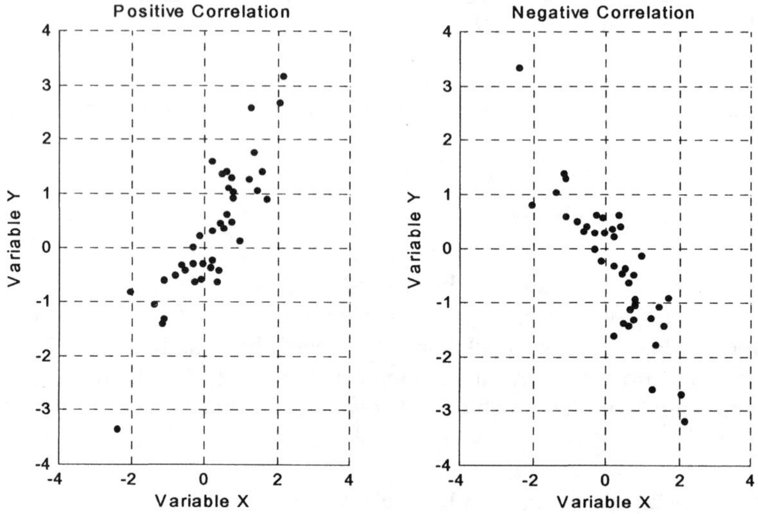

Figure 5.18 Scatter plots.

Figure 5.19 shows the degree of correlation with respect to the scatter plot. As shown in the figure, a weak correlation shows that the points are scattered loosely around a straight line. When the correlation is stronger, the scattered points are tightly bound to the line (not shown). In this case, if a realization of x is known to be x_i, there is less variation in y_i compared to the weak correlation case. When the correlation is perfect (total correlation), the scatter plot just shows samples of a straight line. In that case, when x is known, y will be specified as well.

More precisely, the degree of correlation can be measured quantitatively with a covariance function, $\sigma_{x,y}$, defined as

$$\sigma_{x,y} = E\{(x-m_x)(y-m_y)\} \tag{5.5}$$

where m_x and m_y are the means of x and y, respectively.

Thus, the covariance is given by the expected value of the product of $(x-m_x)$ and $(y-m_y)$. If the covariance is positive, that means that when x is larger than its mean, y will usually be larger than its mean, and the converse is also true. In such case, x and y are positively correlated. On the other hand, if the covariance is negative, then x is larger than its mean usually implies that y is smaller than its mean, and we have a negative correlation. If the covariance was zero, x and y could be independent.

The covariance provides a measure of the correlation of two random variables. However, it is not suitable as a quantitative measure of the strength of correlation. The reason is that the covariance depends on the units used for the measurement. To circumvent this problem, it is convenient to normalize the covariance and create an index that is independent of the unit of measurement. The correlation index, $\rho_{x,y}$ (or correlation coefficient), defined in (5.6) satisfies this requirement:

$$\rho_{x,y} = \frac{\sigma_{x,y}}{\sigma_x \sigma_y} \tag{5.6}$$

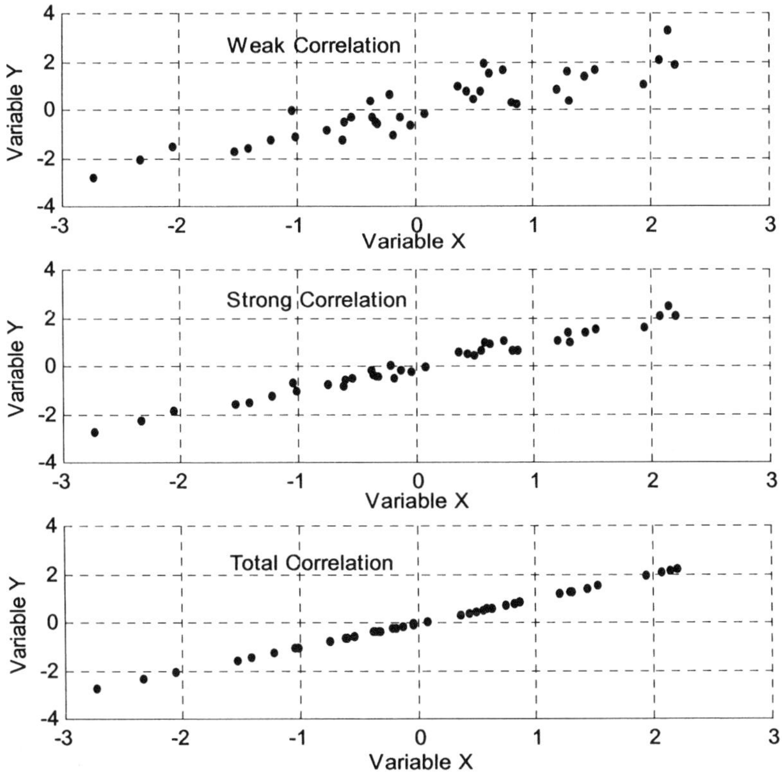

Figure 5.19 Strength of correlation.

It can be shown ([8, 9]) that $\rho_{x,y}$ has all the nice properties of $\sigma_{x,y}$, plus its range is limited to between -1 and $+1$. In general, the closer the correlation index is to $+1$ or to -1, the stronger the correlation. When the correlation index is closer to zero, it indicates very weak correlation. Thus, the correlation index provides a very effective measurement for the strength of correlation between two random variables.

5.6.3 Modeling KPIs and KQIs as Random Processes

Many KPIs and KQIs of interest are best represented as time series of random variables. As such they are best modeled as random processes. A random process (or random signal, or stochastic process) is defined as a sequence of random variables, $x_0, x_1, x_2, \ldots, x_n, \ldots$, where the index n is taken to be the discrete time variable. $X = [x_0, x_1, x_2, \ldots, x_n, \ldots]$ can be considered a random vector of which each element is a random variable. It is also called a discrete random process since x_n can be obtained from sampling a continuous signal. Since each sample x_n is a random variable, the mean and the autocorrelation function of a random process are defined as:

$$x_n = E\,(x_n) \tag{5.7a}$$

$$R(m,n) = E[x(m)x(n)] \tag{5.7b}$$

A random process is wide sense stationary if its mean is a constant and the autocorrelation depends only on the time lag; that is,

$$E\,(x_n) = \text{constant} \tag{5.8a}$$

$$R(\tau) = E[x(n+\tau)x(n)] \tag{5.8b}$$

When the random process is not stationary, it can usually be approximated as a piecewise stationary random process. This means that within a time window, the random process is stationary. One example is that the statistics in busy hours, say from 7 a.m. to 9:30 a.m. and 4:30 p.m. to 7:00 p.m., have different characteristics from those from 9:30 a.m. to 4:30 p.m. Within the busy hours, however, the variables can be considered to be stationary. The assumption about stationary process is very important as most random variables we are interested in modeling are time series, and we can gather a lot of observed samples, which are realizations of the random variables at specific times. We can then use these observed samples to estimate the statistical properties of the random process. For example, if we want to estimate the mean and the standard deviation of a random process X, given the time samples of x_0, x_1, x_2,..., x_n, we use the following estimators:

$$\hat{m}_X = \frac{1}{n}\sum_{i=1}^{n} x_i \tag{5.9a}$$

and

$$\hat{\sigma}_X = \sqrt{\frac{1}{n-1}\sum_{i=1}^{n}(x_i - \hat{m}_X)^2} \tag{5.9b}$$

One can use an average of samples of the random variable, collected over time, to estimate the actual mean of the random variable because one makes an assumption that the random process is ergodic [7]. For simplicity, we will make the assumption of ergodicity throughout this book.

We can use an example of applying the statistical information model given above for the analysis of KPI data. Suppose two common KPIs of the IP network are one-way IP packet latency and IP packet loss. We can model these KPIs as random processes D and P. We further assume that we have a data-collection agent that collects the latency and loss information once every 5 minutes. Thus, the collected data on each of the KPIs constitutes the samples of the random processes. Applying the formula given in (5.9a) and (5.9b), we can then estimate the mean and the standard deviation of one-way latency and packet loss. We can also use (5.5) and (5.6) to see if the measured latency and loss are correlated and quantitatively measure the strength of the correlation.

5.6.4 Hypothesis Testing and Confidence Level

Suppose we want to test a hypothesis (H) of the form

H: There is a performance problem in the service.

How do we convert the above hypothesis into something that can be mathematically tractable and consistently applicable across different services and networks? We have, in fact, described part of the solution, as the service model decomposes a service into components, and each of these components can be represented by three fundamental sets of KPIs, one of which is the set of performance KPIs. For a particular performance KPI, we have collected a set of sampled values. The question becomes, how do we analyze the set of sampled values to determine whether the monitored service has incurred performance problems?

A common way of testing hypothesis H is to set a threshold on the selected performance KPI. When the measurement exceeds the threshold, it is then concluded that H is true and that a performance problem has been detected. So, the key question is, how should one select the threshold? Setting the threshold to a value that is too low causes too many false alarms (false positives) (i.e., alarms are generated without the occurrence of a real performance problem). On the other hand, setting the threshold too high will cause unnecessary false negatives (i.e., the occurrence of a performance problem is not detected).

To pose the problem of how to set the threshold of a performance parameter in a hypothesis testing form, let's assume the KPI is represented as a random process x, and that x is normally distributed with known (or estimated), long-term average m_x and standard deviation σ_x. Let's first make the following substitution of variable:

$$v = \frac{x - m_x}{\sigma_x} \tag{5.10}$$

where x is the original variable and v is the normalized variable.

Since x is normally distributed, it can easily be shown that v is also normal with a mean of 0 and a variance of 1. The normalization given in (5.10) makes the explanation later easier. From the table of standard normal distribution in [7], we can immediately write down the following probabilities:

$$\text{Prob}(v < 1.28) = 0.9$$

$$\text{Prob}(v < 1.65) = 0.95$$

$$\text{Prob}(v < 2.33) = 0.99 \tag{5.11}$$

This means that if v is selected at random, there is a 90% probability that v is less than 1.28, a 95% probability that v is less than 1.65, and a 99% probability that v is less than 2.33. Replacing v back with x, we have, for the 95% case,

$$\text{Prob}\left(\frac{x - m_x}{\sigma_x} < 1.65\right) = 0.95 \tag{5.12a}$$

or

$$\text{Prob}(x < m_x + 1.65\,\sigma_x) = 0.95 \qquad (5.12b)$$

The interpretation of (5.12b) is as follows. If x is random, we can, with 95% confidence, predict that the value of x satisfies the following inequality:

$$x < m_x + 1.65\,\sigma_x \qquad (5.13)$$

When applied to setting a threshold for hypothesis testing, we can make the following interpretation: even if there is no performance problem, since the samples of x that we are collecting are assumed to be normally distributed, 5% of the samples are expected to fall outside the range specified in (5.13). Therefore, if we set the threshold to be $m_x + 1.65\,\sigma_x$, we will draw a wrong conclusion for 5% of the samples when there is no performance problem. This 5% error corresponds to the samples that fall outside of the threshold without the occurrence of a performance problem. We say that the way the threshold is set corresponds to a 95% confidence level. Formally, the hypothesis is written as

$$H0\text{: if } x < m_x + 1.65\,\sigma_x \qquad (5.14a)$$

We conclude that there is no performance problem, and set no alert.

$$H1\text{: if } x \geq m_x + 1.65\,\sigma_x \qquad (5.14b)$$

We conclude that there is a performance problem, and set an alert.

When threshold is set according to (5.13) and the hypothesis is postulated as in (5.14a) and (5.14b), we can formally define two types of errors: a type I error occurs if we reject the hypothesis $H0$ when it is true. In this example, a type I error means we conclude that there is a performance problem (i.e., we accept $H1$) when there is none. This corresponds to the sample falling outside the range (5.13) due to its random nature but produces the conclusion of a problematic performance event, causing the sample to fall outside the range. So, a type I error corresponds to a false positive. In our performance example, the type I error corresponding to (5.13) is 5%, and the confidence level, defined as (1– type I error), is 95%.

A type II error occurs if we accept the $H0$ hypothesis when it is actually false. In our performance example, we say there is no performance problem when there is one. This corresponds to a false negative, which occurs when the sample falls within the range (5.13), but the actual value lies outside. An example illustrating a type II error is shown in Figure 5.20. The normal curve on the left side of Figure

5.20 represents how the threshold of 1.65 is set ($m_x + 1.65\sigma_x = 1.65$). The normal curve on the right represents the distribution of the measured samples of *x* when a performance problem has occurred. The performance problem renders the distribution curve to move from a mean of 0 to a mean of 3, assuming the same $\sigma_x = 1$.

The shaded area in Figure 5.20 represents the situation where the conclusion of accepting *H*0 is wrong. Since the threshold is set at 1.65, a type II error is made when the sample of *x* is less than 1.65 with respect to the curve on the right. The probability of making a type II error is thus the integral of the shaded area in Figure 5.20. In this example, the type I error is 5% and the type II error is about 9%. It is clear from Figure 5.20 that if the threshold is made larger (than the 1.65 in this example), the chance of the sample falling below the threshold is larger, when the actual value is outside the range. Thus, by designing a test whereby the type I error is made smaller, the corresponding type II error will be larger.

Since the type II error depends on the severity of the performance problem, which is usually unknown (unless we know the statistics of the event that causes threshold crossing), type II errors cannot easily be expressed in closed form, thus cannot be controlled directly. For this reason, in almost all the hypothesis tests, the type I error is used as a design criteria.

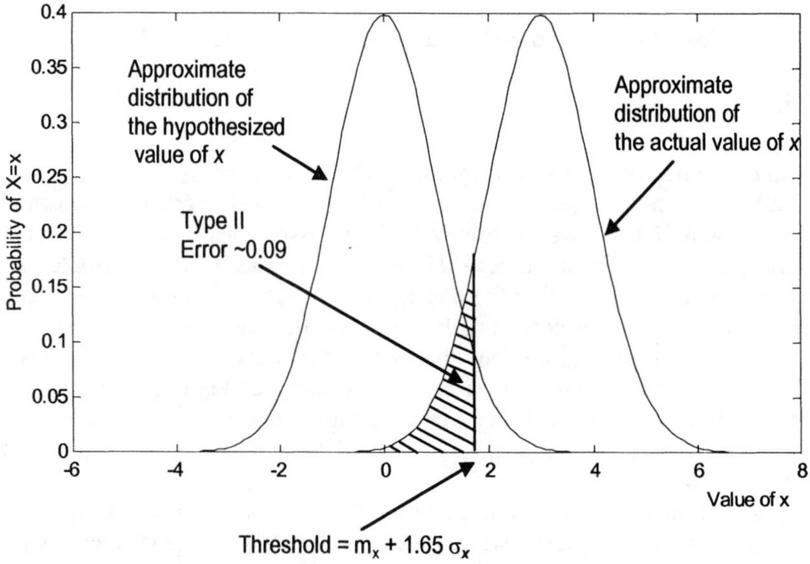

Figure 5.20 Type II error.

Figure 5.21 shows an example of threshold setting. The IP packet-loss ratio was collected from one of the large ISPs. Figure 5.21 shows the monthly average packet-loss ratio for two consecutive years. The 2-year average loss is 5%, and the standard deviation is 6.5%. Both the 95% and 99% confidence levels are shown in the figure. For the 95% case, there are 2 months where performance alarms will be set off, while for the 99% case, only 1 month will have the alarm triggered.

5.6.5 Applications Based on Statistical Relationships Between KPIs

We have described previously how we can set a threshold on a KPI based on the hypothesis theory. That would apply to an independent KPI that does not have a parent component. In some cases, the parent and child KPIs have a well-defined relationship. If the parent KPI's threshold is already defined, it is desirable to discover the appropriate threshold for the child-component KPIs. In the following, we first look at some common mathematical relationships between the parent and the child KPIs. Once the relationship is defined, we proceed to examine how to set the thresholds for the child KPIs, given the threshold of the parent KPI.

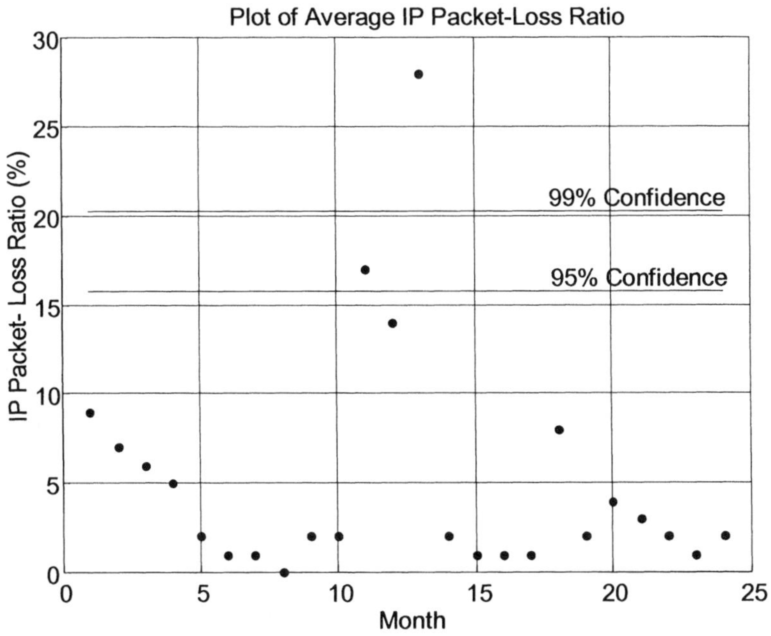

Figure 5.21 IP packet-loss ratio showing threshold setting.

5.6.5.1 Statistical Relationships Between KPIs

With respect to the service model, let us assume that the KPI w of a parent component is related to the KPIs $x, y, z, \ldots$, of the children components such that

$$w = f(x, y, z \ldots) \tag{5.15}$$

We use three KPIs x, y, z for illustration, but the methodology can generally be used to model any number of KPIs. In many practical situations, the functional relationship f is, or can be approximated as, a linear relationship so that we can write

$$w = a + b\,x + c\,y + d\,z + u \tag{5.16}$$

where $a, b, c,$ and d are constants, and u is an unknown variable.

If we can represent the KPI w with the relationship described in (5.16), we have an analytical model that can be used to support various assurance functions. For example, the weights $b, c,$ and d, together with the variances, give the relative importance of the influence of the child KPIs on w. Note that assuming that the child KPI random variables are uncorrelated, the variance of w is given by [7]

$$\sigma_w{}^2 = b^2\,\sigma_x{}^2 + c^2\,\sigma_y{}^2 + d^2\,\sigma_z{}^2 + \sigma_u{}^2 \tag{5.17}$$

By collecting statistical information, such as the mean and the standard deviation of each term in (5.16), we can quantify the impact of the KPIs $x, y,$ and z on w. This can also be used to assist in root-cause analysis of performance problems. Another application of (5.16) is for the interpolation of missing samples in w. This can be particularly important if an SLA is written on w.

Next, we describe how the estimation model of (5.16) can be obtained. This estimation problem can be posed as follows: Given a set of samples of the KPIs w, $x, y,$ and z, we would like to estimate the parameters $a, b, c,$ and d. The assumptions are:

- u is an unknown variable that has zero mean.
- u is uncorrelated to the child KPIs [i.e., $E(ku) = 0$, where $k = x, y,$ or z].

From these two conditions, we can generate the following equations:

$$E(w) = a + b\,E(x) + c\,E(y) + d\,E(z)$$

$$E(xw) = a\,E(x) + b\,E(x^2) + c\,E(xy) + d\,E(xz)$$

$$E(yw) = a\,E(y) + b\,E(yx) + c\,E(y^2) + d\,E(yz)$$

$$E(zw) = a\,E(z) + b\,E(zx) + c\,E(zy) + d\,E(z^2) \tag{5.18}$$

Equation (5.18) is called the normal equation. It can be written in matrix notation as

$$\boldsymbol{M}\,v = C \tag{5.19}$$

where

$$\boldsymbol{M} = \begin{pmatrix} 1 & E(x) & E(y) & E(z) \\ E(x) & E(x^2) & E(xy) & E(xz) \\ E(y) & E(xy) & E(y^2) & E(yz) \\ E(z) & E(xz) & E(yz) & E(z^2) \end{pmatrix} \quad v = \begin{pmatrix} a \\ b \\ c \\ d \end{pmatrix} \quad C = \begin{pmatrix} E(w) \\ E(xw) \\ E(yw) \\ E(zw) \end{pmatrix}$$

The solution of (5.19) is given by

$$v = \boldsymbol{M}^{-1}\,C \tag{5.20}$$

Numerical Example

A KPI w is estimated by the following:

$$w = a + bx + cy + dz + u \tag{5.21a}$$

where a, b, c, and d are constants, x, y, and z are the child KPIs, and u is the error term, which cannot be estimated. We simulate w by the following:

$a = 0.4, b = 0.6, c = 0.8, d = 0.1$

x, y, and z are normal random sequence with zero mean and unity variance ($N(0, 1)$).

$u = 0.2 \times N(0, 1)$

The computer simulated w as given by

$$w = 0.4 + 0.6\,x + 0.8\,y + 0.1\,z + u \tag{5.21b}$$

where x, y, and z are computer-generated time series (300 points) using MATLAB. In this example, we want to show how to estimate the constants a, b, c, and d from the samples of w, x, y, and z. Using the solution given by (5.20), we obtained the estimates of the constants as

$$\hat{a} = 0.42 \qquad \hat{b} = 0.55 \qquad \hat{c} = 0.77 \qquad \hat{d} = 0.08 \tag{5.22}$$

Therefore, the estimated w ($\hat{w}$) is given by

$$\hat{w} = 0.42 + 0.55\,x + 0.77\,y + 0.08\,z \tag{5.23}$$

Both w (KPI) and $\hat{w}$ (estimated KPI) are shown in Figure 5.22. As shown in Figure 5.22, the estimation is quite good. The difference in the two time series is due to the estimation error of the constants a, b, c, and d and the unestimated error (u).

5.6.5.2 Setting the Threshold of a Child KPI with Respect to a Parent KPI

Suppose a parent KPI w depends on child KPIs x and y (Figure 5.23) in a linear fashion such that

$$w = a\,x + b\,y \tag{5.24}$$

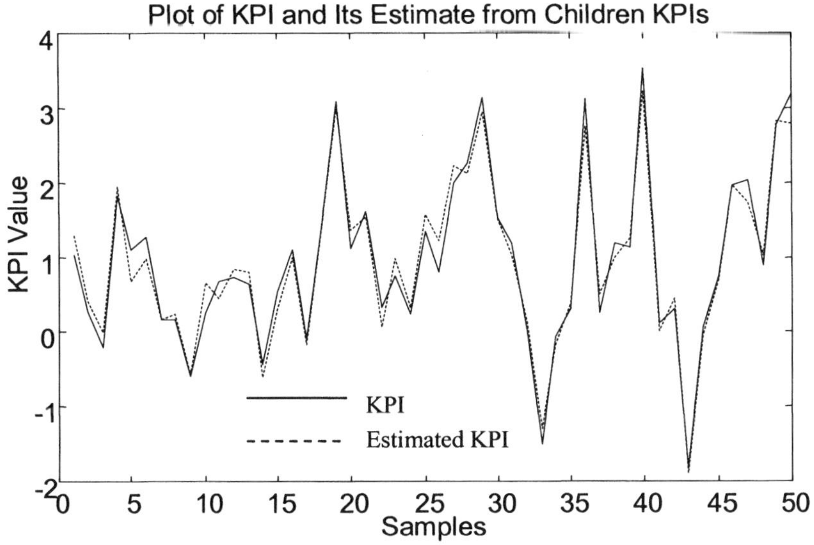

Figure 5.22 Plot of KPI w and estimated KPI $\hat{w}$.

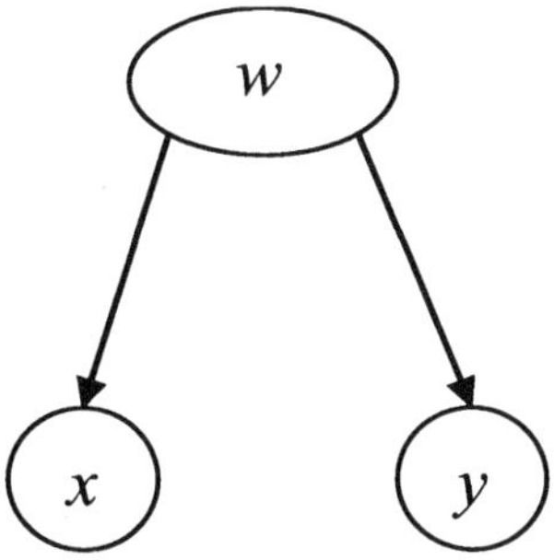

Figure 5.23 Simple parent-child KPI dependence.

Suppose the threshold of $w(w_T)$ is set so that the confidence level is 95%. Thus, $w_T = m_w + 1.65\,\sigma_w$. The question is, what should the threshold be for x and y. Recall that one of the purposes of the dependence relationship of the service model is that when service-affecting events happen, the model provides tools for finding the root cause of the performance problem. Now, if it is detected that w crosses the threshold, we want to identify which of the two component KPIs (x or y) is the most likely cause. Since the threshold of w is known, we would like to assign the thresholds of x and y to be proportional to the weights a and b, so that

$$w_T = a\,x_T + b\,y_T = a\,x_m + b\,y_m + 1.65\,\sigma_w \tag{5.25}$$

Let

$$x_T = x_m + x_t$$

$$y_T = y_m + y_t$$

where x_t and y_t are the threshold above the mean. We have from (5.25)

$$a\,x_t + b\,y_t = 1.65\,\sigma_w \tag{5.26}$$

It is desirable that the threshold be designed to be proportional to the standard deviation, so as to minimize the probability of false alarm. We have another constraint given by

$$x_t / \sigma_x = y_t / \sigma_y \tag{5.27}$$

Solving (5.26) and (5.27), we have

$$x_t = 1.65\,\sigma_w \,/\, [\,a + b\,(\sigma_y / \sigma_x)\,]$$

$$y_t = 1.65\,\sigma_w \,/\, [\,b + a\,(\sigma_x / \sigma_y)\,] \tag{5.28}$$

Equation (5.28) provides a formula for setting the thresholds for the child KPIs, reflecting the weights of each of the KPIs, as well as their statistical characteristics. Although it is demonstrated for a two-child KPI case, the idea can be easily extended for any number of child APIs. In a performance root-cause analysis application, the threshold crossing of the child KPIs defined in (5.28) provides a good indication of the likely problem source.

5.6.5.3 Interpolating Missing Data Using Prediction

Suppose a KPI is continuously monitored and is used to compute an SLA. Occasionally, anomalous situations occur and cause the KPI data-collection process to be interrupted. This may last for a few minutes, hours, or even days. In such a situation, how does one deal with the missing data that directly affects the SLA report? Traditionally, interpolation of the KPI based on previous collected data and the data collected after recovery can be used. However, from an SLA point of view, this may not be desirable. For example, the data before and after the anomaly may temporarily exceed the performance threshold. Interpolation based on such data leads to the conclusion that the KPI also crosses the performance threshold during the period of missing data. This may cause unnecessary and severe penalties.

If the desired measured KPI can be estimated from other random variables, for example, as a linear combination of attributes from dependent components, as indicated in (5.16), and assuming that the dependent variable data is available when the KPI data is missing, the desired KPI can be readily estimated from the other variables with much less chance of making an error in the SLA report.

5.7 APPLICATION OF SERVICE MODEL TO SERVICE ASSURANCE

So far, we have given a lot of effort to explaining the motivation, basic building blocks, and mathematical formulation of the service model. In this section, we will see how the service model is applied to solve the service assurance problem.

5.7.1 Service Model–Based Assurance

The role of a service model for the support of the wireless service assurance function is described in the functional block diagram of Figure 5.24. The diagram illustrates the key building blocks of the OSS functions, which include the business layer, the service fulfillment layer, and the service assurance layer. Our focus in this book is on service assurance, and in particular, in this section, on how a service model is applied to support various assurance functions.

As discussed in Chapter 4, managing services begins with service fulfillment, which includes service creation, service activation, service ordering, and customer management. In a service model–based management platform, the service creation process includes setting up a service model for supporting service configuration and service assurance. This includes defining the components of the service model and assigning KPIs and KQIs to the service model, as well as configuring the service model parameters such as data-collection intervals, thresholds, types of notification, and types of reports. Instantiation of the service model is part of the service activation process, which includes initiating the data-collection process and the service model with respect to a network inventory database system for retrieving topology information.

Once the service model is created, configured, and initiated, an instance of the service model is created. The assurance functions that are supported by the activated service model include SQM/SLM problem identification and resolution, service, capacity, and traffic planning. The service model makes use of a set of lower-level tools to support the above assurance functions.

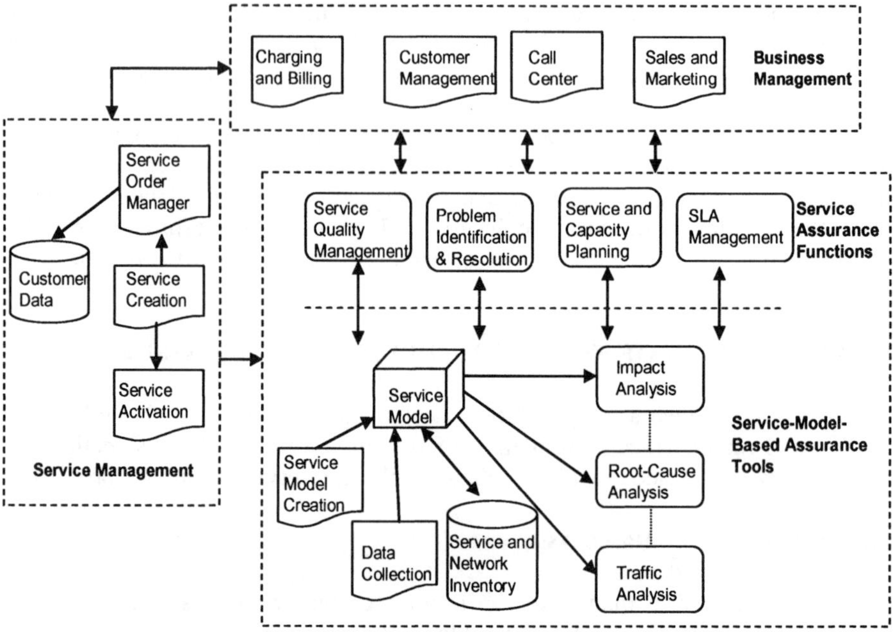

Figure 5.24 Service model applied in service management process.

These lower-level tools include statistical analysis tools from the previous section and generally fall into the categories of impact analysis, root-cause analysis, and traffic analysis. We will go into details about these assurance applications in the following sections.

5.7.2 SLA Management

As described in Chapter 2 (see Figure 2.11), SLAs are contracts between the provider of a service and the user of that service. The purpose of an SLA is to define service performance targets and to put in place measurement mechanisms whereby actual performance can be monitored against the targets. In essence, an SLA reflects the provider's commitment to its customers. An SLA is a key aspect of service management that underscores the service-driven aspect compared to the traditional network-focused performance management paradigm. Note that customers include external customers, both wholesale or retail, and internal customers, who are administrative personnel from the same company but who have different responsibilities. For example, the service management department may be the customer of the network management department, which will be responsible for ensuring the performance of the network defined by the internal SLAs.

SLA management begins with defining and negotiating between the provider and the customer the desirable quality metrics, usually based on the specific business needs of the customer. Examples of these quality metrics are:

- Service availability;
- Maximum service downtime;
- Response time;
- Speed of connection;
- Accuracy.

As shown in Figure 5.25, the SLA management process starts with a customer interaction to negotiate an agreement on the service quality metrics, sometimes referred to as KQIs. The negotiation involves defining the set of KQIs, acceptable range of values, methods of measurement, and clear definitions of the criteria for violations.

Once a set of KQIs and the associated conditions are agreed upon, they need to be converted into a set of KPIs which can be measured and used for planning, engineering, and analysis. Planning involves understanding how much of the resource (transmission facility, radio spectrum, server capacity, traffic load) is needed to meet the requirements of the KQIs and KPIs demanded by the SLA.

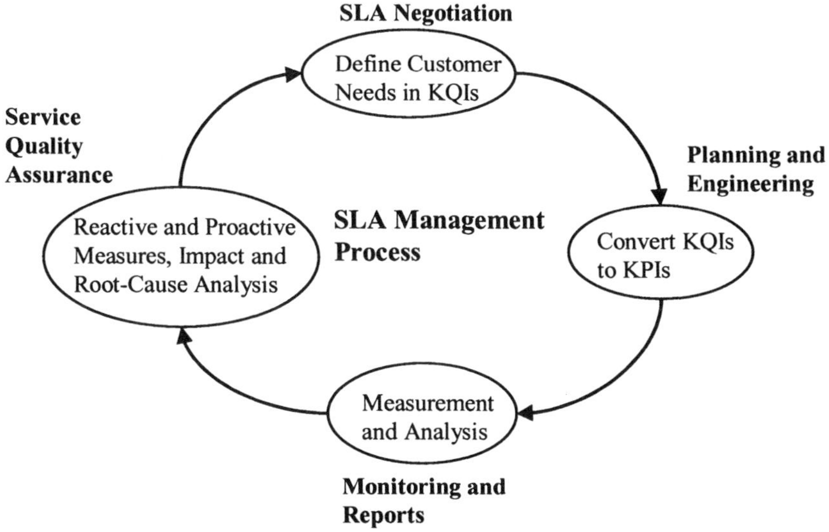

Figure 5.25 SLA management process.

Engineering concerns the network realization to achieve the performance goals. Once the network is engineering and realized, appropriate monitors should be put in place to collect performance measurement information. The measured data is then processed and filtered to provide various reports. The raw data, together with various reports, is used to support the service assurance process, which includes reactive assurance, which reacts to detected problems; proactive assurance, which aims at detecting problems before severe damage is incurred; impact analysis, which assesses the impact of the detected problem; and root-cause analysis, which finds the reason for the observed symptoms.

We use an example to illustrate how SLAs are supported via the service model. Figure 5.26 shows a simple service model of a Web-access application, and Figure 5.27 shows a simple network and a service model of a Web access service. The actual service model is a lot more complicated than the one shown in Figure 5.27, but here we use a simpler model to show how KQIs are decomposed into component KPIs. We define the following three KQIs in the Web access service component:

- Availability;
- Maximum downtime;
- Response time.

Service-level availability can be approximated by the sum of the availability of each of the component-level availabilities. Thus, we have

$$\mathit{Avail}(\text{Web access}) \cong 1 - [(\mathit{Unavail}(\text{DNS}) + \mathit{Unavai}(\text{Web server}) + 2\times \mathit{Unavail}(\text{GPRS}) + 2\times \mathit{Unavail}(\text{IP network})] \qquad (5.29\text{a})$$

A multiplication factor of two is applied to the unavailability of GPRS and the IP networks since those are the same network, which supports both the DNS query and the Web access traffic. The approximation holds if the chance of two or more of the components being simultaneously unavailable is low, a condition that should normally be satisfied. Also, the unavailability definition has taken into account the highly redundant design of server components, such as two out of three servers (with load sharing) dropping below the required performance threshold. Writing (5.29a) slightly differently, we obtain

$$\mathit{Unavail}(\text{Web access}) \cong \mathit{Unavail}(\text{DNS}) + \mathit{Unavail}(\text{Web server}) + 2 \times \mathit{Unavail}(\text{GPRS}) + 2\times \mathit{Unavail}(\text{IP network}) \qquad (5.29\text{b})$$

The second KPI is maximum downtime. Very often, the downtime of a network may include periods with a large error rate, in addition to periods with a complete outage. For example, for IP networks, downtime may be defined as a packet-loss rate larger than 15%. In the above example, maximum downtime can be expressed as

$$Max\_Downtime(\text{Service}) = \underset{i}{Max}\ (Max\_Downtime \text{ of component } i) \quad (5.30)$$

In other words, the maximum downtime of the end-to-end service is given by the highest downtime among GPRS, IP network, Web server, or DNS server. Similarly, the response time KQI is given by

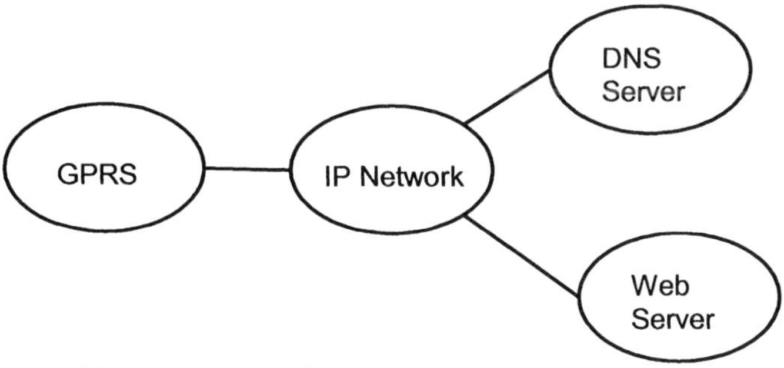

Figure 5.26 Simplified Web access service network.

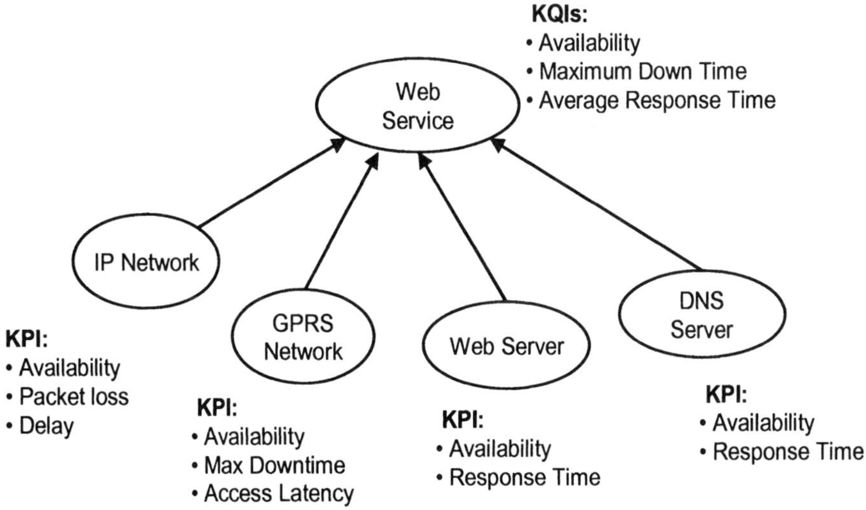

Figure 5.27 Web access service model with KPIs.

$$\textit{Response time}(\text{Web access}) \cong \textit{Response time}(\text{DNS}) + \textit{Response time}(\text{Web server}) + 2 \times \textit{Latency}(\text{GPRS}) + 2 \times \textit{Latency}(\text{IP network}) \qquad (5.31)$$

Equations (5.29) through (5.31) are based on the service model and form the basis for computing the KQIs from the dependent KPIs. If the service model is defined with a further decomposition, a further breakdown of the KPIs is possible. As we can see, both the availability and response-time KQIs depend on child KPIs in a linear relationship. Thus, the discussion in Section 5.6 will be directly applicable.

5.7.3 Solving QoS Problems

Identifying performance problems is traditionally the responsibility of performance engineers or network planners. Their responsibility is to make sure network performance is optimized with respect to the traffic load, given the available network resources. In the mobile world, much of the performance effort concerns the monitoring of the RAN, identifying service-affecting problems, and adjusting radio-related network parameters for meeting certain performance objectives. Identifying and solving performance problems are usually long-term projects. The processes involved usually last for weeks, months, or even longer. It is therefore imperative that a well-thought-out plan and procedure be established with well-defined goals and objectives.

As mobile networks evolve, the proliferation of advanced services such as Multimedia Messaging Service (MMS), IP Multimedia Subsystem (IMS), push-to-talk, VoIP, WiFi, and location-based services, in addition to traditional informational services, places new requirements on how to manage the QoS and customer satisfaction. The performance management issues are no longer limited to the RAN, as other pieces of the network can critically impact the end-to-end QoS. We will describe how the service model makes the management of QoS easier with respect to the large number of new services.

Although services, networks, and technologies may evolve or change rapidly, it is important to note that a management process should be designed to be flexible to deal with technology evolution. In the following, we first describe a service performance management process flow. We then demonstrate how the service model fits in the process. Details of the execution of the process will be described in Chapter 8. Figure 5.28 shows a general service-quality management process. Here, we have focused on the problem detection and resolution functions of service assurance. In general, there are two types of problem handling: Reactive problem handling deals with existing indications of a problem. In other words, a problem has occurred, and there are some indications of the problem in the form of either a customer complaint, a performance alarm or alert, or an SLA violation. The second type of problem handling is proactive problem handling, which deals with the process of actively looking for potential problems that are either already happening but with no obvious symptoms or that could occur if current network and traffic conditions persist. The goal of the proactive problem process is to avoid the occurrence of problems.

As shown in Figure 5.28, reactive problem management includes the handling of customer complaints originating from the NOC, QoS alarms and alerts, and SLA violations. Problem prioritization uses algorithms as described in the following section. The output of problem prioritization is a prioritized list of QoS events or alarms, which feeds into the problem incident management system, which is also known as the trouble ticket manager. In some scenarios, customer complaints and other forms of anecdotal information form the basis of proactive problem management, which also feed into the incident manager.

Once an incident is logged, a trouble ticket is generated, which describes the nature of the problem. The incident report or trouble ticket is assigned to a technician who "owns" this problem incident. This technician will then use available tools to find the root cause of the problem. The tools, including the service model, the CSIs, impact analysis, drill-down capability, access to customer and network configuration data, retrieval of statistics related to the problem, and viewing of past statistics, will all be used by the technician for resolving the QoS problem.

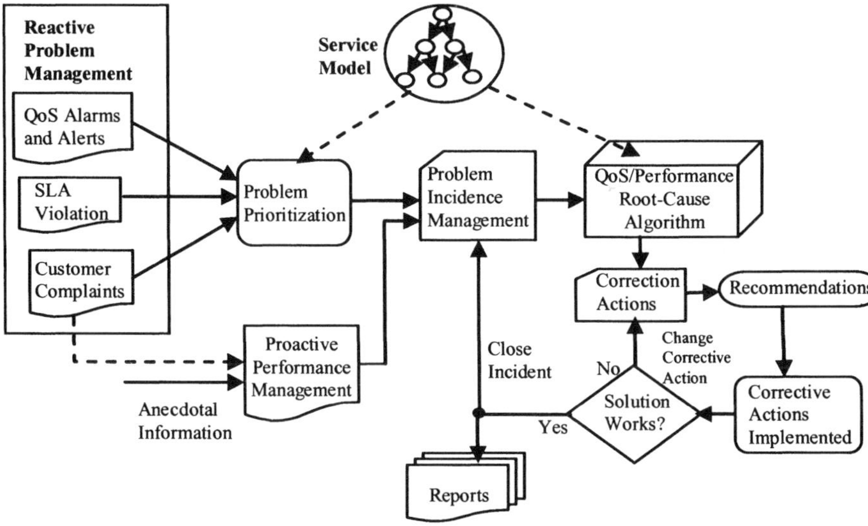

Figure 5.28 QoS problem-resolution process flow. (*From:* [6]. © 2004 Telcordia Technologies. Reprinted with permission.)

When a potential problem is identified, the technician will then be in a good position to suggest a fix. However, before one recommends a certain resolution, it is advisable first to test the solution. The testing instance may emulate clients or an instance supported by the service model. In this way, technicians will be able to emulate the effect of the recommendation before it is actually implemented in the real network.

5.7.4 Service Impact Analysis and Prioritization

A large wireless carrier may encounter tens of thousands of alerts in one day. However, it is nontrivial to sort out which alerts are severe enough to warrant immediate attention from the maintenance personnel, which can be dealt with in a longer time frame, and which are just warnings that have no bearing on service quality. Lacking resources and a strategy to scrutinize the large number of alerts, NOC personnel tend to ignore them, resulting in false negatives and subsequently deferring the necessary actions until customers complain. The proliferation of thousands of alerts is a consequence of the bottom-up approach of traditional network-centric performance management, in which alerts are very often generated without a well-thought-out plan of how to use them. We will show in this section how the service model allows a different approach to solving this problem. This relatively new research area is referred to as impact analysis. The idea is that by taking advantage of the inherent dependence of the subcomponents of a service, the alerts generated can be organized and correlated in such a way that thousands of seemingly unrelated alerts can be quickly and effectively correlated to sort those that are service impacting from those that are inconsequential warnings.

The overall goal of impact analysis is to quantify service-quality degradation with respect to certain predefined service-level criteria. The result of impact analysis can be used to support the following operations:

- Prioritization of service and network alarms, QoS alerts, or other performance impacting events with respect to trouble ticket generation;
- Prioritization for network and service resource expansion;
- Adjustment of SLA for marketing.

At the high level, we would like to take a QoS-related alert, apply an algorithm to it, and associate with the alert a priority index. Prioritization should take into account the following considerations:

- What service(s) is affected by the QoS alert(s)?
- To what extent is the service(s) affected?
- What is the impact on the customer?

The impact analysis and prioritization algorithm is aimed at addressing these questions by collecting information on the following:

1. Identification of affected services;
2. Service-quality impact based on KQI, service index, severity of degradation (total interruption, duration of the interruption, performance degradation, data transfer accuracy);
3. Number of subscribers affected (percentage of premium and regular customers);
4. Usage impact.

With information on the above items, rules will be used to weigh them to create a final priority index. Identifying affected service in (1) above depends on how the service is implemented and the components of the service. It is also highly dependent on the topology and the structure of the service components. On the surface, it may be tempting to conclude that any QoS alarms associated with a service subcomponent (such as a router or a server) imply that the service is impacted. In practice, there is uncertainty as a result of the self-healing or fault-bypassing capabilities of IP networks and many fault tolerant mechanisms that are built into the application layer. A simple example is that the failure of a router interface may be automatically bypassed by the routing algorithm, and subsequently, the router interface failure may manifest itself as just a reduction in capacity, which may or may not impact end-service, depending on the traffic load. Another example is if an application server is load-balanced among multiple computers, each running a copy of the application software. Requests for the service are served by multiple servers according to certain load balancing algorithm, such as DNS round-robin or traffic-based allocation. If one of the servers experiences a crash, that server becomes unavailable, which causes the generation of a severe alarm. However, since other servers are still functioning properly, depending on the load-balancing algorithm (e.g., traffic-based), all the requests may now be directed to the remaining healthy servers. In this scenario, once again, service impact may not be severe if the load is light.

In the above examples, the service model should be intelligent enough to know when to propagate the CSI upward in the graph. This decision is made at each component level as shown in Figure 5.29. Basically, one needs to take into account other information from the subtending children components before the degree of service impact is concluded. As an example of CSI propagation, Figure 5.29 shows that the component *X*1 experiences a local failure (Alarm 1) and causes a CSI (*X*1, *h*1) to be propagated to the next level. The handle *h*1 carries relevant information about the nature of the detected Alarm 1. At *Y*1, further information about the other component (*X*2) is evaluated before the decision to propagate the CSI (*Y*1, *h*1) further upward is made. Note that *h*1 is carried inside the CSI from *Y*1 so that the source of the problem is identified in the alarm-propagation chain. At component *Z*2, although a CSI is received from one of its children components (CSI(*Y*3, *h*3)), no CSI is propagated because the rules in *Z*2 are not satisfied. This suppresses Alarm 3 as it is not service affecting.

If a CSI successfully propagates to the service component, the algorithm concludes that the alarm underlying the CSI is impacting a service. In practice, multiple CSIs from the same or different components may propagate to the service level. The next step of impact analysis is to quantify how much impact these CSIs have. To quantify the impact, we define a set of impact weights as shown in Figure 5.30.

The key weights for prioritization include:

1. Service index (SI). For a specific service, the KQIs that are impacted by the CSI are given as an index, I_{KQI}. Each I_{KQI} is a measure of the severity of the impact due to the CSI. The sum of the impacted KQI indexes constitutes the service index, which has to be computed for each service separately and the results added together to form the total service impact index with respect to the CSIs.
2. The total number of subscribers (*w*1) for each affected service.
3. The ratio of premium versus nonpremium affected customers (*w*2).
4. Current usage (*w*3) for each service.
5. Duration of the outstanding alert (*w*4). All the alerts are defined with respect to the sampling period (e.g., 15 minutes). If the problem is corrected, the alert is expected to be removed. Long outstanding alerts are given more weights than fresh alerts.
6. Number of services. The total impact depends on all the impacted services.

Referring to Figure 5.30, after all the indexes and weights are computed, a single index for a particular CSI is obtained. The index can then be sorted to produce an impact priority with respect to all CSIs.

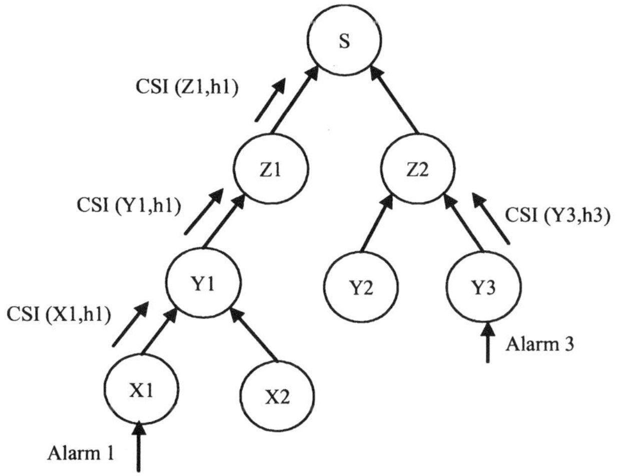

Figure 5.29 CSI propagation.

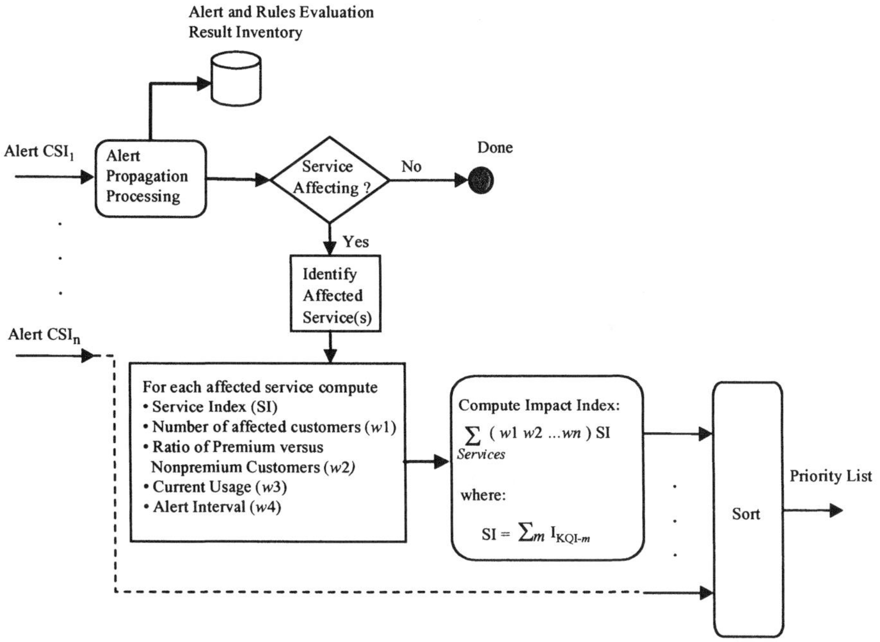

Figure 5.30 Alert prioritization flows. (*From:* [6]. © 2004 Telcordia Technologies. Reprinted with permission.)

5.7.5 Finding the Most Likely Root-Cause KPIs

Once the impact analysis is complete, NOC technicians will need to tackle high-impact problems. This section discusses how the service model and the associated mathematical algorithms can be used to expedite the process by finding the most likely root cause.

Suppose a KPI z depends on two component KPIs, x and y, such that

$$z = x + y \tag{5.32}$$

Assume that the threshold of z is set based on some user requirements (e.g., end-to-end delay is required to be less than 100 ms for voice). Accordingly, the thresholds of the KPIs x and y are set based on the criteria described in Section 5.6.5.2. In the event that the threshold of z is exceeded for a certain period of time, we are interested in finding out which component KPI, x or y, is the more likely KPI responsible for z exceeding the threshold. It should be noted that within a given monitoring period, the value of z may fluctuate around the threshold, thus can cross the threshold multiple times. This is a common phenomenon when dealing with soft faults, and the QoS assurance system should be designed to capture these exceptions.

Recall that in Section 5.6.5.2, we defined the threshold for x and y such that if both of them exceed the threshold, then z should also exceed threshold. However, the converse is not necessarily true. That is, if z exceeds the threshold, it is not necessarily true that both x and y have exceeded the threshold. In fact, in most cases, just one of the KPIs x or y exceeding the threshold can cause z to exceed the threshold, as illustrated in Figure 5.31.

We also notice from Figure 5.31 that the thresholds are crossed intermittently. This suggests that the best way to detect and analyze this type of performance problem should be based on statistical analysis rather than the traditional on-off detection commonly used in fault management systems. We use an example based on simulated data to illustrate the nature of the problem.

Figure 5.32 shows a time series of data for a KPI, $x_1(t)$, with no degradation, and $x\ (t)$ is the result when a degradation $d(t)$ of magnitude 2 is added to $x_1(t)$, such that

$$x(t) = x_1(t) + d(t) \qquad \text{where } d(t) = 2 \text{ for time} = 30 \text{ sec to } 60 \text{ sec} \tag{5.33}$$

The bottom graphs in Figure 5.32 show the resulting signals when the thresholds are applied to $x_1(t)$ and $x(t)$. We observe that even when there is no obvious degradation, there are occasional crossings of the threshold. However, for the degraded signal $x(t)$, the threshold crossings become crowded at around the time when there is degradation (i.e., time = 30–60). At this point, we cannot draw a conclusion regarding the degradation, as we have not seen the impact on the parent KPI, *z*.

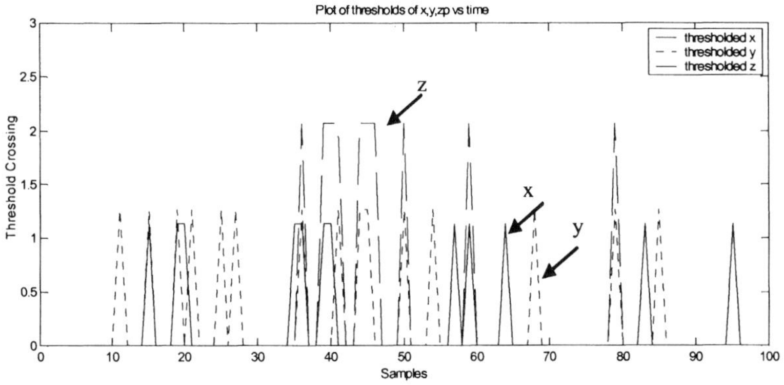

Figure 5.31 Threshold crossing of *x*, *y*, and *z*.

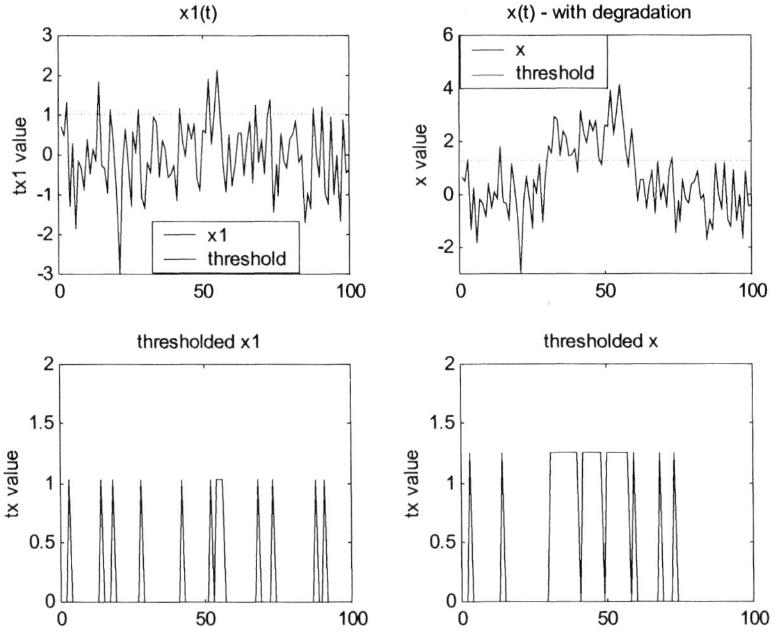

Figure 5.32 Time series of KPI with thresholds.

Next, we show the time series of $y(t)$ and $zp(t)$ in Figure 5.33, where

$$zp\,(t) = x\,(t) + y\,(t) \tag{5.34}$$

The bottom graphs of Figure 5.33 show the corresponding thresholded time series $ty(t)$ and $tzp(t)$. From observing Figures 5.32 and 5.33, it is still not obvious which KPI, $x(t)$ or $y(t)$, is responsible for the threshold violation in $z(t)$. However, it is quite clear, by examining the thresholded graphs, that there are more activities (violations) crowded around the centered area in both the $x(t)$ and $z(t)$, whereas the violations in $y(t)$ are more evenly distributed across time. This suggests that if we apply a correlation operation to the time series between $x(t)$ and $zp(t)$, and between $y(t)$ and $zp(t)$, we can expect to see a clear difference. This difference can be used to distinguish the two cases and thereby identify the likely root-cause component.

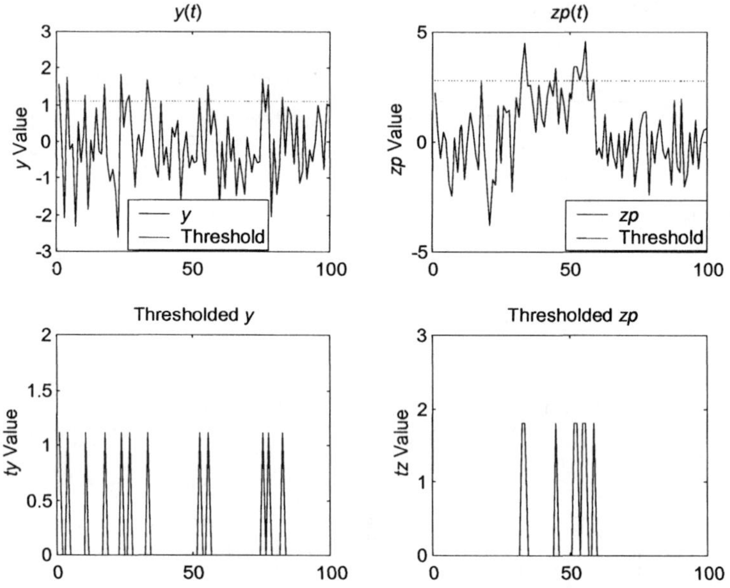

Figure 5.33 Time series of KPI and thresholding of $y(t)$ and $zp(t)$.

Therefore, we proceed to compute the following correlation indexes $\rho(zp, x)$ and $\rho(zp, y)$, and conclude that

$$x \text{ is the likely root cause if: } \rho(zp, x) - \rho(zp, y) > \varepsilon$$

and

$$y \text{ is the likely root cause if: } \rho(zp, y) - \rho(zp, x) > \varepsilon \qquad (5.35)$$

where $\rho(a, b)$ is the correlation index between a and b as defined in (5.6) and ε is a system level parameter. A rule of thumb is to set ε to be 10% of the smaller value of the correlation indexes.

Figure 5.34 shows a computer experiment consisting of 50 runs of a simulation where there is a performance degradation. We see that there is a consistent result since the correlation index of *zp–x* is always larger than that of *zp–y*. It should be mentioned that although the above example is based on two-child KPIs, the problem formulation applies to the general case with any number of dependent KPIs. In fact, the algorithm can be iteratively applied to the lower layers of the service model until the final root cause is identified.

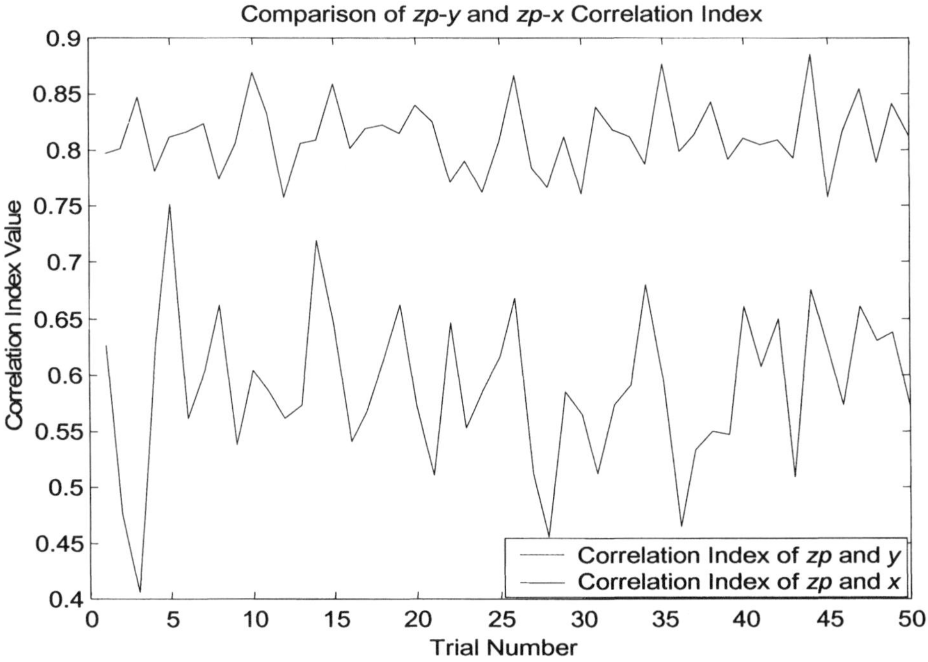

Figure 5.34 Comparing correlation indexes.

Another interesting observation is that the function describing the relationship between the parent and the child KPIs does not even need to be exactly known. Although we have used a simple linear relationship in the above example to illustrate the approach and the algorithm, it can be shown that the algorithm also works well for other functions (even nonlinear ones) or, in some cases, for totally unknown functions. However, it should be pointed out that if the degradation is small compared to the variance of the time series, the subsequent correlation computations may not show any significant difference (i.e., the difference of the correlation indexes is less than ε). In such a case, we will conclude that both child KPIs are equally likely to be responsible.

5.7.6 Corrective Actions

When the technician reaches a conclusion about the root cause of a problem and is confident that a recommendation should be made, the recommendation may call for some corrective measure, which can either be actionable in a short time frame, such as performing a reconfiguration of certain network parameters, or it can be a longer-term solution, involving adding new network capacity or server resources. In such cases, the recommendation will be directed to the appropriate organization. The trouble incident will not be closed until it has been confirmed that the recommended solution and corrective measures have been effective. At that time, the incident will be closed, and the overall process will be completed with the generation of a number of reports to document the findings.

5.7.7 Service and Traffic Planning

A service model provides a convenient structure for bringing together all the relevant data for service and traffic planning. For each component, information regarding availability, performance, and usage is available for analysis. Using the Web access example, where the KQIs and KPIs of each of the components were explicitly monitored, it would be relatively simple to conclude that the service problem was related to traffic overload. The service model then could be used to assist further drill-down analysis. Figure 5.27 is repeated here in Figure 5.35, highlighting the case of a violation of the KQI on average response time. The related child KPIs on response time of the Web access server and the delay of the IP network are also found to be exceeding their thresholds.

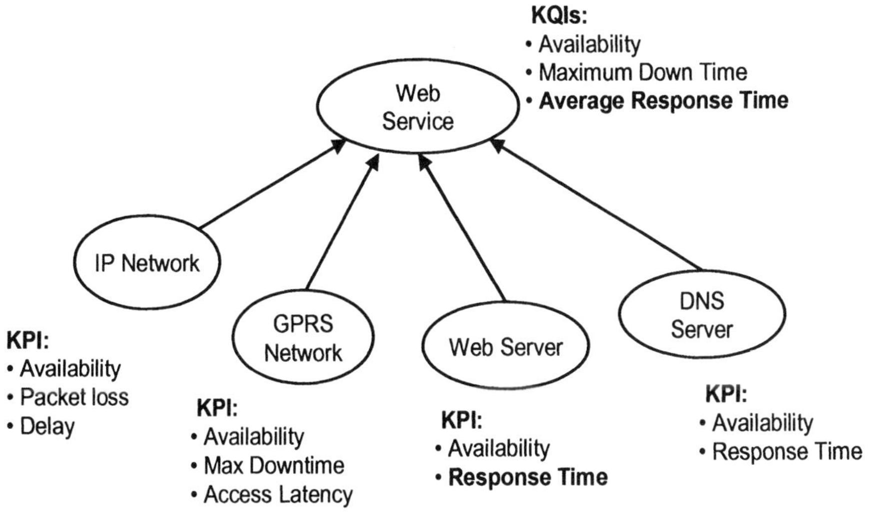

Figure 5.35 Service and traffic planning.

This provides a suggestion of what the next step towards identifying the bottleneck of the traffic and service QoS problem may be. Further drill-down and root-cause analysis may reveal that the Web server response time is most responsible for the service-level QoS violation and more detailed analysis of the KPIs related to the Web server may reveal what resource should be added, and also predict the future impact of the potential problem (e.g., a trending analysis could reveal that with an increase in subscription, the server's capacity could increase to a severe level in 2 months, thereby providing some urgency to the server upgrade).

5.8 SUMMARY

In this chapter, we focused on the design of the service model. We first surveyed the service models in the literature. Various service models have their advantages and disadvantages. For the purpose of supporting large-scale assurance operations, we defined a scalable and modular design following the work of [6], but also incorporated many of the performance analysis tools available from various earlier work. We also gave a summary of statistical analysis and mathematical tools suitable for supporting the service model implementation. These tools take advantage of many existing statistical algorithms and can be used to extract useful knowledge for assurance operations, in particular for root-cause analysis.

The service model and its associated tools will be the basis for the discussion of the following chapters, in which VoWiFi and 3G integration operation is described.

References

[1] Berry, R., and J. Hellerstein, "A Unified Approach to Interpreting Measurement Data in Performance Management Applications," *First IEEE Conference on Systems Management,* University of California, Los Angeles, May 1993.

[2] Hellerstein, J., et al., *GAP: A General Approach to Quantitative Diagnosis of Performance Problems,* IBM Research Report, December 16, 2002.

[3] Smith, M., D. Caswell, and S. Ramanathan, *Modeling of Internet Services*, Agilent Technologies, patent number 6138122, October 2000.

[4] "Automating Root Cause Analysis," SMART white paper, 2004, http://www.smarts.com.

[5] Hasan, M., B. Sugla, and R. Viswanathan, "A Conceptual Framework for Network Management, Event Correlation and Filtering Systems," *Proceedings of the Sixth IFIP/IEEE International Symposium on Integrated Management,* May, 1999.

[6] Lau, R., and R. Khare, "Service Model and Its Application to Impact Analysis," *First International Workshop, SAPIR,* 2004.

[7] Sanders, G., et al., *GPRS Networks*, New York: John Wiley, 2003.

[8] Blommers, P., and E. Lindquist, *Elementary Statistical Methods*, Boston, MA: Houghton Mifflin, 1960.

[9] Kelejian, H., and W. Oates, *Introduction to Econometrics*, New York: Harper & Row, 1981.

[10] Caswell, D., and S. Ramanathan, "Using Service Models for Management of Internet Services," HP Lab, Palo Alto, March 1999.

Chapter 6

Voice over WiFi and Integrated WiFi-3G Networks

This chapter provides a survey of the voice over WiFi (VoWiFi) or WLAN (VoWLAN) and voice over integrated WiFi and 3G networks (VoWiFi/3G), services architectures, and domain definitions. It first reviews basic VoIP technologies, then considers VoWiFi and voice VoWiFi/3G networks. Then, it describes various network domain, technology, and business aspects and VoWiFi-related call flows. User roaming and terminal mobility are considered, and some VoIP QoS and networking issues are summarized. It provides the necessary background for defining the VoWiFi/3G service model to be described in the following chapters. The service model for VoWiFi/3G helps define domain and aggregate KQI requirements. These KQI requirements may then be used to help in the design and use of an approach for VoWiFi/3G service assurance.

6.1 INTRODUCTION

The emergence of VoIP technology in wireline networking presents an opportunity for combining packetized voice and data over one convergent network infrastructure and management. In the 3G wireless spaces, both 3GPP, in its Releases 5 and 6, and 3GPP2 have specified an access-network independent IP multimedia (IM)–core network (CN) domain to support VoIP and other multimedia services. At the same time, the IEEE 802.11 WLAN standard is being accepted widely and rapidly for many different environments. The WiFi market has grown at 40% to 60% over the last few years. Now, capabilities to enable VoWiFi are becoming available. As WiFi use grows, both enterprises and wireless operators are looking for opportunities to leverage their wireless infrastructure for both voice and data traffic over the same backbone and managed IP (M-IP) network support infrastructure. Wireless operators and enterprises are also looking into opportunities for WiFi hotspots and 3GPP network integration for both voice and data services, as well as added services such as presence and location-based services.

VoIP is now a maturing and accepted technology. Many enterprises have used wired VoIP and are considering using VoWiFi, particularly in some vertical markets, such as the health industry and on college campuses. There are several VoIP networks deployed. Now vendors have equipment handling millions of VoIP call minutes, and carriers are using VoIP globally. VoIP is growing faster than other IP multimedia services, such as PTT and instant messaging (IM); in 2006 and beyond, the growth of enterprise VoIP is a key driver for this market segment.

6.2 WIFI AND INTEGRATED WIFI/3G SERVICES

VoIP is a network application that sends voice packets over IP. A voice signal is digitized, compressed, and converted into IP packets. Specialized signaling protocols, such as SIP and H.323, are used to set up and tear down calls or voice sessions, carry information required to locate users, and negotiate capabilities. VoIP is an end-to-end architecture that exploits processing in the end points, unlike the traditional PSTNs, where processing is done inside the network.

The VoWiFi service runs VoIP over WiFi. Besides VoWiFi, it is also called by other names, such as VoWLAN, voice over wireless IP (VoWIP), Wireless Voice over IP (WVoIP), Voice over IP over WiFi (VoIPoWiFi), Mobile VoIP, and WiFi telephony.

VoIP over 3G is a 3GPP Release 5 and 6 IMS application, and a 3GPP2 Multimedia Domain (MMD) application.

Integrated VoWiFi/3G service is an application running on dual-mode mobile host and gaining access VoIP service either via WiFi or 3G networks, and allowing roaming and mobility between WiFi hotspots and 3G cellular networks. Integrated VoWiFi/3G service bridges traditional PSTN, data communications, and mobile technologies. It is functionally equivalent to a wired or mobile phone and is also a WiFi and 3G data-client device using the technology used in wireless laptops and PDAs.

6.2.1 Why Use VoWiFi?

WiFi use is being driven by employees; according to some estimates, there will be over 30 million WiFi users by 2007. Over 50% the major financial institutions are expected to deploy WiFi by 2006. In enterprise environments, the average employee spends about 40% of his or her time away from desks; as a result, about 70% of calls fail to reach the person called, requiring employees to spend additional time listening to voicemail and also increasing cellular phone costs in the office. For enterprises that are planning to use WiFi hotspots for data applications, enabling these WiFi hotspots for VoIP is an attractive opportunity.

WiFi networks use IP technology, which makes them suitable for VoIP applications. The synergy between WiFi and VoIP technologies makes wireless voice application less expensive and easier to install, configure, and maintain. Some of the important reasons VoWiFi is becoming important are

- *Business needs:* In some industries, such as health care, retail, and manufacturing, the need for mobile communications is sufficient to justify installing WiFi hotspots just for voice applications.
- *Acquisition and operational cost savings:* The cost of WiFi APs has been going down for the last few years, and since a single WiFi AP can serve both voice and data for 10 to 15 users, depending on their data bandwidth needs, in enterprises, the cost savings in new wiring, installation, and maintenance, as well as later move, add, and change functions of AP, as compared to the wired network, can easily justify use of VoWiFi.
- *Transport cost savings:* By combining voice and data over the same WiFi access network, the overall transport cost and total cost of ownership are reduced by more efficient network use.
- *Competitive advantage:* In an enterprise environment, VoWiFi has an added advantage over cellular phones in that these can easily integrate with enterprise's internal data and voice networks and can be used as a replacement for desktop wired phones. There are extra opportunities to provide new services and increasing end-user productivity.

6.2.2 Why Use Integration of VoWiFi/3G Networks?

In Section 3.4.1, we discussed reasons for integrating data applications over WiFi and 3G networks. The integration of these data applications with VoWiFi and 3G further increases the value of such integration to business users as well as consumers. VoIP users with integrated WiFi and 3G capabilities, while within their coverage area, can use the WiFi networks at lower cost as compared to using 3G networks, then switch over to the use of 3G networks outside the WiFi coverage area. With the growing popularity of WiFi hotspots, many operators are deploying WiFi as an adjunct to 2.5G or 3G. Integration of the best of both WiFi and 3G networks can provide better coverage, integrated voice services, seamless wireless data services, and common billing. It will make wireless multimedia and other high-data-rate services a reality for a large population.

Since 3G IMS/MMD infrastructures that are needed to support VoIP over 3G networks, they are not yet deployed, and it may take a few years for these infrastructures to become available. Several efforts are now underway to integrate VoIP over WiFi with currently available circuit-switched cellular voice.

In Section 3.4.2, two such recent developments were mentioned. The first is the development of specifications by the SCCAN Forum. Now, SCCAN is being promoted by the IEEE Industrial Standards and Technology Organization (ISTO).

The focus of the SCCAN effort is integration of VoIP over WiFi and circuit-switched voice in GSM/CDMA networks. It is tightly linked with private branch exchange (PBX) architecture and some proprietary systems and extends the PBX experience into the wireless domain for enterprise users.

The second is the UMA architecture and specifications developed by the UMA Consortium formed in September 2004. This group includes major operators and network vendors and promises to let users make mobile calls over WiFi networks in homes and hotspots. The focus of this effort is the integration of both VoIP over WiFi and circuit-switched voice over GSM/CDMA and data over WiFi and GPRS/CDMA2000 networks. UMA is tightly linked to the core network, which is used for routing, authentication, and billing. The focus of UMA is to extend the cellular experience into the WiFi domain and is intended for mobile carriers.

We will further discuss the SCCAN and UMA architectures in Section 6.10.2.

6.3 EVOLUTION OF VOIP SERVICES

Over the last decade, VoIP services have evolved in the following steps:

1. *Low-cost/low-quality VoIP:* Voice conversation became available between users with PCs equipped with telephony software. The software provides data compression and translation to IP packets. VoIP began in 1995 by some hobbyists when only PC-to-PC communication was available. Later, in the same year, Vocaltec released Internet Phone Software to run on a home PC (486/33 MHz) with sound cards, speakers, microphone, and modem. The software compressed the voice signal, translated it into voice packets, and shipped it out over the Internet. The technology worked when both the caller and the receiver had the same equipment and software. Although the sound quality was poor, this effort represented the first IP phone. VoIP was originally offered as a toll-bypass service by ISPs and Internet carriers. These companies used VoIP to avoid high international tolls. As the QoS improved, VoIP was adopted by smaller businesses as well, mostly for intersite international calling. In many developing countries, VoIP providers were able to win a large market share from incumbents, which led to regulatory reforms aimed at liberalizing the market and readapting tariffs.
2. *Toll-quality VoIP:* Development of PSTN interface protocols and the use of phone numbers for IP address mapping enabled a user with an Internet connection to talk to any PSTN user. The development of phone gateways, which provide a two-way interface between the PSTN and the Internet, allowed conversation between two users with no computers and Internet access. With voice telephony still generating about 80% of telecoms' profits, toll-grade VoIP becomes a major revenue earner for

new entrants. But, VoIP has allowed operators, already offering data services, to reduce costs by moving all their traffic onto a single network. Incumbents started to shift to VoIP over time, integrating their dial-up Internet traffic with voice and realizing cost savings in the process. Within an enterprise, there are advantages of combining voice, fax, data, and multimedia traffic onto a single multipurpose network. It reduces recurring transmission charges and long-term network ownership costs and provides the ability to implement a host of new and powerful voice-enabled applications. Large enterprise customers started to explore the use of VoIP to reduce future costs or bring about new application opportunities. The small to medium enterprises (SMEs) started using VoIP initially to save costs on intersite calls. With the widespread availability of broadband Internet access via cable and DSL modems, VoIP service providers are offering services to residential customers. This is all wired VoIP, where enterprises started replacing their legacy voice PBX with IP PBX, SMEs started replacing their legacy Centrex with IP Centrex, and operators started using VoIP in the middle of the network and offering wired VoIP from customer premises.

3. *Value-added data services:* Declining voice revenues have created interest in service offerings that bundle VoIP with value-added services, such as integrated voice and data services for enterprise contact centers. Currently, the development of these applications is driven by the relationship between corporate clients and systems integrators. But as their benefits become more widely recognized, they could be tailored to meet the requirements of organizations in the SME sector. Home users are first likely to face them when providers develop multimedia instant messaging applications that use voice alongside text and eventually offer videophone capabilities.
4. *VoIP on wireless technologies:* Development of 3GPP IMS and 3GPP2 MMD architectures allow VoIP in wireless networks. Now there is a wide deployment of 802.11 WiFi and VoWiFi in selected vertical markets. VoIP interworking among WiFi, 3G wireless networks, and wired PSTN, and multimedia services that include voice, as well as text and video, offer mobile operators a means of driving up use, and a premium can be charged for such services. Now, the widespread adoption of SIP allows 3G mobile customers to bypass the operator and set up their own phone calls, using data services. This would have the same impact on mobile operators as basic VoIP had on wireline operators, driving down voice tariffs. With wireless VoIP, enterprises are replacing their wired IP PBX and IP Centrex with WiFi, and SIP-based phones. The operators are deploying VoIP to WiFi hotspots, introducing IMS/MMD and gateways for VoIP in PLMNs, and planning integration of VoIP in WiFi and 3G networks, and end-to-end native VoIP.

The data and voice convergence over IP has led to growing deployment and use of VoIP. With VoIP, voice becomes another data application running over the IP network, and the enterprise PBX becomes a large server. VoIP offers the possibility of increased separation between network ownership and service provision, which will enable existing and new competitors, including ISPs and systems integrators, to enter telecom markets and begin offering services. The result may be that the incumbents are simply transporting bits, a service which will rapidly become commoditized with a corresponding impact on revenues.

6.4 BASIC VOIP TECHNOLOGY

Legacy telephony, or PSTN, is TDM/SS7-based infrastructure and uses legacy class 5/class 4 switches for call control, signaling, bearer/media, and other features. VoIP is an IP-based packet infrastructure that replaces PSTN voice transport and requires new elements that collectively perform traditional PSTN roles.

Since VoIP is just another IP application, it requires an IP infrastructure that can provide service dependability. While IP connectivity is available, it is independent of the underlying transport bearer. However, voice has the following distinct requirements that set it apart:

- Voice is a real-time interactive application. This means data packets must be processed as they happen in the real world. They do not have the same tolerance for delay and packet loss as other data applications.
- Data traffic sometimes happens at unpredictable intervals, or bursts. Voice, on the other hand, follows a consistent flow and is more predictable.
- Data transmissions are asymmetrical because file transfers are typically much larger on the download side than they are on the upload side. Conversely, voice transmissions are typically symmetrical because the rate of transfer is almost the same in both directions.
- An industry concern is the reliability and quality in a migration from traditional PSTN-based voice to VoIP. It is a critical application, and users have come to expect near 100% reliability from any voice network of uninterrupted service. Voice service must be accessible and predictable to provide the continuity human conversation depends on. In a digital voice network, this continuity requires that digital voice packets arrive in the same order in which they were transmitted and be assembled in the proper order. Otherwise, it would not be an accurate representation of the source analog input. This means that the underlying IP bearer for VoIP must meet voice QoS requirements. These are:

- *Delay intolerance:* ITU-T Recommendation G.144: 0–150 ms; acceptable for most applications; 150–400 ms, acceptable for limited applications; 400 ms, unacceptable for general network purpose.
- *Jitter and packet loss:* 5%, acceptable without noticing packet-loss noise; 10%, acceptable but noticing packet-loss noise.

Additionally, there are security and CoS issues.

WiFi and public Internet generally do not meet these voice QoS requirements and need improvements to the IP network and additional enabling capabilities and QoS-managed IP network to support VoIP.

There are three VoIP infrastructure components: Media, transport, and gateways; gateway control and signaling; and services. In this chapter, we will consider only media, transport, and gateways, gateway control, and signaling. Services here refer to applications that use VoIP platforms; examples are PTT, instant messaging and Location Based Services (LBS). In PSTN, media can be analog or digital [uncompressed at 64 Kbps (DS0) or compressed at constant 8, 16, 24, 32, or 40 Kbps and variable 6.3/5.3 Kbps rates]. Signaling can be analog, ISDN Q.930/Q.931, and SS7. In VoIP, signaling, as discussed later, is H.323 or SIP; however, regardless of which of these signaling protocols is used, the voice information is carried over a protocol called Real-Time Protocol (RTP).

6.4.1 VoIP Service Alternatives and Protocols

At present, there are five major alternatives for VoIP services:

- Conferencing industry led effort that uses signaling concepts from the traditional PSTN—H.323;
- An Internet-centric protocol, such as SIP;
- Control concepts from the traditional CS PSTN that uses softswitches or media gateways and controllers;
- Enhancements of legacy PBX to IP-PBXs that use some concepts from softswitches and can support both H.323, and SIP-based signaling;
- Convergence ITU standards using bearer independent call control (BICC).

As depicted in Figure 6.1, there are three signaling protocols that are commonly used for IP telephony: H.323, SIP, and media gateway control (MEGACO). The H.323 and SIP both take the approach of putting the most intelligence at the IP telephones. In both cases, a small-scale IP telephony system (about 10 to 20 phones) can be constructed without any servers or gatekeepers. However, servers are necessary to support supplementary services and bandwidth management features and for large-scale IP telephony systems. Irrespective of the signaling protocol used, RTP and related protocols are used for media transfer. A brief description of these protocols follows:

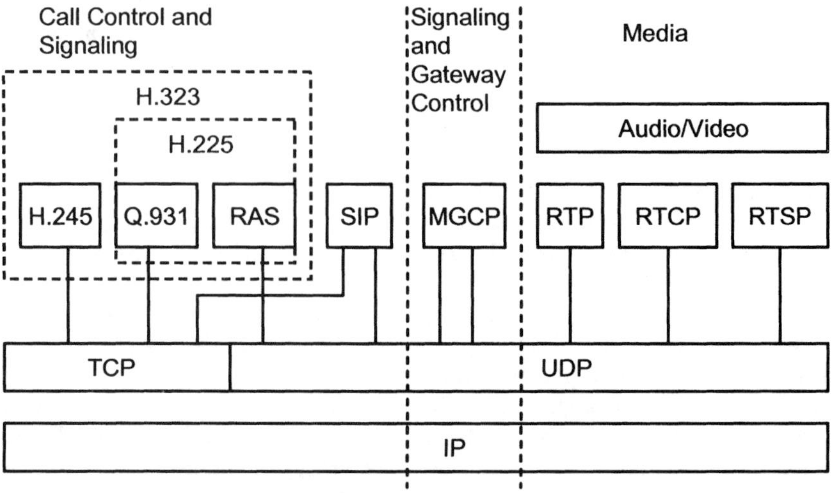

Figure 6.1 VoIP signaling and media protocols.

6.4.2 H.323 and VoIP

ITU-T recommendation H.323 is a flexible set of umbrella recommendations that specify the components, protocols, and procedures used to enable voice, video, and data conferencing over a packet-based network. This standard was originally created for multimedia applications. By 1997, H.323 was accepted as the prevailing VoIP network standard, a position now being overtaken by SIP. Since H.323 is still widely deployed, particularly in enterprise environments, in the following, we will briefly describe components of the H.323 network and the VoIP call setup process with H.323.

6.4.2.1 H.323 Network Components

The H.323 standard defines four components, which when networked together provide both point-to-point and point-to-multipoint multimedia services. As depicted in Figure 6.2, these components are:

1. *Terminals:* These are the endpoints (LAN clients) of the H.323 session. A multimedia PC with an H.323 compliant protocol stack can act as a terminal.
2. *Gateway (H.323 to H.320/H.324/POTS):* This is needed only when conferencing needs to done between different H.32X-based clients.

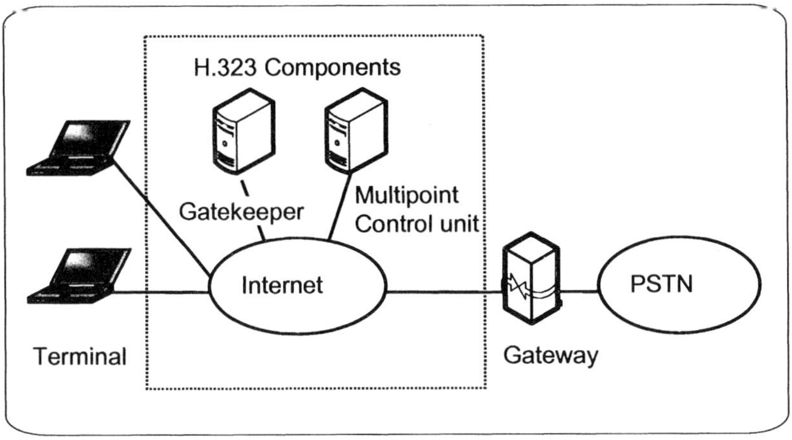

Figure 6.2 H.323 components.

3. *Gatekeeper:* This provides functions such as admission control and bandwidth management. H.323 terminals get permission from the gatekeeper to place any call within a given domain.
4. *Multipoint control unit (MCU):* This is an optional component that provides point-to-multipoint conferencing capability to an H.323 network. It consists of a multipoint controller (MC) that handles control and signaling for multipoint conference support and MCUs that support multipoint conferences where more than one terminal or gateway joins in.

6.4.2.2 VoIP Call Setup with H.323

Establishing communication using H.323 may occur in five steps:

1. Call setup;
2. Initial communication and capabilities exchange;
3. Audio/video communication establishment;
4. Call services;
5. Call termination.

If both endpoints have previously registered with the gatekeeper, a simple VoIP call flow using H.323 is set up (Figure 6.3). Terminal A initiates the call to the gatekeeper [registration admission status (RAS) messages]. The gatekeeper provides information for Terminal A to contact Terminal B. Terminal A sends a setup message to Terminal B. Terminal B responds with a call proceeding message and also contacts the gatekeeper for permission. Terminal B sends an alerting and connect message. Terminal B and Terminal A exchange H.245

messages to determine master-slave relationships, terminal capabilities, and open logical channels. These two terminals then establish RTP media paths.

H.323 has been used in the VoIP system widely, but the original H.323 has issues on large call setup time and poor functionality and scalability. The SIP protocol has been proposed to solve some of these issues in IETF. However, later versions of H.323 are improving themselves by solving the above issues, and the differences between H.323 and SIP are diminishing with their new versions as they influence one another.

6.4.3 SIP and VoIP

The IETF's SIP (RFC 2543) is a relatively new standard. SIP is an application-layer control protocol that makes up for some of H.323's inherent faults. SIP addresses call setup and teardown, error handling, and interprocess signaling, which are functions of every point-to-point connection. It also changes and ends multimedia sessions, including conferences, Internet telephony, distance learning, and other applications. SIP enables VoIP gateways, client endpoints, PBXs, and other systems to communicate over packet networks from an equipment perspective. It is independent of the underlying transport [IP, User Datagram Protocol (UDP), or Stream Control Transmission Protocol (SCTP)].

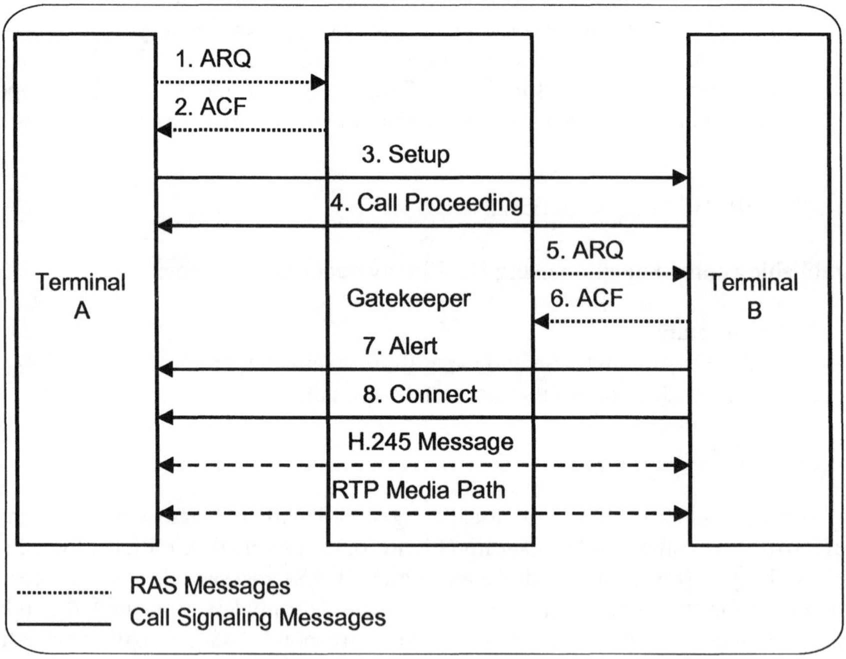

Figure 6.3 VoIP call setup with H.323.

Compared with H.323, SIP is a simpler protocol with less overhead. SIP locates the recipient of a call, ensures that the equipment is congruent with the caller's equipment, then allows other protocols to take care of other functions, such as data transfer and security. SIP distributes much of the call management and routing among different areas of the network. Like H.323, SIP has expanded to include several related protocols. Furthermore, SIP is a text-based and largely free-formatted protocol, making it easy to debug protocol implementations and to add new features to the protocol to meet new industrial demands.

SIP is becoming for person-to-person IP communications what HTTP is for the Internet, and while much of its early development was focused on fixed-line Web services, the attention of many developers has now switched toward mobile networks. Making SIP work with wireless devices will allow next generation mobile users to access multifunction IP-based services, which can combine voice, messaging, and e-mail on their cellular handset or PDA.

SIP has won full backing in the mobile industry; the 3GPP decided to base 3G mobile call setup on SIP. SIP is used in packet-based IP communications networks, so in the mobile field, it will come into play with the launch of new high-speed 3G systems and through existing technologies such as WLAN and GPRS. Bringing SIP into the mobile arena means that mobile services can converge with other IP services to provide a wider range and a broader reach of applications. Some of the services that use SIP include

- *Instant messaging:* This is the natural successor to mobile text messaging. SIP plays an important role in the signaling side of mobile instant messaging.
- *LBS:* Specific services can be triggered depending on the location of a user. SIP enables real-time transmission between the service provider and the user.
- *Unified messaging:* This enables a single user to be contacted on several different devices. SIP simplifies the integration of different media types such as voice- and e-mail.

The key SIP network components are described next.

6.4.3.1 SIP Network Components

Broadly speaking, SIP user agents running on UE communicate with several network servers, in a distributed environment to provide peer-to-peer signaling and VoIP service. Figure 6.4 depicts SIP components and their distributed environment. A brief description of these components follows.

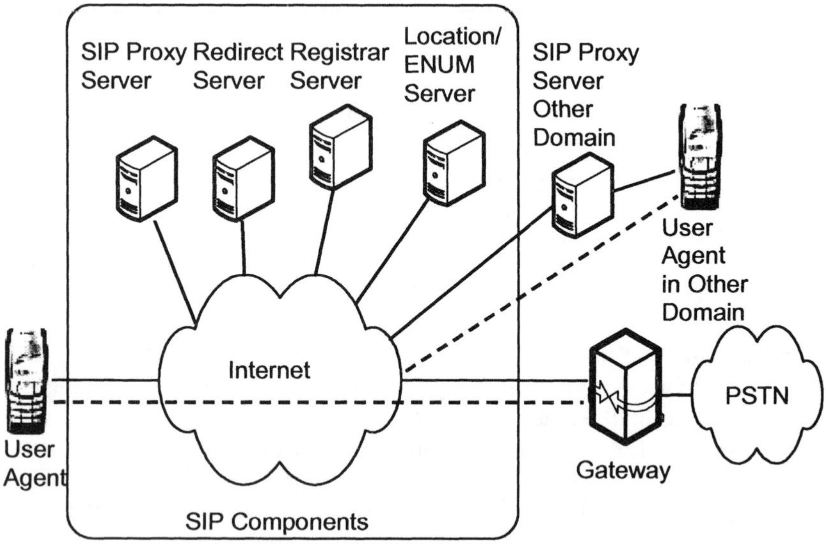

Figure 6.4 SIP network components.

1. *User agents:* A user agent (UA) is an application that acts for a user. It can act both as a user agent client (UAC) and user agent server (UAS), as the user probably wishes to be able both to call and to be called. UAC is used to start a SIP request. The UAS receives requests and returns responses to the user. The response accepts, rejects, or redirects the request. These user agents contain the full SIP state machine and can be used without intermediate servers. Thus, a UAC-1 can call another UAC-2 or UAS-2, if the IP address of the UAC-2 or UAS-2 is known, without the use of any intermediate servers. However, the IP address of the callee may not be known or may change with time, requiring use of a network proxy server as described below.
2. *Proxy server:* A SIP proxy server receives SIP messages and forwards them to the next server after deciding which it should be. A proxy server interprets and, if necessary, rewrites a request message before forwarding it. This next server could be any kind of SIP server; the proxy does not know and does not have to know. Before the request has reached the UAS, it may have traversed several servers. As a proxy server issues both requests and responses, it contains both a client and a server. A proxy server can either be stateful or stateless. When stateful, a proxy remembers both the incoming and outgoing requests. The incoming requests generate outgoing requests. A stateless proxy forgets all information once an outgoing request is generated. A proxy server can fork the incoming request to multiple locations if the callee has multiple

location registrations with the server. A forking proxy is always stateful because it needs to remember the states of all the branches to which the incoming SIP request was forked. The SIP proxy server controls calls in each domain. Proxy servers can also provide functions such as authentication, authorization, network access control, routing, reliable request retransmission, and security. The SIP proxy server acts on behalf of and provides services to all clients in the access network or the administrative domain. A SIP proxy server serves only one IP domain; however, more than one proxy server may be used in one IP domain to handle traffic load. Clients requesting call setup first have to be registered with the SIP server before obtaining authorization for supported calls. After registration with the SIP server, the server may handle all call requests to and from that client. This does not exclude, however, direct client-client call setup without the benefits of SIP server. Such direct client-client call setups can be faster and may be desirable for special services, such as the equivalent of the direct hot line. Clients that are not registered and authorized for direct calling cannot have the QoS benefit via support from the SIP and policy servers.

3. *Redirect server:* The redirect server does not forward requests to the next server. It accepts a SIP request and maps the address to zero or newer addresses and returns these addresses to the client, and then the client can contact the server directly. Unlike a proxy server, it does not start its own SIP request. Unlike a UAS, it does not accept calls.
4. *Registrar server:* A registrar is a server that accepts register requests and maintains the availability details of various servers and clients. A registrar is typically collocated with a proxy or redirect server and may sometimes offer location services also.
5. *Location server:* The location server locates the next hop for an incoming session request. It is a generic term for a database. Sometimes it is combined with an E.164 Numbering (ENUM) server as well.

Besides the above-described network servers, SIP servers also use the following servers:

- Directory server, such as Lightweight Directory Access Protocol (LDAP), to map names to user domain;
- DNS for mapping uniform resource identifiers (URIs) to ENUM and to callee's domain;
- AAA server for UAC authentication and authorization;
- Application servers for other value-added services.

Application servers typically bundle a predefined suite of enhanced services (special-purpose application server), but many also include a service creation environment where new or customized services can be deployed (general purpose application server).

LDAP, DNS, AAA, and even application servers are intrinsic to any IP network and are needed for its proper functioning. These are not exclusive to VoIP and will not be discussed in this chapter.

Since a SIP phone (SIP UAC/UAS) can only signal to another SIP phone, various gateways, described later, are required to signal from a SIP phone to other PSTN phones or even other SIP phones connected via PLMN or via IP PBX.

6.4.3.2 Naming and Addressing

IP URIs are used to identify uniquely SIP UAs. These URIs have the same basic form as e-mail addresses—user@domain.

ENUM (IETF RFC 2916) employs the principles of DNS to simplify the convergence of PSTN and IP networks by using familiar telephone numbers to be used as universal IDs to reach Internet services. It allows mapping of telephone numbers to URLs and has powerful application in VoIP, e-mail service, instant message, and so forth.

The mapping between URI or ENUM and a specific host address or address resolution can use DNS server lookup, ENUM, or location server lookup.

6.4.3.3 VoIP Call Setup with SIP

As depicted in Figure 6.5, basic SIP call operations include:

1. Pair of UAs in different domains set up a session using a pair of proxy servers, one in each domain.
2. Each user agent is configured with a default outbound SIP proxy server, to which it sends its requests.
3. The proxy server typically authenticates the user agent.
4. For interdomain sessions, the proxy server in one domain locates the proxy server in the other domain, sometimes called the inbound proxy server, using the DNS server. It may use directory service (e.g., LDAP) to map the name to user@domain (the LDAP part is not depicted in the figure).
5. The called server may use the location server to locate the callee, then routes the request to the callee.
6. Callee accepts, rejects, or forwards (to new address) the call.
7. If new address, go to step 4.
8. A conversation happens.
9. The caller or callee sends BYE.

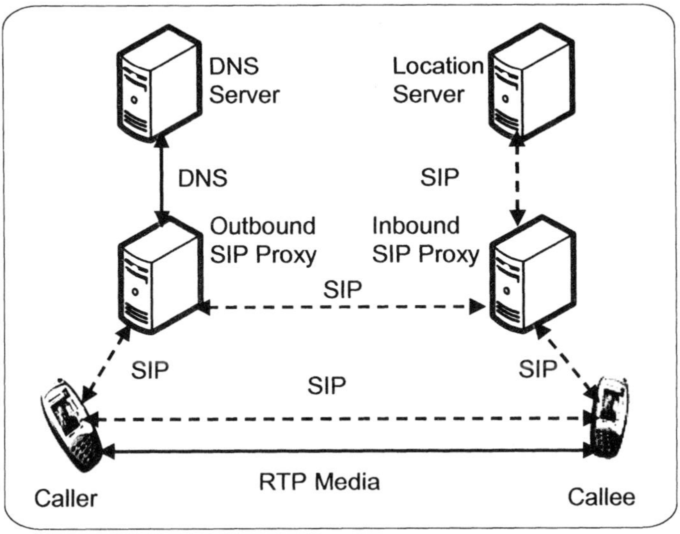

Figure 6.5 Basic SIP call setup.

SIP is a peer-to-peer protocol; thus, a proxy cannot issue a BYE, and only UAs can. Two approaches for call control are used. Either a proxy, after passing an invite on, stays in the signaling path, or it uses the refer command to start third-party control (the third party is no longer in the signaling path). Call control is required for PTT, automatic call distribution (ACD), and Web call-center applications.

6.4.4 Other VoIP-Related Protocols

Many other protocols are part of a VoIP network. Brief remarks about some of these follow.

6.4.4.1 Bearer Control Protocols: RTP/RTCP

VoIP stream control is done by RTP (RFC 1889, 1996), which provides end-to-end network transport functions suitable for applications transmitting real-time data, such as audio, video, or simulation data, over multicast or unicast network services. RTP does not address resource reservation and does not guarantee QoS for real-time services. In RTP, voice is packetized at certain periods (20 ms, 30 ms), and the packetized voice frame is sent to the network with RTP, UDP, and an IP header. The main issue with RTP is its large header overhead. RTP services include payload type identification, sequence numbering, time stamping, and delivery monitoring. Since RTP was originally developed for use in the wired IP network, header overhead was not an issue, but for wireless networks, the main issue with RTP is its large header overhead of 40 bytes (discussed further in Section 7.3.2.2). Due to limited bandwidth, IP, UDP, RTP, TCP packets (the size of voice payload can be as low as 15 to 20 bytes) sent over wireless links can benefit considerably from header compression. Another consequence of low-bandwidth wireless links is the long session setup delays when text-based signaling protocols, such as SIP and Session Description Protocol (SDP), are used. These delays can be significantly reduced by compressing not only the headers but also the signaling information. Therefore, in wireless networks, header compression is needed to improve the performance of RTP. For mobile wireless environments, such header compression schemes (RFC 2507, 2508, and 3095) can, under favorable circumstances, reduce the combined 40-byte IP, UDP, RTP, or TCP header to as little as 4 bytes.

RTSP (Real-Time Streaming Protocol) (RFC 1998) is a client-server multimedia presentation control protocol, designed to address the needs for efficient delivery of streamed multimedia over IP networks.

Real-Time Control Protocol (RTCP) (RFC 3550) is used to allow monitoring of the data delivery in a manner scalable to large multicast networks and to provide minimal control and identification functionality. The protocol is based on the periodic transmission of control packets to all participants in the session, using the same distribution mechanism as the data packets. In Section 7.3.2.3, we discuss further how RTCP is used in monitoring the quality of VoWiFi/3G.

6.4.4.2 Gateway Controller Protocols: MGCP/MEGACO

Media Gateway Control Protocol (MGCP): Created by the IETF, MGCP is a standard to convert audio signals on the PSTN to data packets that traverse the Internet. The protocol is based on an architecture that moves call-control intelligence away from the gateway for processing by external called media gateway controllers (MGC) or call agents. MGCP allows media gateways to communicate. The gateways (e.g., RGW and TWG) provide physical interfaces between VoIP networks and residences. The goal of separating control intelligence from media transport is to achieve scalable gateways between IP telephony and PSTN. Currently, two variants of MGCP exist: one, per RFC 2705, under control of the IETF, and another under the auspices of the International Softswitch Consortium (ISC). It does not, as H.323 does, specify a complete end-to-end communication.

Megaco/H.248: Megaco/H.248 is a new protocol developed as a joint effort between the ITU and the IETF (RFC 3015). Functionally, it enables control of media gateways. Megaco/H.248 is designed to succeed MGCP, adding peer-to-peer interoperability and ensuring a way to control IP telephone devices operating in a master-slave manner. The standard breaks the H.323 gateway function into separate subcomponents. It also determines the protocols employed by each communication component. The main thrust of Megaco is to permit greater scalability than allowed by H.323 and to address the technical requirements of multimedia conferencing. At present H.248/Megaco is not nearly as widely supported as H.323 or MGCP, but some vendors plan to implement it in their products.

Telephony Routing over IP (TRIP) (IETF RFC 3219): TRIP is a policy-driven inter-administrative domain protocol. It advertises reachability of telephony destinations and route attributes to the destinations. It is independent of the signaling protocol and is modeled after border gateway protocol version 4 (BGP-4).

Signaling Transport (SIGTRAN): SIGTRAN addresses transport of PSTN signaling over IP networks by allowing SS7 signaling to be carried between the signaling gateway and IP signaling points [MGC or IP service control points (SCPs)]. This allows carriers to maintain their existing SS7 infrastructure. SIGTRAN protocol suite consists of stream control transmission protocol (SCTP), and user adaptation (UA) layers.

6.4.4.3 SIP Extensions

Session Description Protocol (SDP) (IETF RFC 2327): SDP is used by SIP in the context of QoS to describe the capabilities of the client and other properties of the session ([SDP-QoS). It is a format for describing multimedia sessions. It conveys session setup information to the participants for session announcement and session invitation. The information includes session name, session duration, media type, and information like address, port, and format to receive the media. SDP consists of one session-level description and, optionally, several media-level descriptions.

Session Announcement Protocol (SAP) (IETF RFC 2974): SAP is an announcement protocol used for multicast conference sessions.

SIP-T (SIP-Telephony) (RFC 3204): SIP-T is used at SIP-PSTN boundary gateways. It is relevant in the following scenarios:

- PSTN origination, IP termination;
- IP origination, PSTN termination;
- PSTN origination, PSTN termination with IP transit.

SIP-T is not required for native-IP-originated and IP-terminated calls. The gateway between the SIP and the PSTN networks looks like a SIP user agent to other SIP entities and like a terminating telephone switch to the PSTN. It provides

ISDN User Part (ISUP) transparency (by carrying ISUP messages as multipart multipurpose Internet mail extensions (MIME) messages in the SIP messages between SIP-T gateways).

SIP-Business (SIP-B): This initiative is bringing the requirement and implementation of features needed for voice services in an enterprise environment, over and above making and receiving calls. For example, SIP-B will address how to create PBX-type features, such as bridge line appearances and line sharing.

6.4.5 Softswitch, Media Gateways and Controllers, and VoIP

Softswitch solutions offer the most comprehensive form of VoIP. Based on a layered architecture that separates call logic from voice switching, these solutions provide a distributed packet-switched alternative that is more flexible and cost-effective than the centralized circuit-switched solutions of the traditional PSTN. Different vendors use different concepts and definitions of "softswitch" and emphasize their products' gateways and session border controller (SBC) capabilities. In technical terms, the central attributes of a softswitch are an open architecture with standardized protocols and application programming interfaces (APIs), along with the decoupling of applications, call control, and bearer control, providing greater flexibility for service creation, provisioning, and network management. One way of defining softswitch is in terms of the following logical components:

- Provides call-control functionality;
- Supports all existing telephony functionality, along with new ways of communicating;
- Uses standard protocols (such as SIP, H.323, MGCP, MEGACO/H.248, SIGTRAN);
- Offers interoperability between different vendors' equipment;
- Works together with one or several of the following elements: media gateway, signaling gateway, application server, media server, and management, provisioning, and charging/billing interfaces.

The essential idea is to decompose legacy switches first into call control from media control and then decompose these further in media control (MC), bearer control (BC), session or call control (CC), service control (SC), service execution (SE), and media device (MD) functions. These functions may then be grouped and built in various combinations as required.

The *call agent* or *call control agent* controls calls in softswitch environments and functions similarly to SIP servers described earlier. The softswitch or call agent is called the media gateway controller (MGWC) in 3GPP IMS architecture. In this chapter, we will not go into further details of softswitch but rather use SIP and related components to illustrated VoIP functions and flows from a service assurance perspective. Various gateways and controllers are used in VoIP architecture; the key ones are as follows:

1. The *MGWC* mediates call control (including call setup and teardown) between the gateways described next (SGW, MGW, and others) and controls access from the IP world to and from the PSTN. Of all the components, the MGWC has most of the intelligence and is sometimes called a softswitch. MGWC = SC + CC + BC layers mentioned earlier.
2. The *MGW* connects the SIP network to the PSTN by converting RTP media to pulse code modulation (PCM) traffic and vice versa. MGW = MC + MD.
   - The SGW interfaces to the SS7 network and passes signaling messages to the IP nodes. An SGW can relay, translate, or terminate SS7 signaling. SGW = SC = CC. SGW is sometimes further decomposed into the transport signaling gateway (T-SGW).
   - The T-SGW maps call-related signaling from the PSTN or PLMN on an IP bearer and sends it to the MGCF. It provides the PSTN or PLMN to IP transport-level address mapping and connects the SIP network to an SS7 network by converting SIP signaling to SS7, ISDN primary rate interface (PRI), or channel associated signaling (CAS) and vice versa.
3. *Roaming signaling gateway (R-SGW)* provides communication with a 2G/Release 99 MSC/VLR. It connects the SIP network to SS7/PLMN networks by converting SIP signaling to SS7/MAP signaling.
4. The *VoIP gateway (VoIP GW)* connects the SIP network to the PSTN network. It has functions of both MGWs and SGWs.

Figure 6.6 illustrates functional components of gateways. The VoIP gateway includes functions of both the MGW and MGWC.

6.4.6 IP PBX/Call Manager and VoIP

The IP PBX, sometime called the call manager, leverages VoIP technology by replacing legacy PBXs. A key benefit is reduced toll calls, since they bypass long-distance PSTN calls by using the Internet, while keeping call control required in enterprise environments.

There are two categories of IP PBX systems: converged and client-server. A converged PBX system, also called an IP-enabled PBX, is a circuit-switched system with optional IP capabilities, such as IP station or trunk interfaces, or dispersed port carrier interface equipment supported by IP control signaling over a LAN/WAN infrastructure.

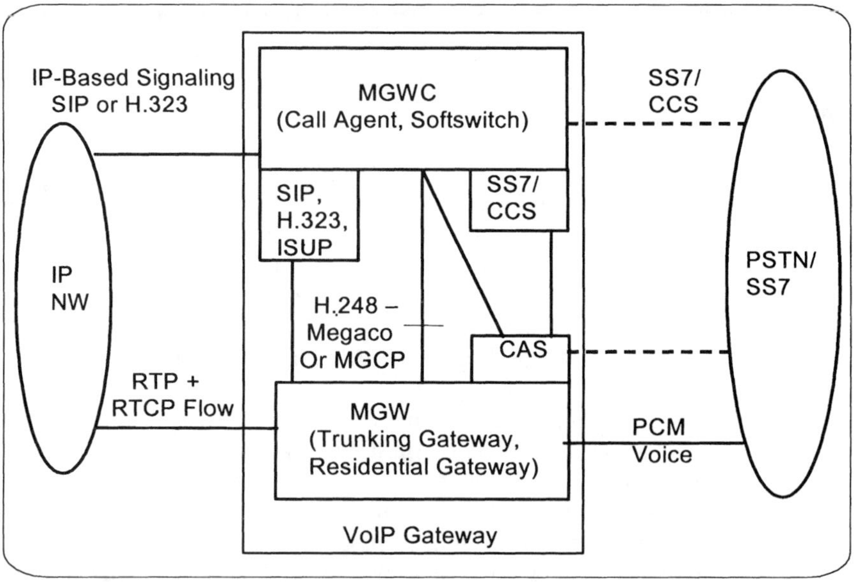

Figure 6.6 Gateway functional components.

IP PBXs are call-processing servers that run digitized voice and call control over the IP network, provide most or all the features of their legacy PBX predecessors, and connect over the LAN or WAN with IP-enabled phone handsets (softphones).

Figure 6.7 compares IP PBX with the SIP components described earlier. IP PBX acts as the outbound proxy and the registrar server of all extensions.

IP PBX is client-server architecture as compared to SIP, which is a peer-to-peer distributed architecture. It requires a numbering plan and use of an ENUM server, rather than URI. Call control and services are centralized in IP PBX architecture, but these are decentralized in end devices in SIP architecture. It can support network address translation (NAT)/firewall traversal, whereas additional techniques and processes are required for such support in a pure SIP environment. IP PBX can collect usages and billing data but, unlike pure SIP, requires centralized control.

6.4.7 Controller-Controller Protocols: BICC

BICC development was triggered by a need for a packet-based PSTN replacement. It is based on functional separation of call and bearer signaling protocols in a broadband network. Besides TDM bearer, it includes support for IP/ATM bearers and uses SS7 signaling (with extensions to ISUP), and binding information allows correlation between call control and bearer. BICC defines three capability sets:

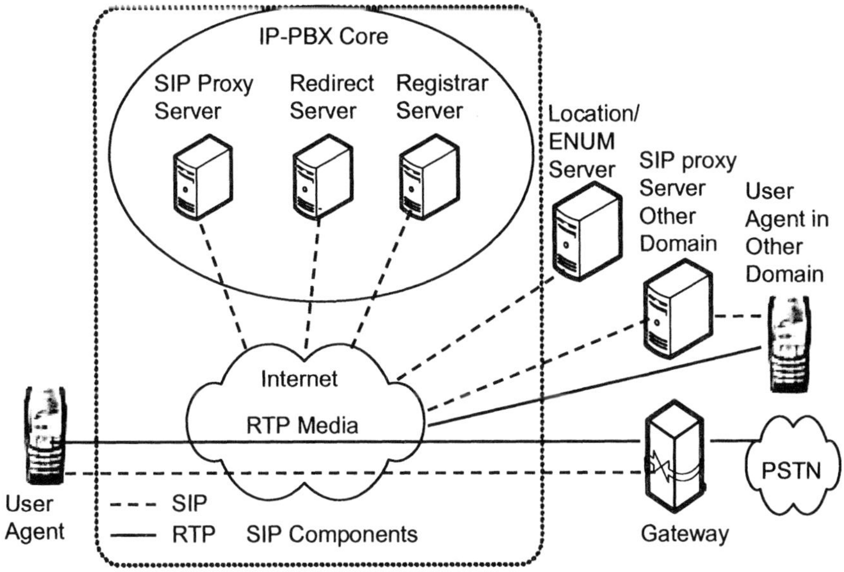

Figure 6.7 IP PBX using SIP and its comparison with SIP components.

- CS1: Supports ATM-based (AAL1/AAL2) bearer;
- CS2: Supports IP-based bearer;
- CS3: still in works to support advanced services and interoperability with SIP.

We will not go into details about BICC here.

6.4.8 H.323/SIP Interworking

At present, H.323 has been widely deployed throughout the world, both in enterprises (IP PBXs) and in PSTN (VoIP gateways). The trend in the industry appears to be to migrate toward a SIP-based network infrastructure in the future rather than continue to expand H.323 networks. Legacy H.323 networks will likely remain in place and be somewhat extended, and some new H.323 networking will be used by certain providers. Therefore, interworking between SIP- and H.323-based networks is an important issue.

Although H.323 and SIP address similar requirements, the mechanics of how they perform call setup, media negotiation, and call teardown makes them incompatible and prevents direct connectivity between SIP and H.323 endpoints. A session controller (described further in Section 6.9.2) provides interworking function (IWF) services between H.323 and SIP endpoints. In effect, this hardware-software combination operates simultaneously as an H.323 gatekeeper and SIP proxy server. IP PBXs or VoIP gateways may perform this interworking function.

There are several problems with H.323-SIP interworking; in particular:

- There are no agreed on specifications for requirements or mechanisms of IWF.
- There is a need for H.323 to use H.245 Fast Connect to enable any handling of media descriptions in SIP. The media description in SIP/SDP is dynamically chosen from listed modes, but H.323 and H.245 need to declare exact modes.
- There is context dependence of SIP messages in relation to ISUP.
- There are no standards or requirements for 3G networks to provide H.323 gateways.

Additionally, today's VoIP products do not scale well, and today's VoIP solutions do not interface with value-added feature databases or signaling control points (SCPs). Voice feature support requires interaction with existing and future SCPs for functions such as local number portability (LNP) and 800 services. Challenges are mainly due to the number of devices used in the network, and call setup time and reliability. Some of issues with SIP are:

- SIP is still evolving. There are some 25 request for comments (RFCs) (over 800 pages) relating to SIP, and there are several extensions.
- SIP was originally developed for wired Internet and is not optimized for wireless environments.
- ASCII coding increases the signaling overhead in radio access.
- QoS, particularly SIP latency issues remain.
- Security is an issue.
- In a pure SIP environment, there is no call control after the initial call setup. Parties can release the call session, but since they have obtained each other's IP-addresses, they can continue sending media streams to each other.
- Response messages (e.g., 180) are not reliably delivered. This may cause teardown of the call if it was started from integrated services digital network (ISDN).
- Legacy service assurance support systems are not SIP-friendly.

6.5 WIFI AND VOIP

In the previous section, we discussed how using the capabilities of SIP servers and H.323 or SIP-based IP PBXs can provide VoIP services to Internet-connected end devices with SIP clients or H.323 terminals. Although these capabilities of SIP servers or IP PBXs are independent of how the end devices are connected to or access the Internet, the access network, whether wired, or wireless, WiFi or 3G based, has an effect on how end-to-end VoIP service is provided and what service quality can be achieved. In this and the following section, we described how VoIP service is provided over WiFi and 3G wireless networks.

The basic WiFi technology and architectures were discussed in Section 3.2.6. WiFi was originally designed as an access network domain to reach the Internet for data services and not for voice traffic. To provide VoIP service, which is very sensitive to latency and jitter, over WiFi, the basic WiFi architecture for Internet data services is enhanced with new VoIP-specific capabilities, such as network and service components to support QoS and security. Specifically, the capabilities required to enable WiFi to support VoIP are:

1. Addition of VoIP-specific call control and signaling support by using
   - SIP servers;
   - IP PBXs (sometime called call managers).
2. Addition of VoIP Gateways to PSTN;
3. Addition of application servers for supporting other service capabilities beyond simple voice calls, such as voice mail, conferencing, and other back-office OSS or network management (NM) servers.

These capabilities can be centralized or distributed in various WiFi hotspots. Additionally, a server farm with IP utility servers such as DNS, DHCP, and AAA, and an M-IP network to control QoS over IP is required.

We will consider some of these components in further detail in Section 6.7 dealing with VoIP in specific business environments but will not cover applications servers or back-office OSS servers.

Figure 6.8 illustrates the SIP-server-based WiFi architecture to support VoWiFi. The major components are one or more APs connected via Ethernet to an AR, which acts as a gateway to an ISP network. It also includes local access control and security functions.

In case of broadband (BB) access, a DSL or cable modem is also a part of the access network (AN). What is different here from the common architecture described earlier in Section 3.2.6.2 is the deployment of a firewall for enterprise security and NAT for use of private IP numbers. Use of a firewall and NAT requires use of an RTP proxy to transverse SIP messages, as discussed later in Section 8.3.4. SIP and RTP servers are deployed in the DMZ area, and the SIP server itself is a part of the M-IP network supporting the VoIP service infrastructure domain.

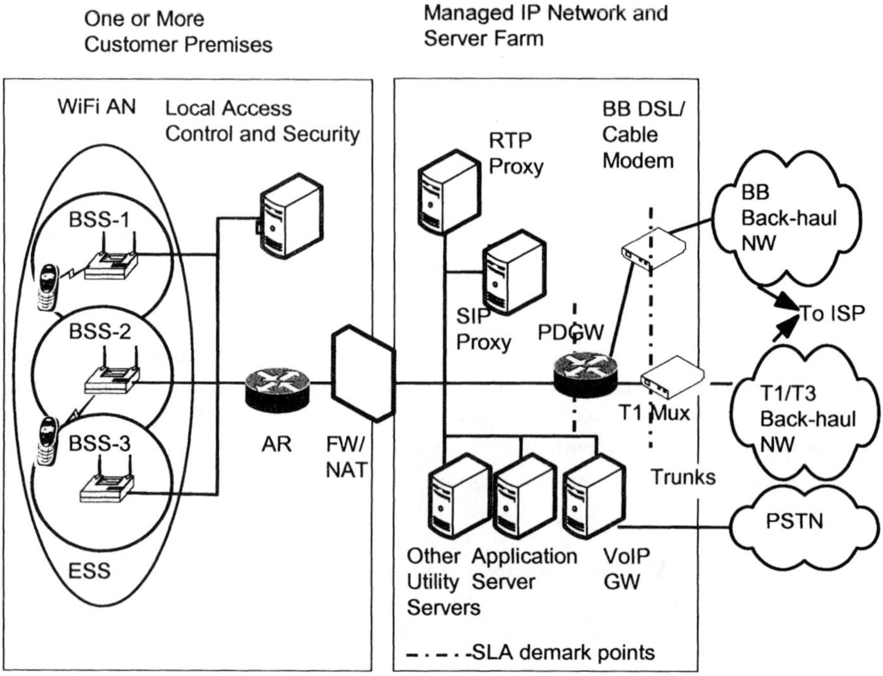

Figure 6.8 VoIP-specific components to support SIP-based VoWiFi.

Figure 6.9 illustrates the IP PBX or call manager-based WiFi architecture to support VoWiFi. Since the same architecture can also support wired VoIP access, it is also depicted in the figure to illustrate how enterprises can migrate from wired to wireless VoIP.

The key elements of this architecture are:

- The IP PBX/call manager functions similarly to the SIP server.
- The media gateway, which acts as the gateway to PSTN, is used to translate between IP and the TDM scheme used by legacy PBXs and the PSTN. Gateways provide the translation necessary to add IP phones to a legacy PBX, to connect two legacy PBXs over an IP WAN, or to provide an IP PBX with trunks to the PSTN.
- The media server provides conferencing and other features to support compressed RTP for end-to-end VoIP calls, and application servers provide enterprise service features like voice mail.

As depicted in Figure 6.9, IP PBX, gateway, and application servers are components of VoIP service infrastructure domain.

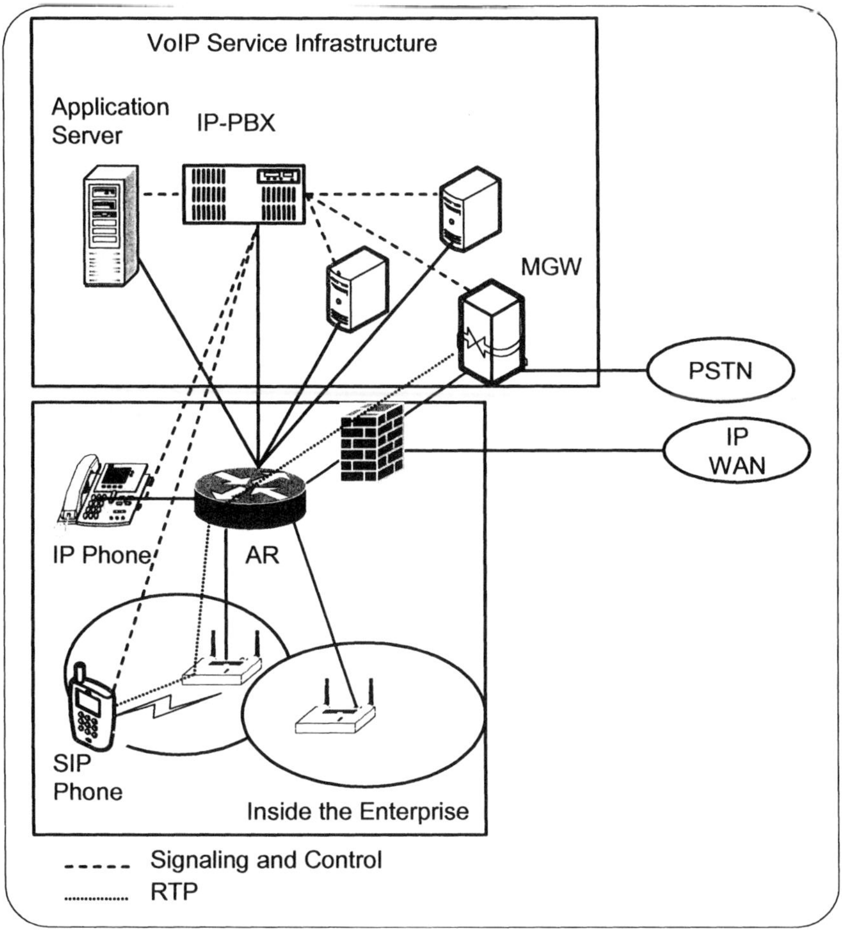

Figure 6.9 VoIP-specific components to support IP-PBX-based VoWiFi.

Most IP telephony systems support H.323. More and more VoIP vendors have started to support SIP. Although most vendors are still struggling to provide popular PSTN features with maturing standards and broad industry support, clearly enterprises are taking VoIP seriously. These IP-PBX-based services can be implemented at customer premises or can be hosted by a service provider.

6.6 3G NETWORKS AND VOIP

The basic 3G technologies and architectures were discussed earlier in Section 3.3. The 3G network architecture grew from the 2G GSM architecture, which was originally designed as PLMN for voice services and not for data services. To provide data services, 3GPP developed the 2.5G GPRS network as an overlay PS domain, which was later enhanced to a 3G all-IP network to support VoIP and other multimedia services. The basic GPRS architecture for Internet data services is enhanced with new VoIP-specific capabilities, as an access-independent IMS domain, which defines several network components.

Similarly, 3GPP2 enhanced CDMA2000-based networks by specifying MMD to support VoIP and other multimedia services. Although these two IMS and MMD architectures are similar in their direction to support VoIP, there are important differences as well.

VoIP-specific elements introduced in the 3GPP IMS domain and 3GPP2 MMD are covered in the following sections, where their similarities and differences will be pointed out. In later sections, we will only focus on 3GPP-based 3G networks and their integration with WiFi networks to support VoIP.

6.6.1 3GPP IMS Domain and VoIP

In Section 3.3.1.2, we discussed 3GPP-based 3G networks, including the IMS domain that supports VoIP and other multimedia services. The IMS uses SIP as primary signaling protocol and the IMS core network is based on IPv6. It is independent of circuit-switched networks, uses packet-switched transport for signaling and bearer traffic, and uses existing radio infrastructure—UTRAN radio network and 3G (WCDMA) core. In Release 5, services' control is in the home network (by S-CSCF), but services can be executed anywhere (home, visited, or external network) and delivered anywhere. As depicted in Figure 6.10, a WiFi access network can interface with an IMS core and share the VoIP service infrastructure.

The details of 3G PS-domain components were also described earlier in Section 3.3.1.1, and will not be repeated here. 3GPP Release 5 allows mobiles operating in packet mode to establish VoIP calls using the signaling protocol and IMS domain. A brief description of key components of the WiFi and IMS domain integrated architecture is depicted in Figure 6.10. This figure depicts only VoIP-related components and is a simplified version of Figure 3.8.

In WiFi network domain, we consider a WiFi packet gateway (WiFi PGW) [also called packed data gateway (PDGW)] that consolidates WiFi traffic from several hotspots. It is functionally equivalent to GGSN in GPRS/UMTS networks. The VoIP traffic consists of two components; SIP signaling and RTP voice packets.

First, both the RTP voice traffic and SIP signaling from UE in the WiFi domain go to the WiFi PGW, which is one of the key interfaces with the WiFi network domain from external networks. The VoIP-related interfaces from the WiFi domain to the IMS are as follows:

- The RTP traffic from the WiFi PGW goes to the Internet.

- The RTP traffic from the WiFi PGW intended for either PSTN or PLMN terminations goes to the IM-MGW. The T-SGW is a signaling gateway that converts SIP signaling into SS7 signaling for use by PSTN, and the R-SGW converts SIP signaling into GSM MAP for use by PLMN.
- The SIP traffic from the PGW goes to the P-CSCF.
- For authentication and authorization of users with their home network in 3G, the AAA in the WiFi domain interfaces with the HSS in the 3G domain.

Similarly both the RTP voice traffic and SIP signaling from the UE in the 3G domain go to the GGSN, which is one of the key interfaces with the WiFi network domain from external networks. The VoIP-related interfaces from the 3G PS-CN domain to the IMS are as follows:

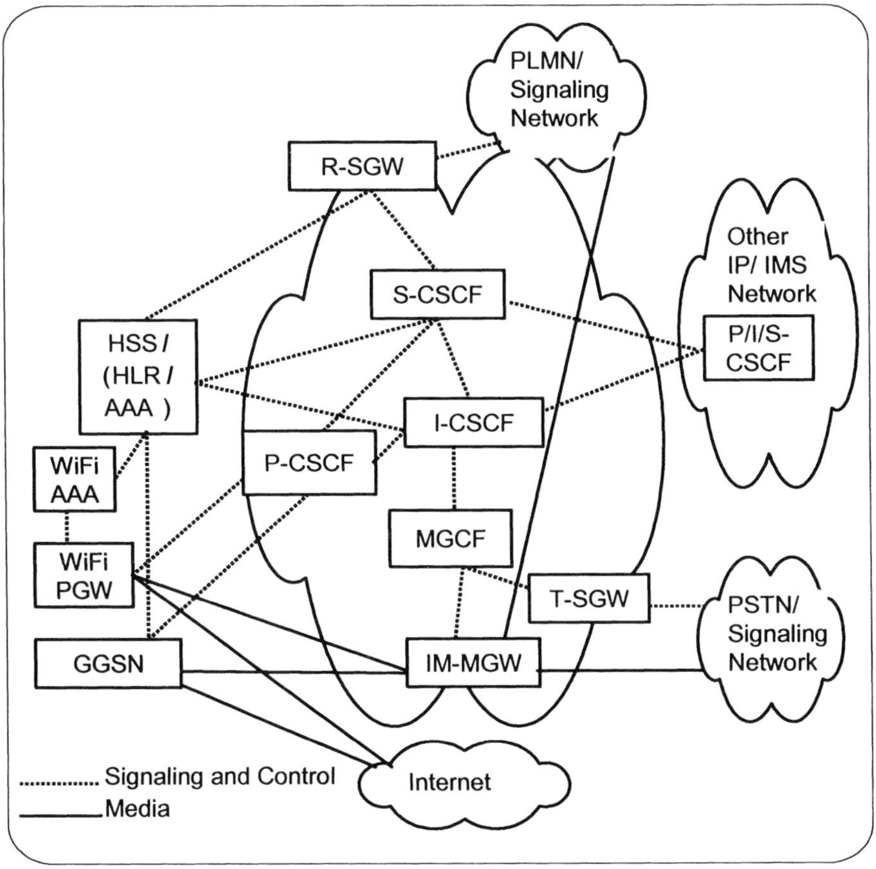

Figure 6.10 VoIP-3G and 3GPP IMS architecture. (*After:* [1].)

- The RTP voice traffic from the GGSN goes to the Internet.
- The RTP traffic from the GGSN intended for either PSTN or PLMN termination goes to the IM-MGW. Again, as before, the T-SGW is a signaling gateway that converts SIP signaling into SS7 signaling for use by the PSTN, and the R-SGW converts the SIP signaling into GSM MAP for use by the PLMN.
- The SIP traffic from the GGSN goes to the P-CSCF.
- For authentication and authorization of users with their home network in WiFi, the GGSN interfaces with the HSS, which in turn interfaces with the AAA in the WiFi domain.

The functions of each element in the IMS domain were described earlier in Section 3.3.1.2 and will not be repeated here. Various cases of how various IMS domain network elements support VoIP call flow are described later in Section 6.12.

6.6.2 3GPP2 MMD Domain and VoIP

Again, the details of 3GPP2 CDMA2000 network architectures were discussed earlier in Section 3.3.2.2 and will not be repeated here. As mentioned in Chapter 3, the 3GPP2 MMD architecture has followed 3GPP IMS architecture closely, and the use of 3GPP MMD for VoIP is similar to that described for 3GPP IMS. Figure 6.11 depicts how the WiFi domain interfaces with the MMD core and shares VoIP service infrastructure. This figure depicts only VoIP-related components and is a simplified version of Figure 3.17.

In MMD architecture, signaling is based on SIP; however, roaming and mobility are based on mobile IP. Also, MMD has tried to adjust its architecture to use IP standards as far as possible, whereas IMS has tried to adapt IP standards to meet its goals. In CDMA2000 architecture, PDSN/FA performs a function similar to GGSN in the UMTS architecture, and again, the PGW in the WiFi domain is functionally equivalent to the PDSN/FA in the CDMA2000 domain.

A brief description of the interfaces and the how these differ from the WiFi/IMS interfaces follows.

Again, both the RTP voice traffic and SIP signaling from the UE in the WiFi domain goes to WGW. The VoIP-related interfaces from the WiFi domain to the MMD are as follows:

- The RTP traffic from the PGW goes to the Internet directly or via a BR.
- The RTP traffic from the PGW intended for either PSTN or PLMN termination goes to the MGW.
- The SIP traffic from the PGW goes to the CSCF.
- For authentication and authorization of users with their home network in 3G, the AAA in the WiFi domain interfaces with AAA in the CDMA2000 domain.

Similarly, both the RTP voice traffic and SIP signaling from the UE in the CDMA2000 domain go to the PDSN/FA. The VoIP-related interfaces from the CDMA2000 domain to the MMD are as follows:

- The RTP voice traffic from the PDSN/FA goes to the Internet via a BR.
- The RTP traffic from the PDSN/FA intended for either PSTN or PLMN termination goes to the MGW.
- The SIP traffic from the PDSN goes to the CSCF.
- For authentication and authorization of users with their home network in WiFi, the PDSN interfaces via the PDF and its AAA, this in turn interfaces with the AAA in the WiFi domain.

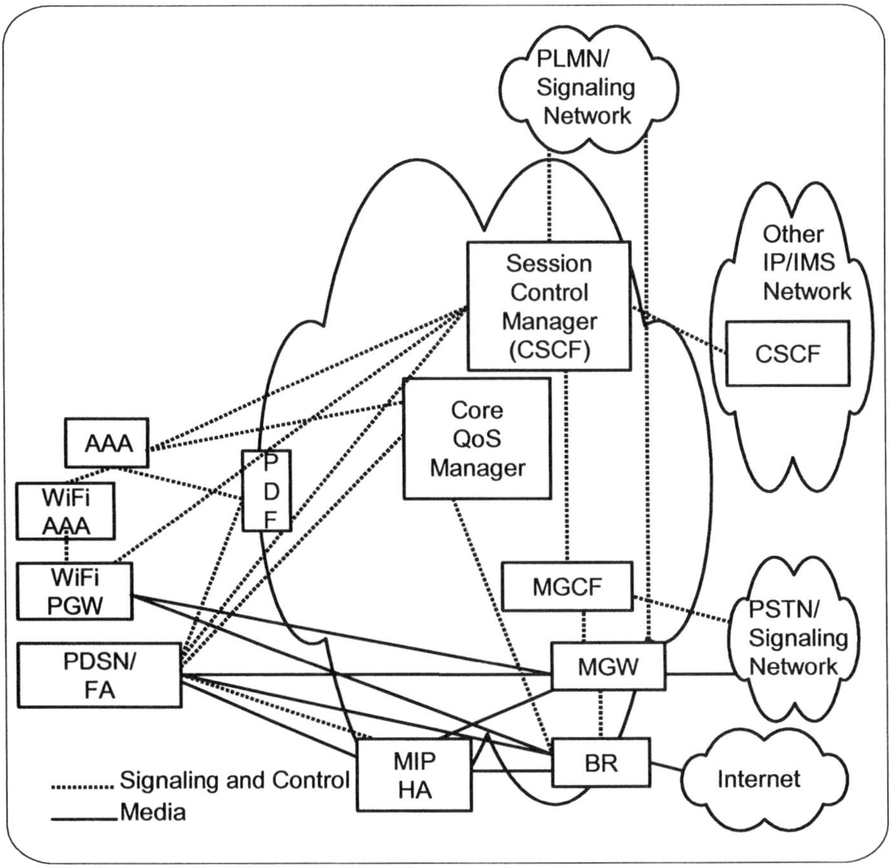

Figure 6.11 VoIP-3G and 3GPP2 MMD architecture. (*After:* [2].)

- Also, the PDSN/FA interfaces with the core QoS manager for controlling QoS.

The functions of each element in the MMD domain, and MIP-based roaming were described earlier in Section 3.3.2.2 and will not be repeated here. We will not go into details about CDMA2000 MMD-specific flows here since these are functionally similar to those for IMS.

6.7 VOWIFI AND INTEGRATED WIFI/3G NETWORK ARCHITECTURE

In this section, we will describe VoWiFi and VoWiFi/3G networks. Although our focus will be SIP-based solutions, we will look at the interworking of SIP with other H.323-based legacy systems as well as SIP-to-PSTN and legacy PLMN terminations. At the functional level, a simple view is that SIP-based VoIP is provided over UE with SIP clients connected by QoS-supported Internet. However, underlying network technologies, end-to-end network connectivity, and organizations responsible for providing and managing different aspects of VoIP service have a large impact on how the service is actually deployed and managed and service quality is assured.

We consider VoWiFi architecture from three different points of view:

1. *Network connectivity or domains:* Includes the necessary network domain to provide end-to-end transport and signaling functions;
2. *Technology:* Includes technologies involved such as VoWiFi, VoB, Vo3GPP-IMS, Vo3GPP-2 IMD, and VoWiFi/3G;
3. *Business:* Includes enterprises, public hotspot operators, broadband operators, voice over broadband operators, and mobile cellular operators.

We consider these three aspects in the following sections.

6.7.1 Network Domain View

Figure 6.12 depicts network domains or connectivity required to support WoWiFi/3G service. IP networks support SIP signaling and RTP media transport and IP bearer in general.

A brief discussion of the VoIP network domains and components follows.

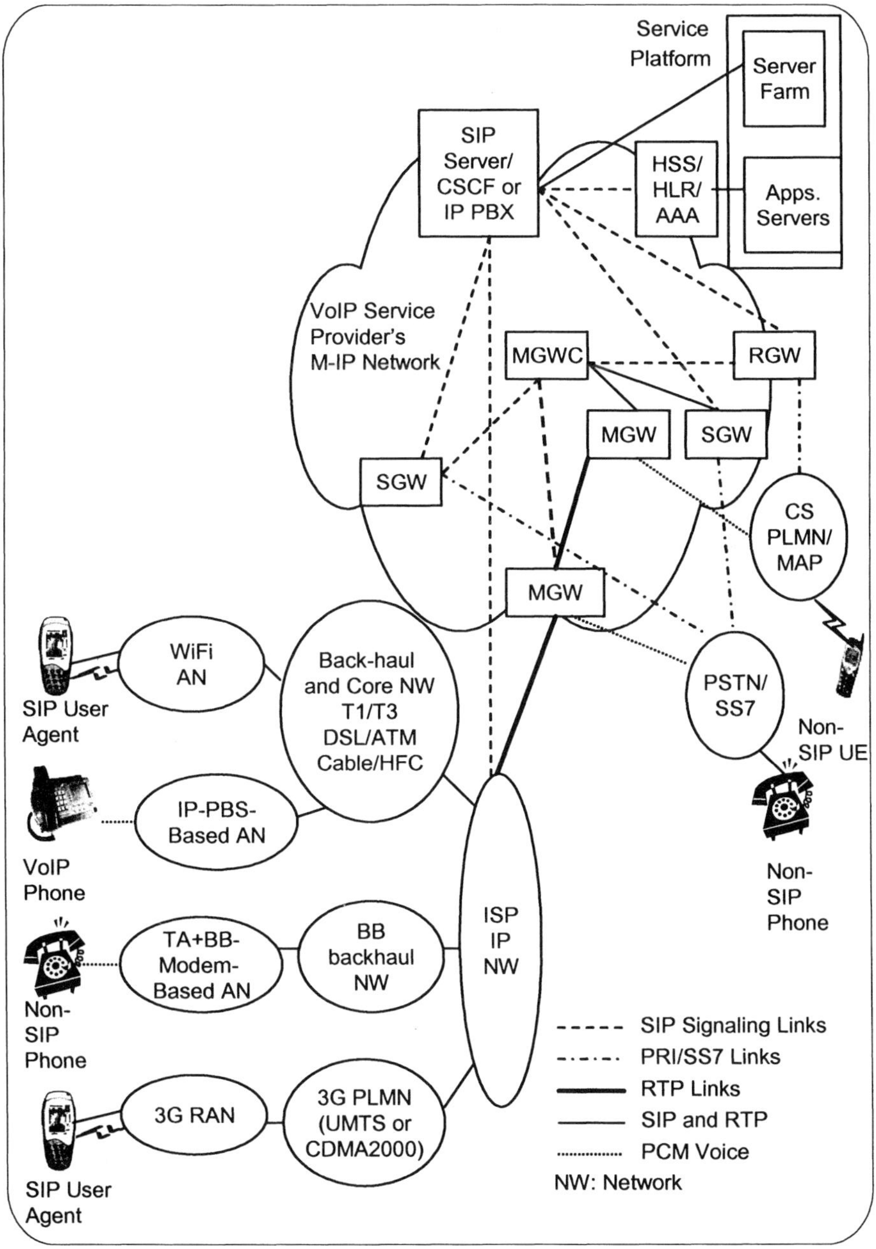

Figure 6.12 VoIP network architecture: domain view.

1. Mobile hosts with SIP UA clients—dual mode to support both WiFi and 3G;
2. Hotspots or ANs—WiFi, 3G RAN;
3. Back-haul network and CNs—T1/T3, DSL/ATM/cable/HFC and ISP's IP network, if enterprise and hotspot operators, and CN if 3G PLMN;
4. M-IP network, server farm, and IMS/MMD VoIP service infrastructure, which include:
   - SIP servers and CSCF or IP PBX provides signaling and control.
   - MGW provides bearer or media conversions, for example from RTP to PCM voice.
   - SGW provides interface between SIP and PRI/SS7 signaling.
   - MGWC controls calls. It performs the session control function, signaling, and resource management.
   - Media server (not depicted in the figure) supports functions such as interactive voice response (IVR), conferencing, fax, speech recognition, and text-to-speech conversions.
   - Application platform/server provides supplementary and value-added features, and service execution, service management, and service creation platform.
   - Subscription server function performs AAA and user and application profiles and includes HSS/HLR/AAA functions.

The VoIP service infrastructure usually supports all hotspot sites from one or more locations; it may also be outsourced and located at an operators' central office.

The main difference in enterprise and other business environments is the responsibility for these domains; for example:

- One operator may be responsible for WiFi ANs (hotspots), another for the VoIP service infrastructure, and a third (or more) for the back-haul network.
- One operator may be responsible for both WiFi ANs (hotspots) and the VoIP service infrastructure, and another (or more) may be responsible for the back-haul network.
- Broadband operators may be responsible for all three: WiFi ANs, back-haul network, and VoIP service infrastructure.
- VoB operators may be responsible for only the VoIP service infrastructure.

We will go into further details about these network domains in later sections.

6.7.2 Technology View

SIP-based VoIP services can be accessed using different technologies, such as broadband access, WiFi, and 3G, both 3GPP-IMS-based and 3GPP2-MMD-based. These different technologies use different network components in the various domains discussed above. The basic technologies of WiFi, 3GPP, and 3GPP2 networks have been covered earlier in Section 3.3, and will not be discussed here.

In the case of voice over broadband (VoB), access is based on DSL/ATM or cable/HFC technology, and the VoIP service infrastructure is similar to that used for enterprises. To understand how SIP processes will be supported by the underlying technologies, Figure 6.13 illustrates the rough similarity between the WiFi- and 3GPP- and 3GPP2-based network components involved in supporting VoIP service. This figure illustrates the following four user cases:

- *VoB users:* UE's connect to the Internet via a media terminal adapter (MTA) and DSL/ATM or cable/HFC access network, and from there to the VoB service provider's VoIP service infrastructure.
- *WiFi users:* UE's connect to the Internet via WiFi AP and then using either T1 or higher links or broadband (BB) connects, and from there to the service provider's M-IP network and VoIP infrastructure.
- *3GPP users:* UE's connect to the Internet via NodeB in the UTRAN and through SGSN/GGSN, and from there to the service provider's IMS network.
- *3GPP2 users:* UE's connect to the Internet via BTS in RAN and through PDSN, and from there to the service provider's MMD network.

VoIP interworking among these four cases is either at the VoIP service infrastructure level or via PSTN.

6.7.3 Business View

The business view of VoIP is concerned with the grouping of technologies and domains to provide VoIP services in a particular business environment. This is also concerned with the organizational responsibilities for managing and providing service assurance to users and other partners who may be responsible for some of the domains to provide end-to-end VoIP service to users.

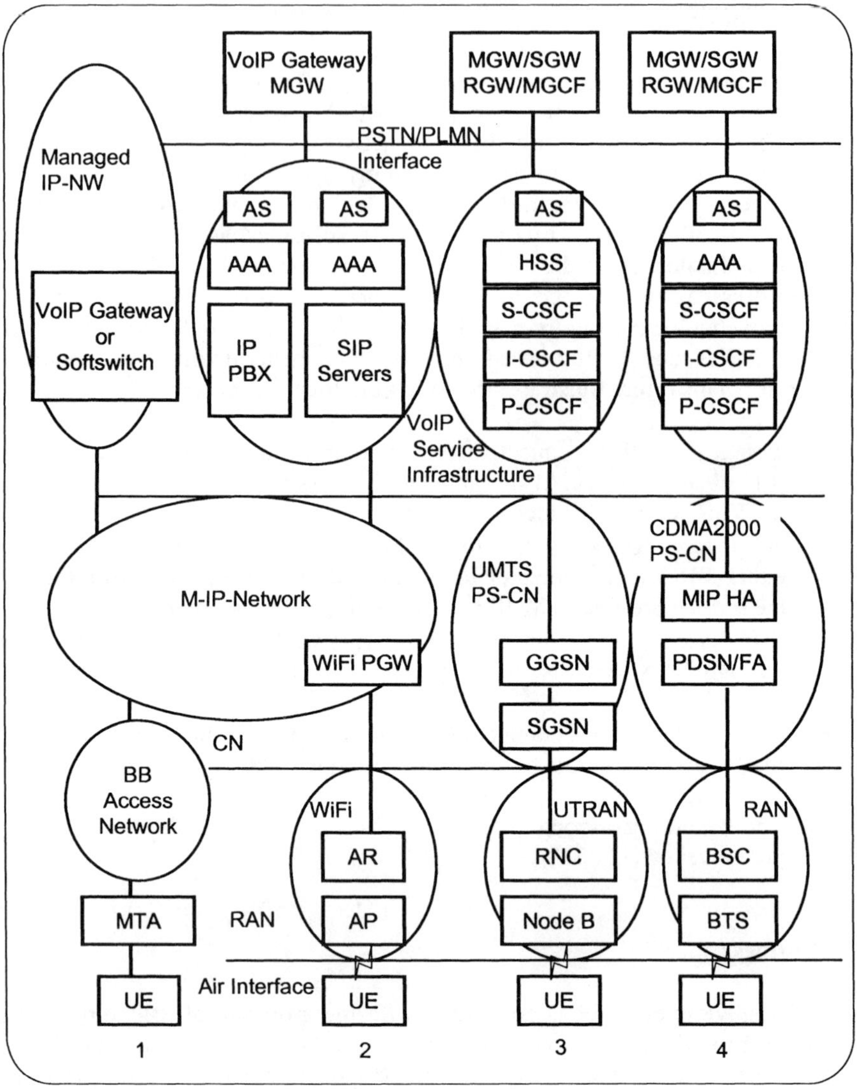

Figure 6.13 VoIP network architecture: comparison of technology and domain views.

Figure 6.14 illustrates four business views and user cases for VoWiFi, 3G, and integrated WiFi plus 3G architectures.

These business views are:

- *Enterprise perspective:* Three options, WiFi-based with and without IP PBX, and wired IP-PBX-based. User cases considered are VoIP calls from:
    - VoWiFi to VoWiFi (intra- and intersite);
    - VoWiFi to IP PBX (intersite);
    - VoWiFi/IP-PBX to PSTN;
    - Calls to legacy or 3G PLMNs or other public hotspots via PSTN, not considered as a separate case.
- *Hotspot providers' perspective:* Use of VoWiFi with or without IP PBX; User cases considered are VoIP calls from:
    - VoWiFi to VoWiFi (intra- and inter-hotspots);
    - VoWiFi to PSTN;
    - Calls to legacy or 3G PLMNs or other public hotspots via PSTN, not considered as a separate case.
- *BB operators' perspective:* Hotspot plus VoWiFi or VoIP service only. User cases considered are VoIP calls from:
    - VoB to VoB (on the Internet);
    - VoB to PSTN;
    - Calls to legacy or 3G PLMNs or other public hotspots via PSTN, not considered as a separate case.
- *Mobile cellular operators' perspective:* VoIP over integrated WiFi and 3G. User cases considered are VoIP calls from:
    - VoWiFi to VoWiFi;
    - Vo3G to Vo3G;
    - VoWiFi to Vo3G;
    - VoWiFi/3G to PSTN;
    - VoWiFi/3G to legacy PLMN.

In the following sections, these various business views are explored further.

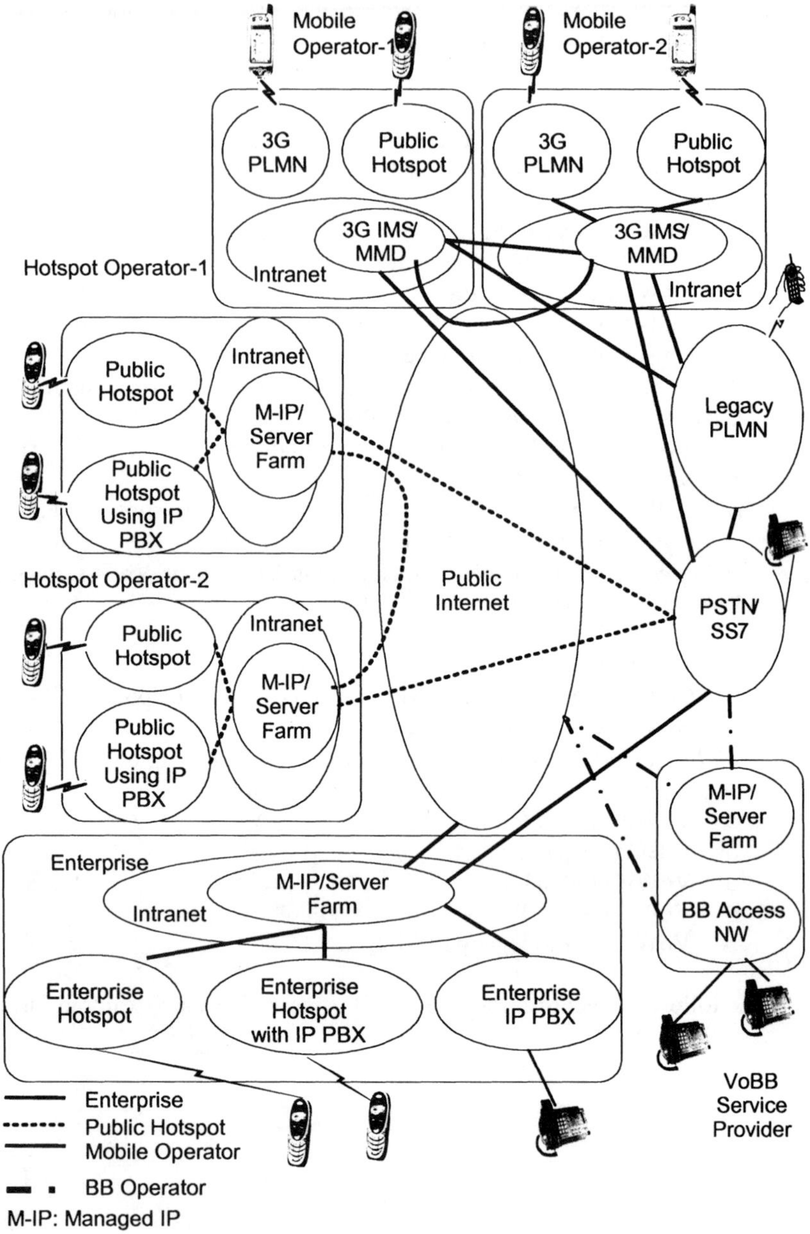

Figure 6.14 VoIP network architecture: business and various user cases.

6.8 VOWIFI FROM ENTERPRISE, HOTSPOT, AND BROADBAND OPERATOR PERSPECTIVES

In this section, first we briefly review the evolution of VoIP developments in enterprise environments, then describe enterprise VoWiFi architectures. We consider VoWiFi architectures in enterprise, public hotspot, and BB operator environments together, since these architectures are very similar, particularly for the access network.

6.8.1 Evolution of VoWiFi in Enterprises

Figure 6.15 illustrates the three-step evolution of VoIP in enterprise environments:

(a) The starting point was the legacy PBXs.
(b) Enterprises started to move from legacy PBXs to IP PBXs/call manager for on-net intersite communications using an M-IP network, and PSTN for off-net communications.
(c) With the growing deployment of WiFi hotspots, enterprises are considering using WiFi for VoIP as well. There are two alternatives here: to use PSTN in the middle or to use end-to-end VoIP for intersite voice.

In the enterprise environment, those who have moved from legacy PBX to IP PBX environments will continue to use these legacy deployments, at least until VoWiFi proves reliable. Therefore, we need to consider not only multisite enterprises deploying WiFi hotspots at multiple sites but also requiring these hotspots to interwork with their existing wired VoIP systems. In the following section, we consider the VoWiFi architecture, combining both wired and wireless environments in an enterprise.

6.8.2 Enterprise Perspective on VoWiFi Architecture

Figure 6.16 illustrates a common VoWiFi architecture for enterprise, public hotspot, and BB operators.

In this figure, three different configurations for hotspot architecture are depicted. The first (left-top block) depicts a WiFi access network supported by centralized SIP-based VoIP servers, VoIP gateway and server farms, and an M-IP network. The second (left-middle block) depicts a WiFi access network, supported by IP PBX and a centralized VoIP gateway and an M-IP network. The third is a wired IP-PBX-based VoIP, where intersite voice traffic can be transported over an IP network, and PSTN traffic can be sent either directly using a local VoIP gateway or via a centralized VoIP gateway located in an M-IP network and server farm.

The back-haul nctwork can be a private line T1/T3 or BB links. In the case of public hotspots, the back-haul network can be a WiFi-based or WiMax-based mesh network (not depicted in the figure).

The main difference in architecture for enterprise, hotspot, and BB operators is in the areas of responsibilities and size of networks. For example, enterprises will deploy WiFi hotspots with many more APs per site and will have relatively fewer sites as compared to hotspot operators, who may have only a few APs per site but will have an order of magnitude more sites. Also, enterprises usually deploy their hotspots behind firewalls and also use NAT for private IP numbers.

In the following, we review access network architectures for enterprises, then point out differences for others, such as hotspot operators, BB operators, and VoB operators. These differences are pointed out when the architectural aspects of each are individually discussed.

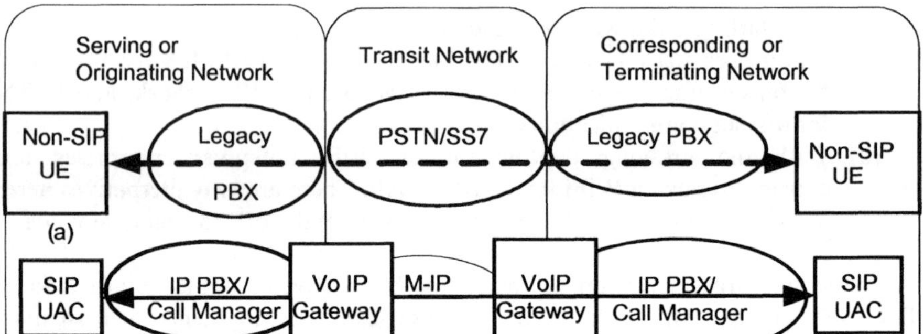

Figure 6.15 Evolution of VoIP in an enterprise environment.

Besides cost savings, the enterprise interest in VoWiFi is to use indoor mobile voice as an extension of corporate telephony (i.e., to use WiFi mobile voice handsets within the enterprise as an extension of the VoIP communications system and a replacement for desk phones).

As illustrated in Figure 6.16 the main components of this architecture are:

- One or more WiFi hotspots at various enterprise locations;
- One or more remote sites with IP-PBX-based VoIP deployments;
- These hotspots and remote sites connected via a back-haul network to the enterprise's M-IP network;
- Enterprise's VoIP service infrastructure and utility server farm.

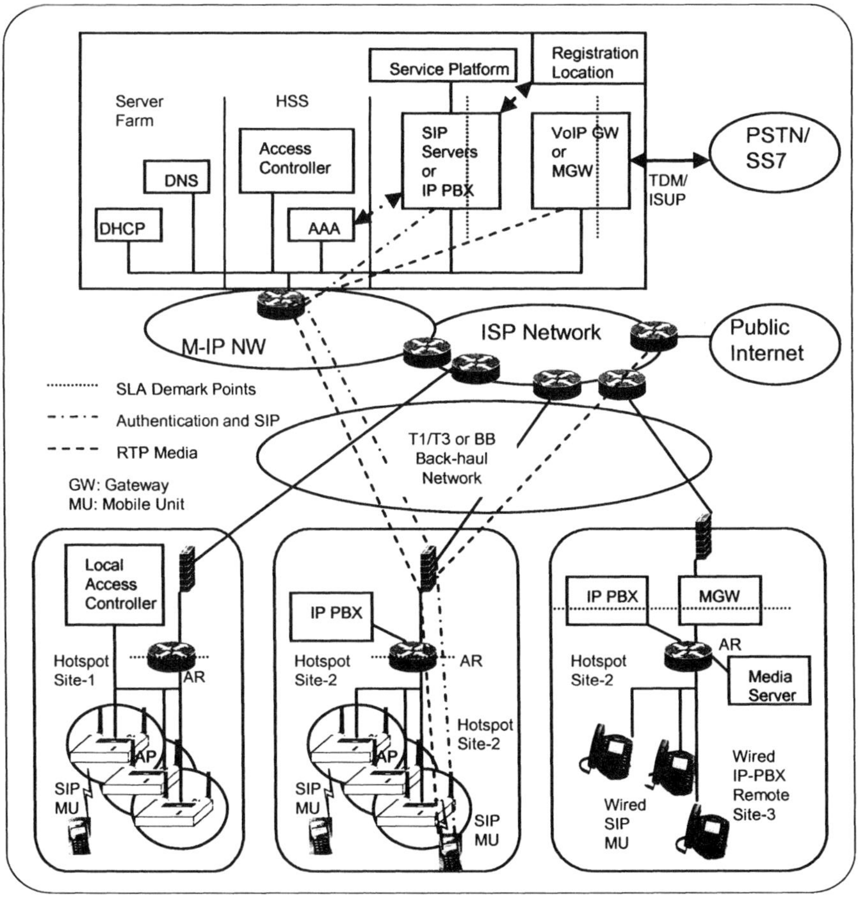

Figure 6.16 VoWiFi architecture from enterprise, hotspot, and BB operator perspectives.

As mentioned earlier, from the enterprise service assurance perspective, the following three cases are of interest:

- VoWiFi to VoWiFi (intra- and intersite) calls;
- VoWiFi to IP-PBX (intersite) calls;
- VoWiFi/IP-PBX to PSTN calls including any calls to legacy or 3G PLMNs or other public hotspots via PSTN.

The SIP signaling and RTP media paths, as depicted in Figure 6.16, are similar to the generic SIP calls described earlier.

6.8.2.1 Access Network Domain

The access network domain for enterprise, as illustrated in Figure 6.16, can be considered in three parts.

- Using WiFi and basic SIP server infrastructure;
- Using WiFi plus IP PBX instead of SIP servers;
- Using IP PBX and wired IP phones.

6.8.2.2 Back-Haul and CN Domain

For VoIP services, the back-haul network options to connect WiFi hotspots to the ISP's IP network or to an M-IP network are the same, as described in Section 3.2.6 for data services.

6.8.2.3 Managed-IP and Server Farm Domain

The operator or VoIP service providers usually deploy their own M-IP network to meet QoS control, security, and other requirements for VoIP not available on the public Internet. This M-IP network, plus associated SIP servers or IP PBX/call manager servers, constitutes the VoIP service infrastructure. As depicted in Figure 6.16, this includes:

- SIP servers and an associated utility server farm: DNS, DHCP, AAA, VoIP gateways;
- IP PBX or call manager and associated gateways.

6.8.2.4 Applications/Service Platform or Domain

For VoIP services, the applications servers or service platform is the same as described in Section 3.3.1.2 for data services.

6.8.3 Hotspot Operator Perspective on VoWiFi

The VoWiFi architecture from the hotspot operator perspective is similar to the architecture for enterprise hotspots. The main differences are that there are no wired IP-PBX-based VoIP sites. The number of hotspot sites is much larger than for enterprises, and the different operator hotspot services need to interwork to provide roaming capability for their customers. In enterprise environments, inter-hotspot roaming is easier since it all takes place within the domains of an enterprise. Hotspot interworking and roaming are covered in Sections 6.9.4 and 6.10.

6.8.4 Broadband Operator Perspective on VoWiFi

The VoWiFi architecture from the broadband operator perspective is also very similar to that of the hotspot operator, with the difference that these operators may control all components of the architecture, hotspot access network, broadband back-haul (which in case of the enterprise and hotspot operators may also use other back-haul transport such as T1 or higher links), ISP network, M-IP network, and VoIP infrastructure and server farms.

6.8.5 Service Provider Perspective on VoB

The architecture for VoB is a subset of that for VoWiFi for BB operators. Here, a VoIP service provider is only responsible for the M-IP network, VoIP service infrastructure, and related server farm, and offers VoB service. Such VoB services are currently offered by several companies.

Although the initial focus of these services is on VoB only, the architectural focus is access-independent service hosting; therefore, in principle, it can also be used to reach such services over WiFi. Recently, Vonage has announced such support for VoWiFi. These services are equivalent to hosted VoIP services. More recently, Skype has started offering such VoIP services, but its architecture and user clients are proprietary, and details are not yet in public domain.

Figure 6.17 depicts a simplified view of the VoB service architecture. At customer premises, a caller's regular telephone is connected to an analog terminal adapter (ATA), which then connects to a BB modem (either DSL or cable). This BB modem connects to an ISP network over a BB back-haul network. ATA converts analog voice to RTP and signaling into H.323 or SIP and sends it to the service provider's VoIP gateway via the ISP's DSL access multiplexer (DSLAM) or cable modem termination system (CMTS). Gateways located at suitable sites within the service provider's M-IP network convert and connect the call and signaling to PSTN for delivery at the callee end.

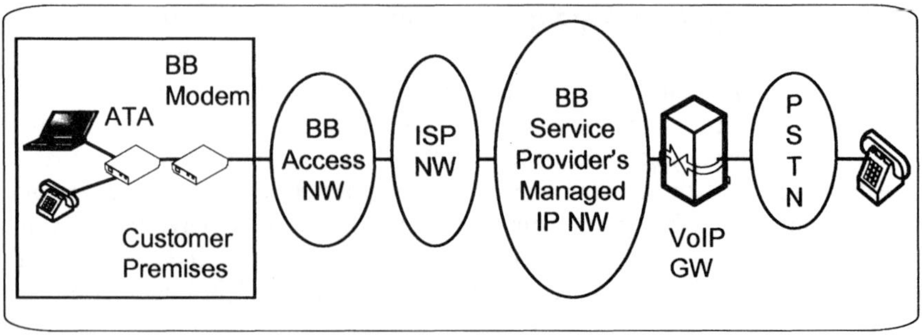

Figure 6.17 Simplified view of VoB service architecture.

6.9 MOBILE OPERATOR PERSPECTIVE ON VOWIFI

Mobile operators would like to take advantage of the growing deployment of WiFi hotspots and their complementary nature to 3G networks. Mobile operators' interest is in integrated VoIP over WiFi and 3G network to integrate indoor VoWiFi with outdoor 3G mobile voice for seamless communications between in-building WiFi and WAN, enabled with dual-band, dual-mode handsets. Thus, callers have seamless handoffs on exiting and entering the building.

In this section, after a review of evolution of VoIP developments in the operator environment, integrated voice over WiFi and 3G architectures are described.

6.9.1 Evolution of VoIP in Operator Environments

Figure 6.18 illustrates the evolution of VoIP in the enterprise environment in the following three steps.

1. *PSTN to IP:* In this configuration, a call begins at the enterprise as TDM and then is converted in the service provider's core as IP. The TDM-to-VoIP conversion is done by VoIP servers or gateways. Phone-to-phone calls between regular telephones on both PSTN ends are bypassed by the Internet through VoIP gateways.
2. *IP to PTSN/PLMN:* In this configuration, calls begin at the enterprise as VoIP and are converted to TDM via a customer-based gateway or MGW within a service provider point of presence (POP). The media gateways (MGW/SGW) handle IP-to-TDM conversion and vice versa. Access and egress networks include WiFi and 3G access. In this configuration [depicted in the middle of case (2) in the figure], conversion from IP to TDM and back creates delays and adds to transport costs. Although this alternative could be extended [depicted in the top of case (2) in the

figure] to include end-to-end VoIP by using the public Internet for transit instead of PSTN, the QoS in the public Internet cannot be guaranteed to support carrier-grade voice; this option is not used by operators.

3. *IP to IP:* In an IP-to-IP configuration, originating VoIP calls bypass the PSTN and peer to another service provider in a pure VoIP format. SBCs typically manage the handoff of IP traffic between enterprise and service provider networks or between different service provider networks.

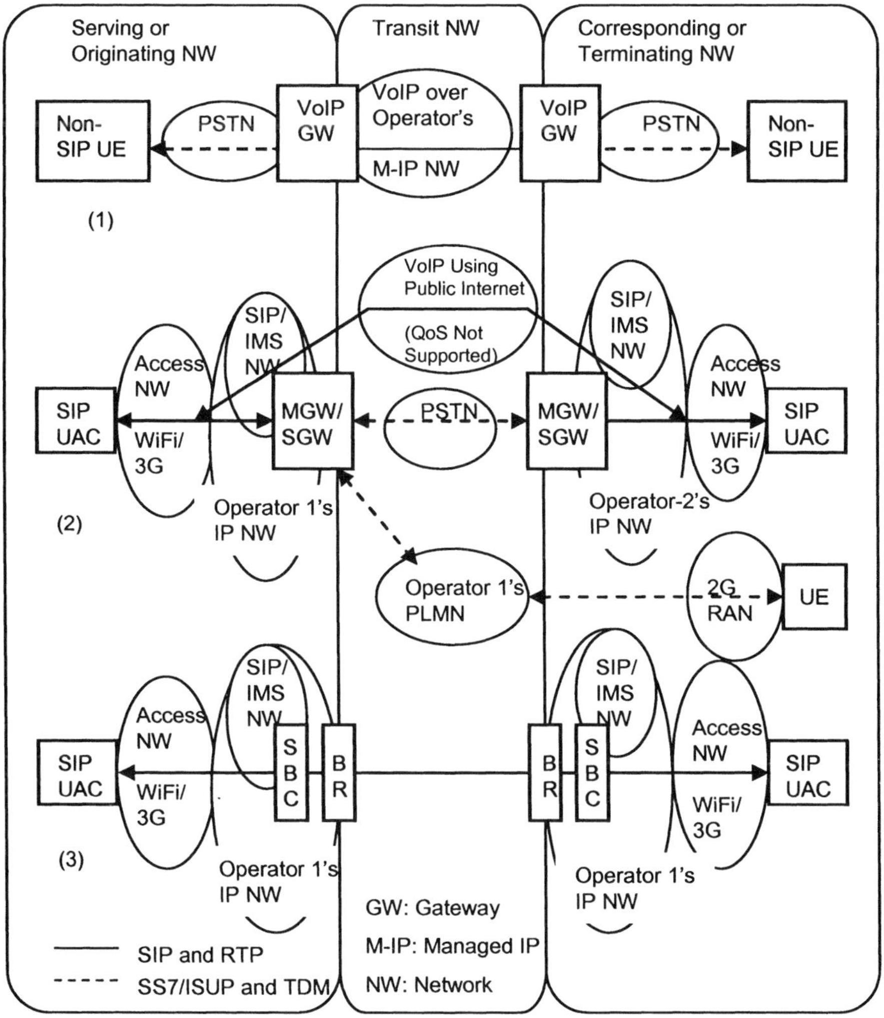

Figure 6.18 Evolution of VoIP in operator environment.

Since SIP servers only support services in one domain, there are issues of inter-operator/inter-domain operations. At present, although the VoIP marketplace is booming, interoperability issues continue to hinder its growth.

6.9.2 New Requirements at the Network Edge

Connecting two IP networks, such as an enterprise and a provider's network, introduces new network-edge requirements in three major areas: security, service assurance, and law enforcement. Since these requirements cannot be satisfied by existing products, a new product category called SBC has been introduced. SBCs sit at the edge of the provider's network and complement existing routers. They perform required control functions by tightly integrating session signaling and media control.

Session admission control is one way to guarantee SLAs for interactive communications over thin pipes. By controlling the number of real-time sessions allowed through network choke points and ensuring their priority, SBCs provide the tools required by service providers to guarantee both call capacity and quality end to end. Multiprotocol support with interworking is another function performed by some SBCs, maximizing network reach and revenues, while minimizing costs. By supporting SIP-H.323 interworking, for example, a provider can build one SIP service backbone, yet support both SIP and H.323 customers.

As illustrated in Figure 6.18, with the availability of SBCs, a direct IP-to-IP interoperation between different operators or domains is now possible. VoIP services are becoming widespread enough that service providers want to do direct IP handoffs of their traffic without having to go back through the TDM-based PSTN, so MGWs and SBCs have, in effect, become competing products. By minimizing packet-to-TDM conversions, providers will be able to reduce sharply costs associated with traffic connections, while they maintain quality and boost SLAs. With the deployment of SBCs, native VoIP interconnect provides dramatic savings for both enterprises and carriers, removes fragmented IP island topologies, and ensures network security and QoS at critical points.

To support their customers, operators need to support VoIP over WiFi, 3G, and integrated WiFi and 3G networks. VoIP over WiFi was covered earlier. Mobile operators may own, operate, or interwork with WiFi hotspots or interwork with enterprises WiFi hotspots. In the following, we consider how VoIP is supported in 3G networks, then in integrated WiFi and 3G networks.

6.9.3 WiFi and 3G Integration Requirements for VoIP

The WiFi/3G integration requirements and integration scenarios described for data services in Section 3.4.3 are applicable to VoIP-related RTP data flows. Additionally, there are some VoIP-specific support requirements for SIP signaling and control flows. In this section, we cover these additional integration requirements and integration scenarios needed to support VoIP-specific signaling and control messages.

In the following, we consider integration of WiFi and 3G networks to provide VoIP services, while the user is in a WiFi or 3G network coverage area, and when the user roams from the WiFi to the 3G coverage area, or vice versa, as well as when the user terminal moves from attachment to one WiFi AP to another or moves from attachment to a WiFi AP to a 3G SGSN or vice versa.

To support VoIP access in public, enterprise and broadband environments require a stepwise integration, starting with roaming, handover, and then seamless mobility across networks. Initially, standardized loose WiFi/3G interworking is preferred for its simplicity.

The key WiFi/3G interworking requirements are the following:

- End systems have SIP clients and support dual-mode radio access and SIP clients.
- Roaming agreements are made between 3G-network operators and WiFi operators. Operators give the user the same benefits as if the interworking were handled within one network.
- Subscriber billing and accounting is handled between roaming partners.
- Subscriber identification is done in a manner that it can be used both in a WiFi/3G environment and in WiFi/3G integrated environment.
- The subscriber database can either be shared or separate for the two networks, but sharing the subscriber's authentication and security information. The subscriber database could be an HLR/HSS (3G terminology) or an AAA server (IETF terminology).
- The user should be alerted of any possible degradation of the provided QoS due to change of access network.

In the following sections, first we review roaming and mobility and then consider integrated scenarios for VoWiFi and 3G.

6.9.4 Roaming and Mobility

Different aspects of roaming and mobility for data services were discussed earlier in Sections 3.2, 3.3, and 3.4. With few exceptions, the problems of roaming and mobility support for VoIP are similar to those for data services. The main difference between data and voice services is that for data services mostly the caller is roaming and mobile and the callee host is fixed, but for the voice case, both caller and callee may be roaming and mobile. In this section, we briefly describe the VoIP-specific roaming and mobility aspects and how any service-assurance-related problems are addressed.

6.9.4.1 User Roaming

For the VoIP user, roaming is concerned with three capabilities: (1) the system acquisition, (2) transport bearer-level authentication, and (3) service registration, when the user is in a visited network. The first two capabilities are the same as those for data services. Service registration is a new VoIP-specific capability.

Again, as for data services, roaming requires the home service provider to offer service using the infrastructure of another operator, but the home service provider does not have good visibility into the operational status of another provider's network. As VoIP services with additional QoS needs are built and offered, there is a requirement for common agreements between service providers on policies for network service levels and network performance information.

Currently, standards and business models for roaming are being developed by different companies and industry organizations.

Roaming support enables a user to make a VoIP call after obtaining service registration from the user's home network (whether WiFi or 3G), while the user is visiting a network (whether WiFi or 3G). Here we need to consider three cases:

- VoIP roaming between WiFi using AAA-based service registration. The architecture for AAA-based VoIP roaming is similar to the roaming for data services between WiFi and 3G discussed in Sections 3.2, 3.3, and 3.4 and will not be repeated here. The processes of user authentication by WiFi and service registration with SIP servers are covered in Section 6.11 on VoIP SIP call flows. These processes support user roaming between WiFi hotspots.
- VoIP roaming between 3G networks using SIP-based service registration for 3GPP-based IMS networks and using AAA-SIP service registration for 3GPP2-based MMD networks.
- VoIP roaming between WiFi and 3G using AAA, and SIP-based service registration. Standards-based AAA implementations allow users the flexibility to use wireless networks anytime, anywhere. A SIP-AAA coupling is used to authenticate the user from WiFi to 3G networks, and the service registration process from WiFi to 3G networks is required to support roaming between WiFi and 3G networks.

The processes of user authentication by 3G networks from users in WiFi and service registration with the SIP servers are covered in Section 6.12 on VoIP SIP call flows. These processes support users roaming between WiFi and 3G networks.

Some of the activities currently underway to address interoperable roaming are as follows:

- IEEE is preparing adoption of SIM-authentication-based extensions to 802.11 standards in Work Group 802.11i.

- 3GPP Release 6 is considering WiFi interfaces in UMTS service architectures (USIM/UICC).

For user roaming and terminal mobility, the mobile nodes (MN) must have dual-mode capabilities.

6.9.4.2 Terminal Mobility

Terminal mobility allows a terminal to maintain connectivity while changing network attachment points. Depending on the level of transparency desired, terminal mobility can be built at either the network or application layer.

- *Mobile IP:* At the network layer, mobile IP (MIP) is the standard solution, with open issues primarily having to do with AAA and low-latency handoffs. As discussed in Section 3.3.2.5, the MIP solution is independent of underlying transmission technology. It is also independent of application, whether data services or VoIP. MIP is of interest in three areas:

    - WiFi-to-WiFi mobility between subnets in large-scale enterprise site deployments, and between public hotspots;
    - WiFi-to-3G mobility to unify mobility between unlicensed WiFi;
    - 3GPP2 technologies.

    MIP is a mandatory part of the CDMA2000 standard and used for VoIP in 3GPP2-based MMD networks.

- *SIP-based mobility:* At the application layer, SIP mobility is used. SIP mobility is covered under the SIP-based 3GPP proposal, where SIP, with minor extensions, better works with low-bandwidth, high-latency wireless networks, such as SIP compression specifications and additional codecs, which allow:

    - Users can move their SIP phones anywhere in network without any additional administrative work.
    - Users can register from anywhere (i.e., the SIP server becomes a "virtual" PBX for both local and remote users, and the address belongs to the user, not to devices—using one address, users register multiple contacts and can be reached at preferred devices).
    - Application-layer mobility redirects existing streams to the new address (e.g., using SIP invite requests or RTSP redirect requests).

Three types of SIP mobility can be considered as follows:

- Precall mobility where
    - Mobile host (MH) or caller can find SIP server via multicast register.
    - MH acquires IP address via DHCP.
    - MH updates home SIP server.
- Mid-call mobility where
    - MH to corresponding host (CH) or callee involves new invite with contact and updated session description protocol (SDP).
    - MH reregisters with home registrar.
- Vertical handoff scenarios where
    - Moving-out (MO): WiFi to 3G;
    - Moving-in (MI): 3G to WiFi;
    - Moving-through (MT): 3G to WiFi to 3G or WiFi to 3G to WiFi.

6.9.4.3 VoIP Roaming, Mobility, and Service Assurance

User roaming is an internetwork or domain and UE capability, and network registration and service registration call flows support user roaming functions. The performance of roaming capabilities and associated service assurance is measured by monitoring, reporting, and analyzing user-network-authorization and service-registration-related performance parameters for each intertechnology case, such as:

- Successful network authorization ratio;
- Successful service registration ratio.

Mobility is a network and UE capability, and there are no mobility-specific call flows. The performance of mobility capabilities and associated service assurance is measured by monitoring, reporting, and analyzing handoff-related performance parameters for each technology, such as:

- Successful horizontal handoff (within a technology) ratio;
- Successful vertical handoff (intertechnology) ratio.

6.10 WIFI AND 3G INTEGRATION SCENARIOS FOR VOIP

We consider WiFi and 3G integration for VoIP service in two parts. The first is what we call, here, a single-mode integration, where UE uses VoIP over both WiFi and 3G networks and is able to roam between these two networks. As mentioned earlier, WiFi networks were initially developed for data services only. However, now these networks are being improved to provide the necessary QoS, roaming, and mobility between these networks for VoIP services. 3G networks grew out of 2G networks, which were developed for circuit-switched voice services and later enhanced to support data services. With the development of

IMS and MMD architectures, 3G networks can support VoIP and other multimedia services; however, these enhanced networks are not yet deployed. It will be a few years before this single-mode WiFi/3G integration will become generally available.

The second is WiFi/cellular integration for VoIP. This approach takes advantage of what is now available and integrates VoIP over WiFi and circuit-switched voice over cellular, both 2G (GSM) and 3G (UMTS/CDMA2000).

The following sections describe both these approaches.

6.10.1 Single-Mode Integration (VoIP on Both WiFi and 3G Networks)

Broadly speaking the two approaches for WiFi and 3G network integration for VoIP services are extensions of the integration scenarios for data services, discussed earlier in Sections 3.4.3.1 and 3.4.3.2. The integration scenarios for data services need to be enhanced to support VoIP-related signaling and control over integrated WiFi and 3G networks. In the following sections, various aspects of these two types of integration are described.

6.10.1.1 Loosely Coupled WiFi and 3G

Loose coupling between WiFi and 3G allows for independent deployment and traffic engineering of WiFi and 3G networks. Loose integration makes the most sense because it allows enterprise WiFi, public hotspot WiFi, and operator WiFi access. Two examples of loose coupling are illustrated in Figure 6.19. A loose coupling at the common authentication and CBS levels for data services was discussed earlier in Section 3.4.3.1.

To support VoIP service, additional VoIP-specific elements, as depicted in Figure 6.19, such SIP servers and IP PBXs in a WiFi operator's M-IP network, and S-CSCF and MGW in 3G's IMS core network (IMS-CN) are also required. This option is better suitable for SIP-to-PSTN terminations. Although a SIP termination via the public Internet is possible, it is not suitable for ensuring QoS.

6.10.1.2 Tightly Coupled WiFi and 3G

Again, the tight-coupling case for data services was described in Section 3.4.3.2. For VoIP, it is an extension of the tight-coupling case for data services that extends this coupling at the common IMS-CN and the Gi interface level, as illustrated in Figure 6.20.

In this option, besides WiFi interfaces with 3G at the Gi interface level, WiFi, instead of using its own AAA and SIP servers, also uses the 3G IMS-CN as its VoIP service infrastructure. This requires changes in the HSS to handle WiFi and specific NICs in UEs.

Two other cases include (1) tight coupling at the SGSN routing area management level, and (2) very tight coupling at the RNC cell management-level described in Section 3.4.3.2 for data services, and impact only the data service part and not the VoIP signaling and control messages. Although these tight-coupling cases may help improve terminal mobility, they are not considered here.

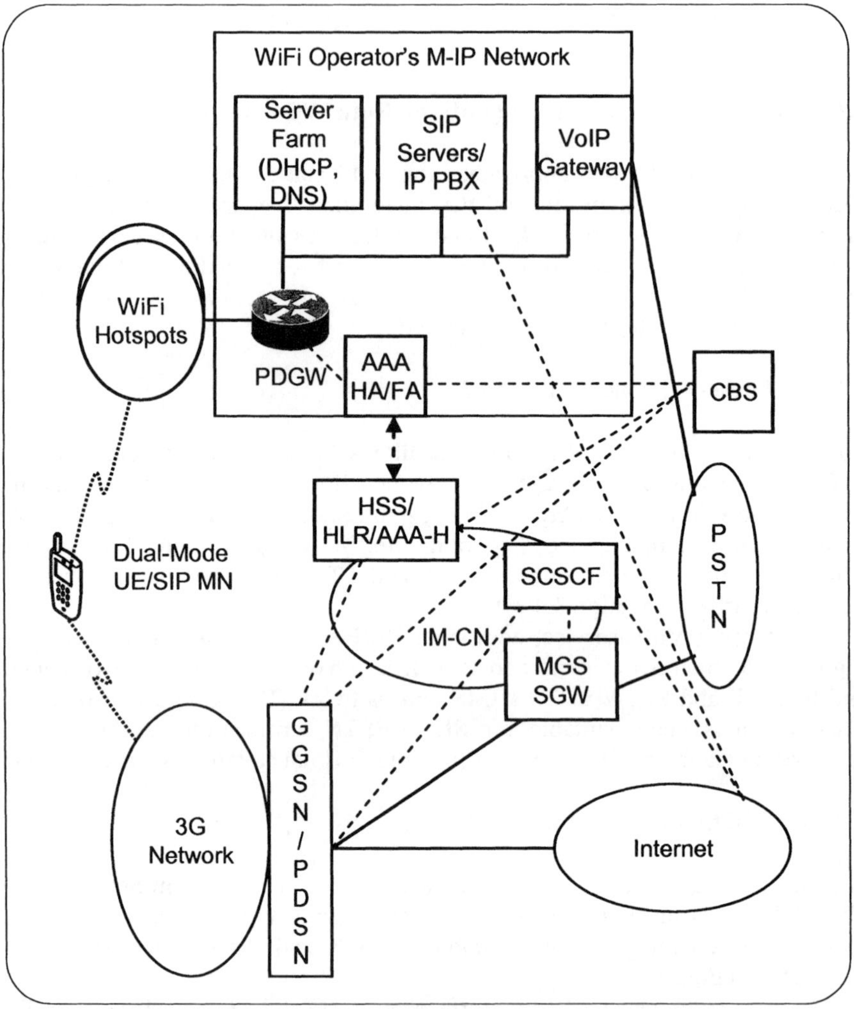

Figure 6.19 WiFi and 3G integration for VoIP: loose coupling.

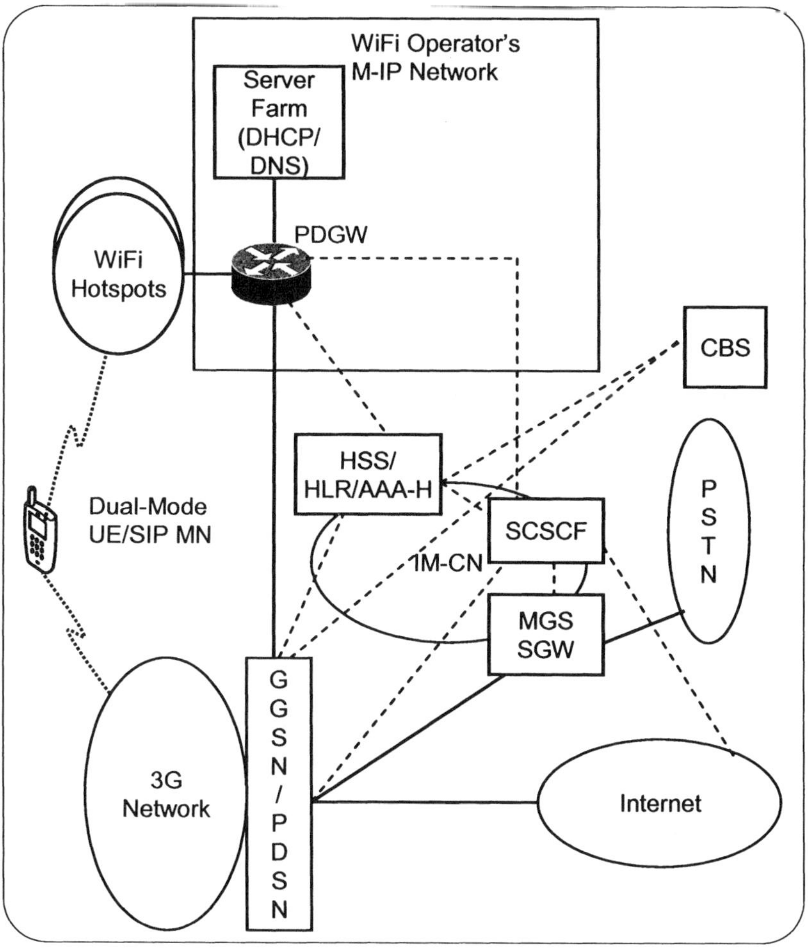

Figure 6.20 WiFi and 3G integration for VoIP: tight coupling at the IMS level.

6.10.2 Dual-Mode Integration (VoIP on WiFi and Circuit-Switched Voice on 3G Networks)

In this section, we describe two promising developments for dual-mode WiFi and cellular integration: the first SCCAN Forum's work and the UMA Consortium's work. There is another recently announced effort called the Fixed-Mobile Convergence Alliance (FMCA) led by the United Kingdom's BT. We will not discuss FMCA here since its initial focus is on Bluetooth and fixed phone integration, and also since at present not much information is available about it.

6.10.2.1 WiFi and Cellular Integration: SCCAN

Initially, the SCCAN effort was a collaboration between Motorola, Avaya, and Proxim. Now, there is a SCCAN consortium, and SCCAN is also being promoted by the IEEE standards group 802.21.

Figure 6.21 illustrates the SCCAN-based WiFi and cellular roaming architecture for enterprise users.

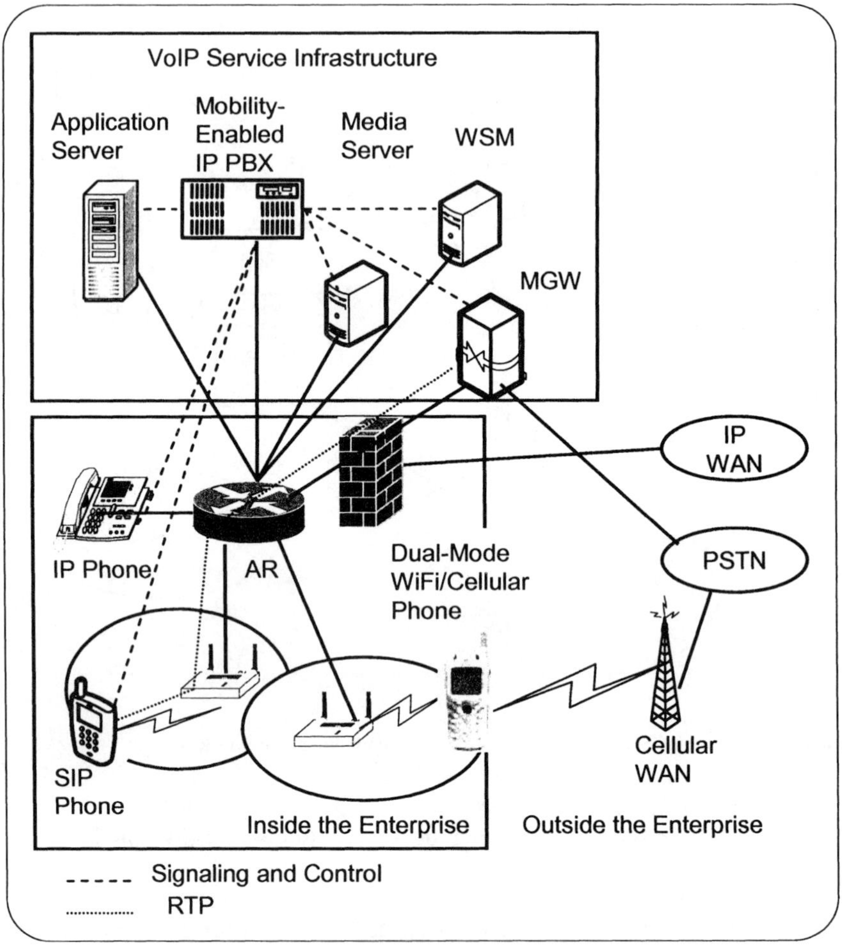

Figure 6.21 SCCAN-based WiFi-cellular roaming architecture. (*After:* [3].)

As mentioned earlier, the focus of SCCAN is the integration of VoIP and circuit-switched GSM and CDMA networks. It is tightly linked with IP-PBX-based VoIP architecture, discussed in Section 6.5 (Figure 6.9). It extends the PBX experience into the wireless domain for enterprise users. As depicted in Figure 6.21, this architecture has two new elements. One is an enhancement of IP PBX called mobility-enhanced IP PBX, and the other is the wireless service manager (WSM). Note, that the only cellular calls that can enter the WiFi domain are via cellular to PSTN to IP PBX. Motorola has announced its CN620 GSM/WiFi dual-mode phone for such use. At present, the WSM and mobility-enhanced IP PBX required for this architecture are proprietary products, and the solution is intended for enterprises only.

6.10.2.2 WiFi and Cellular Integration: UMA Developments

The UMA Consortium's focus is the integration of VoIP over WiFi and circuit-switched voice over GSM and CDMA2000 and data over WiFi and GPRS and CDMA2000 networks.

Figure 6.22 illustrates the UMA-based WiFi and cellular roaming architecture for mobile operators. This architecture introduces a new network element in the WiFi domain called the UMA network controller. This controller interfaces with both CS-CN and PS-CN of 2.5G and 3G visited networks by 3GPP-defined interfaces, such as A or lu-CS for CS-CN and Gb or Iu-ps for PS-CN. Also, although not depicted in Figure 6.22, this controller includes the functions of a service gate that will interface with the AAA server of the visited 3G network. The UMA architecture is tightly linked to the core network of the visited cellular network, which is used for routing, authentication and billing, and supporting roaming. As compared to the SCCAN architecture, where the focus is on enterprise users, the focus of UMA architecture is on cellular carriers, and the goal is to extend the cellular experience into the WiFi domain.

Now UMA client software and suites of UMA-oriented products and services are available that enable dual-mode WiFi and cellular phones to access voice, data, and IMS services over IP broadband and WiFi.

The UMA Consortium released its specifications in September 2004. Later in February 2005, these specifications were incorporated in the 3GPP Release 6 standards, and in May 2005, the UMA completed its work and disbanded itself.

6.10.2.3 Multimode WiFi and 3G Mobile Phones

For integrated VoWiFi/3G networks, multimode phones that work on WiFi, 3GPP IMS, and 3GPP2 MMD networks are required. Currently, most of the available VoIP mobile phones are dual mode and operate on both—over WiFi hotspots using VoIP and over cellular networks using circuit-switched voice. These dual-mode WiFi/3G mobile handsets give users a seamless voice and rich data experience as they move from home to enterprise to public hotspots.

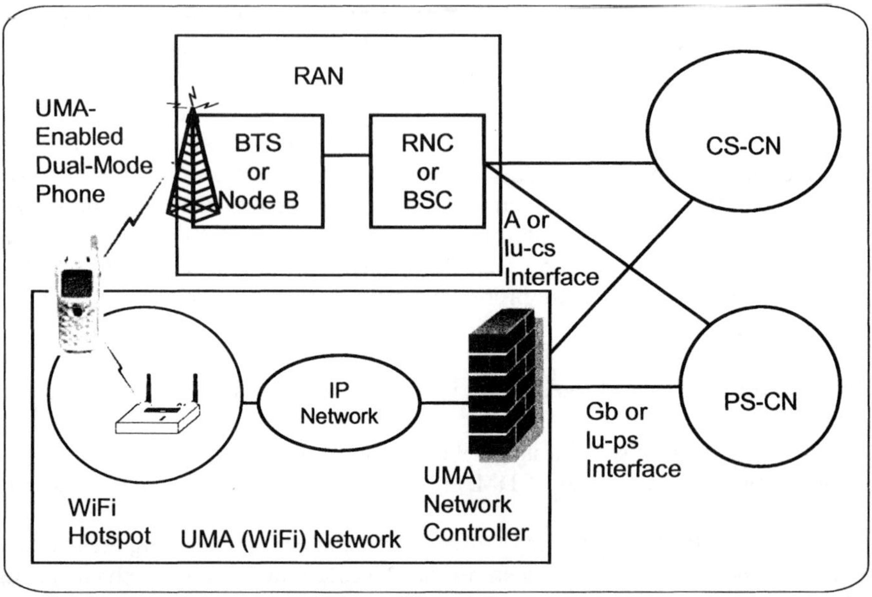

Figure 6.22 UMA-based WiFi-cellular roaming architecture. (*After:* [4].)

According to some estimates [5] nearly 90% of all handsets, handhelds, and laptops will be WiFi enabled by 2007, for a total of more than 700 million devices; at least half of these will include VoWiFi support.

Some of these handsets support WPA for secure authentication and encryption and have proprietary technologies such as fast inter-AP and inter-WiFi-cellular handoff to support roaming, voice flow classification, and control technologies to admit only SIP traffic from these handsets, which prevent the use of unauthorized devices or protocols on the network and also provides the required QoS. To help standards-based VoIP roaming and security, the 802.11 Fast Roaming Working Group is developing the 802.11r standard for inter-AP handoffs that minimize latency and optimize security.

Dual-mode phones are improving in terms of voice quality, support for seamless roaming and billing, reduced power consumption, security, and advanced PBX telephone functionality that goes with the user wherever the user may go. Currently, their interoperability capabilities vary; in the enterprise environment, the WiFi-based phone should interoperate with the existing PBX and a variety of AP vendors offer customers these options when building their WiFi networks.

6.11 OVERVIEW OF VOIP SIP CALL FLOWS

The VoIP SIP call flows for various situations have been defined in detail in various RFCs. The 3GPP-related SIP processes described in [6] are more than 800 pages long. In the following, only some of the basic SIP call flows are covered. Since SIP has been adopted by both WiFi and 3G communities, these flows are generally applicable in both these environments. However, sometimes modifications or enhancements are required to address specific architectures and terminologies (particularly in 3G IMS/MMD) and authentication and authorization needs in these environments. For example, with VoIP using IP PBX, both SIP and H.323, particularly in older legacy deployments, may be in place and require interworking. Also, in WiFi, there is need for AAA and SIP interactions for authentication, and 3G IMS has its own additions to SIP functions in CSCF. Therefore, in the following, we will focus on pure SIP-based VoIP call flows and modify them as appropriate.

As illustrated in Figure 6.23, in SIP call flows between caller (UE-A) to callee (UA-B), up to four network domains may be involved in providing services to UEs. One may consider call flows for the following combinations:

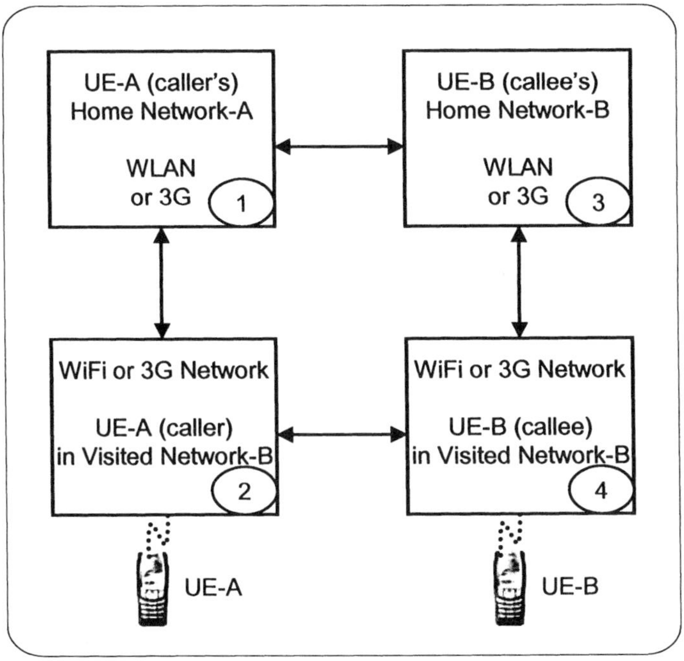

Figure 6.23 Network domains involved in providing services to UEs.

1. UE-A and UE-B in the same home network domain (collapse 2, 3, and 4 in 1);
2. UE-A in its home network, and UE-B in its home network (collapse 2 in 1 and 4 in 3);
3. UE-A in its home network, and UE-B in its visited network, (collapse 2 in 1);
4. UE-A in its visited network, and UE-B in its visited network.

In the rest of the section, we will consider only case 4, which, in effect, includes aspects of the other three cases.

We use a combination of 3G and mobile IP terminology. Here the caller is represented by a user equipment A (UE-A) or a mobile node (MN). The callee is represented by a user equipment-B (UE-B) or a corresponding node (CN).

Since SIP assumes that end systems and SIP servers are connected to the Internet, before a SIP call can be made, the UE or MN needs to perform two non-SIP processes:

1. System acquisition that attaches (registers) UE with the wireless network;
2. Data connection setup or network bearer-level registration for authentication and authorization to gain access to IP networks.

In wired VoIP, this authentication is not required, since the location of the UE and user responsible for service charges is known to the network. Besides system acquisition and setup of data connection, VoIP requires two other sets of process flows:

3. Service registration, authentication, and authorization;
4. Signaling and call setup, transport bearer setup, and call terminations.

Additionally, in a mobile and multiaccess technology (WiFi and 3G) environment, we need to understand two more functional flows for:

- User roaming support (common authentication, user profile access, and billing);
- Terminal mobility support (horizontal and vertical handoffs).

User roaming is based on the user's ability to perform the above 1 to 4 processes from a visited network. Thus, users' roaming-related processes are covered as part of these four processes. Terminal mobility is based on the network and user terminal's abilities to maintain user voice sessions, while the user terminal moves from the coverage of one network attach point to another.

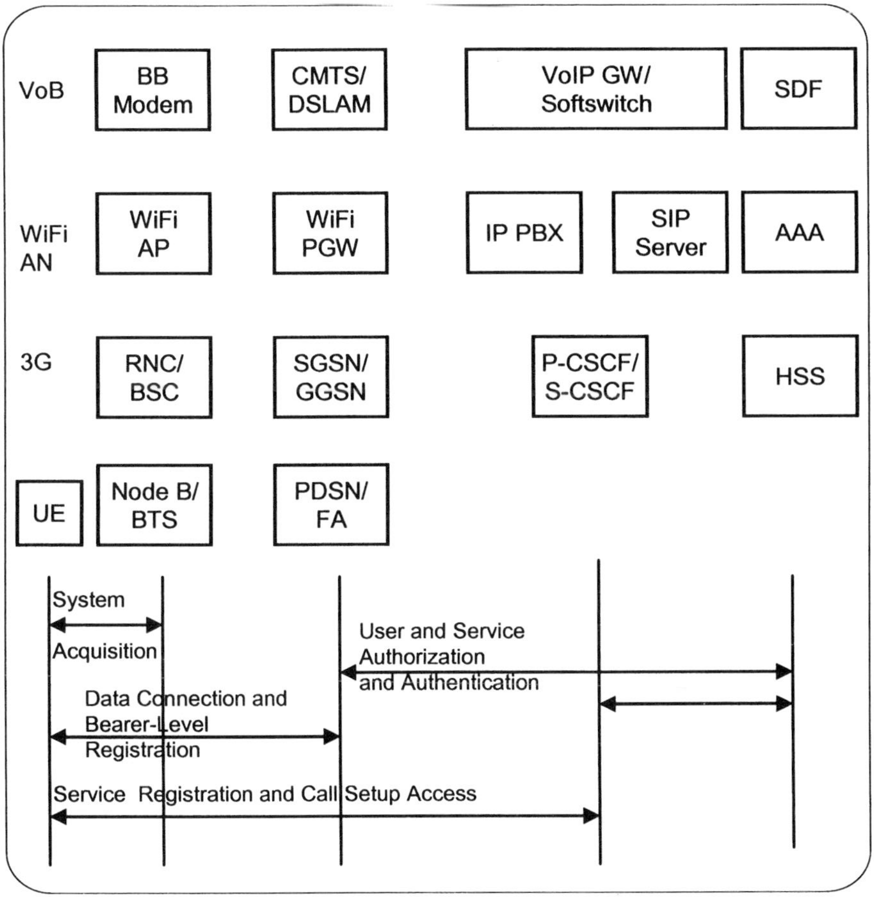

Figure 6.24 Comparison of VoWiFi-, Vo3G-, and VoB-related call flows.

Here no mobility-specific processes are involved; it is the network capability that allows maintaining the existing processes.

Figure 6.24 illustrates the similarity between the VoWiFi, Vo3GPP, and Vo3GPP2 networks, and VoB components involved in supporting VoIP SIP call flows.

Specifically, in this section, we consider one example of VoIP call flows for integrated WiFi hotspots and 3G networks—when the caller is in his or her visited WiFi/3G network and the callee is in his or her visited 3G/WiFi network. The performance of these call flows directly impacts users' experience of service quality and is therefore important for service assurance.

6.12 VOIP CALL FLOWS IN INTEGRATED WIFI AND 3G NETWORKS

Here, the call-originating UE is either visiting a WiFi or 3G network. Since for VoIP call flows in integrated WiFi and 3G networks the system acquisition and data connection setup is the same as described earlier for data services in Sections 3.3.1.4, 3.3.1.6, and 3.3.2.4, these will not be repeated here.

We consider call flows for two cases: (1) calls originating in visiting 3G networks and terminating in 3G networks (Figure 6.25), and (2) calls originating in visiting WiFi and terminating in visiting 3G networks (Figure 6.26).

6.12.1 Calls Originating in V-3G and Terminating in V-3G Networks

VoIP-related call flows for the first case—where UE-A originates a call from a visiting 3G network and terminates it at UE-B, which is also in another visiting 3G network—involve the following steps illustrated in a simplified manner in Figure 6.25:

1. *Service registration:* For service registration, (1) the UE sends a register request to the visiting network's P-CSCF, (2) which sends it to the user's home network's I-CSCF, and (3) the I-CSCF sends a register request to the home network's S-CSCF. To complete service registration, (4 and 5) the I-CSCF and S-CSCF obtains the user's profile from the user's home HSS.
2. *Call setup invite:* (6) UE-A sends an invite request to the P-CSCF, (7) which sends it to the user's home network's S-CSCF, (8 and 10) which sends it to S-CSCF via I-CSCF. Call control is with the I-CSCF of UE-B's home network. As before, (9 and 10) UE-B's I-SCCF and S-CSCF obtains UE-B's user profile and current registration information from HSS-B; S-CSCF then invites UE-B via P-SCSF (11–13).
3. *RTP sessions:* These are set up between UE-A and UE-B. The call control is maintained by UE-A's home network S-CSCF-A.

6.12.2 Calls Originating in V-WiFi and Terminating in V-3G Networks

As illustrated in Figure 6.26, for routing mobile calls from visiting WiFi to visiting 3G networks, the main differences from Figure 6.25 is that 3G RAN and PS-CN are replaced by the WiFi AN components of AP, AR, and packet data gateway (PDGW), and P-CSCF is replaced by local AAA and SIP proxy servers. With these replacements, the call flows for service registration, call setup, and media transport are very similar.

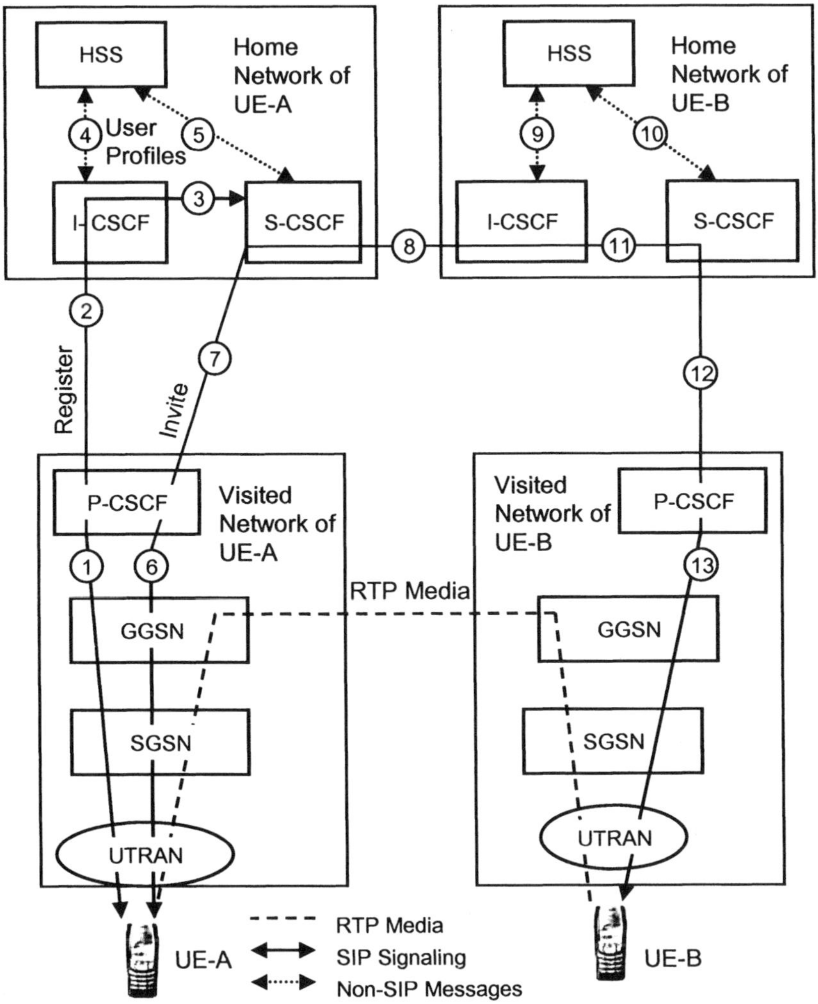

Figure 6.25 VoIP in 3G: originating UE in V-3G and terminating UE in V-3G.

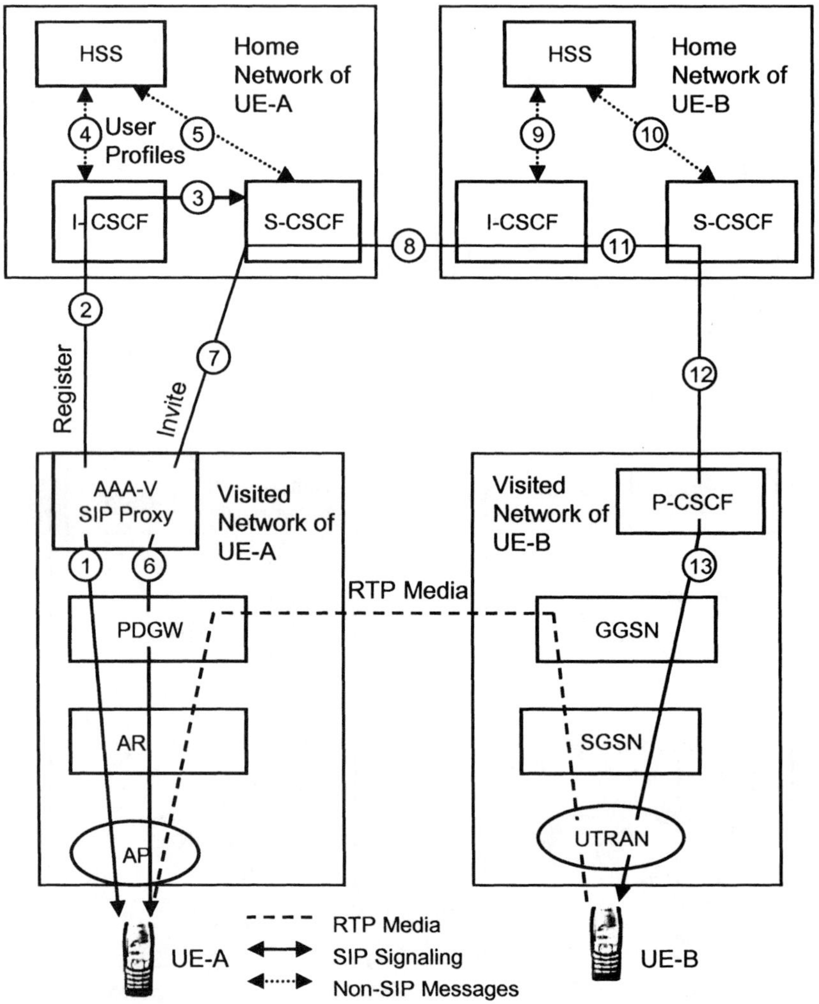

Figure 6.26 VoWiFi/3G: originating UE in V-WiFi and terminating UE in V-3G.

6.13 CONCLUSION

This chapter has surveyed the VoWiFi and voice over integrated WiFi and 3G networks, and also WiFi and cellular integration. We have described basic VoIP technology, network architectures, and domain definitions to support such integration. This chapter has provided the necessary background for defining the VoWiFi/3G service models and service assurance for these services described in Chapters 7 and 8.

References

[1] 3GPP, TS 23.228, "IP Multimedia Subsystem (IMS)," (Release 5), 2003.

[2] 3GPP2: 3GPP2.X.S0013-000-0, "All-IP Core Network Multimedia Domain, Overview," 2005.

[3] http://www.sccan.org.

[4] http://www.umatechnology.org/specifications/index.htm.

[5] VoWLAN, "The Rapidly Evolving Voice over WiFi Landscape," September 22, 2003, http://www.onworld.com.

[6] 3GPP, TS 24.228, "Signaling Flows for the IP Multimedia Call Control Based on Session Initiation Protocol (SIP) and Session Description Protocol (SDP)," (Release 5), 2005.

Selected Bibliography

3GPP, Tdoc S2-00505, "A Comparison of H.323v4 and SIP," Nortel Networks, January 2000.

Abrahams, J. R., and M. Lollo, *Centrex or PBX: The Impact of IP*, Norwood, MA: Artech House, 2003.

Alcatel, "Wireless LAN for Mobile Operators, WLAN Beyond the Enterprise," White Paper, http://cnscenter.future.co.kr/resource/rsc-center/vendor-wp/alcatel/T0210-Wireless_LAN-EN.pdf.

IETF, SIP: IETF RFC 3261, "SIP: Session Initiation Protocol," June 2002.

IETF, RTCP XR: "RFC3611 RTP Control Protocol Reporting Extensions (RTCP XR)," November 2003.

Johnston, A. B., *SIP: Understanding the Session Initiation Protocol*, 2nd ed., Norwood, MA: Artech House, 2004.

Ohrtman, F., *Voice over 802.11*, Norwood, MA: Artech House, 2004.

Struhsaker, P., "VoIP over WLAN in a Seamless Next-Generation Wireless Environment," Texas Instrument, White Paper SPLZ001, June 2003, http://focus.ti.com/pdfs/vf/bband_80211_wp_voipoverwlan.pdf.

Chapter 7

Service Model of Voice over Integrated WiFi and 3G Networks

7.1 INTRODUCTION

As described in the last chapter, the VoWiFi/3G service concept evolved from the VoIP idea, which itself followed a logical evolutionary path from the POTS. As we have seen, replacing the traditional circuit-switched telephone network with an all-digital-packet voice network will be a long-term process. In the meantime, there will be a coexistence of different technologies carrying both packet and circuit-switched voice. Managing all the variations of PSTN, 2G mobile, 3G UMTS, WiFi (or WiMax), and broadband access networks is the challenge many wireless and wireline operators will face in the next decade.

In this chapter, we continue the discussion of carrying packet voice over various types of networks, but now the focus is on defining the service model for VoWiFi/3G and laying down the foundation for the discussion of service assurance operations in Chapter 8. As discussed in the last chapter, there are various forms of packet voice service, which are related to different business arrangements. From the modeling viewpoint, these arrangements can be conveniently represented by a general service model structure. By combining different underlying service and network subcomponents as discussed in Chapter 5, a VoWiFi/3G service model aligned with a particular business arrangement can be constructed.

With the service model framework in place, we will next describe various service performance metrics and their measurements. To address the issue of service quality properly requires a thorough understanding of the critical metrics at various levels of the service model. This is a very important topic since a well-structured set of metrics and its associated SLA can be a differentiator of a service offering. In addition, a quality metric is not very useful if it cannot be readily observed. The monitoring structure is therefore an intimate part of the service model. We will see that the new packet voice technology places new requirements on the monitoring structure, but it also provides many powerful tools that can be used to collect service-quality data. Finally, we will provide an example of a set of critical KQIs and KPIs for an important business case. The business model, the monitoring structure, and the KQIs and KPIs are critical basic building blocks of the service model, which we will use for further discussion of the assurance functions and flow-through operations processes in the next chapter.

Before we get into the details of the service model, it is useful first to present the following views of the targeted service model for VoWiFi/3G integrated service. All of these views need to be considered as the service model is constructed:

1. *Administrative boundary view:* VoWiFi and 3G integration will attract various providers and administrations including WiFi hotspot providers, cellular providers, traditional PSTN carriers, enterprise network administrators, as well as other alternative wireline VoIP providers including cable companies, or Competitive Local Exchange Carriers (CLECs). By constructing the service model to reflect clearly the boundaries and roles of these administrations, the resulting model will be more flexible and useful for supporting a broad range of business arrangements. For example, a 3G cellular carrier offering a dual-mode VoWiFi/cellular service may be partnering with a broadband provider who provides the WiFi and broadband access. The relationship between the end customer and the broadband/WiFi provider, and the relationship between the cellular and broadband/WiFi provider should be clearly reflected in the service model.
2. *End-user perspective view:* The service model needs to be constructed so that user perspectives are reflected. User perspectives are best captured with respect to how the voice is originated and terminated and the roaming perspectives. A user making a voice call between a WiFi home network and an enterprise WiFi network via a broadband VoIP network is likely to have a different QoS compared to a WiFi call placed within an enterprise network. The service model should be defined in such a way that various termination scenarios are represented.
3. *Technology grouping view:* It will be useful to define the service model in such a way that clear boundaries between technology domains can be identified. For example, voice KQI and other performance metrics are well defined in the PSTN world, but not so clearly defined in the packet and new access networks such as WiFi or broadband access. By separating these different networks and their corresponding components, one can direct the focus to assurance issues regarding the new technology.

4. *Performance measurement view:* Since one cannot measure something that is not observable, it is important to structure the service model with alignment to the monitoring architecture. In this view, the set of KQIs and KPIs and how the data is collected help to structure the model.

As a service is constructed, it is desirable to go back to each of the above views and make sure that each criterion is properly addressed.

7.2 BUSINESS-DRIVEN SERVICE MODEL

There are many ways a VoWiFi service model can be formulated. Based on the guidelines given above, we structure the layers of the service model as illustrated in Figure 7.1 (the operations layer is omitted to simplify the picture). The service decomposition is based on the following considerations:

- Service perspective;
- Business perspective;
- Network perspective;
- Operations perspective.

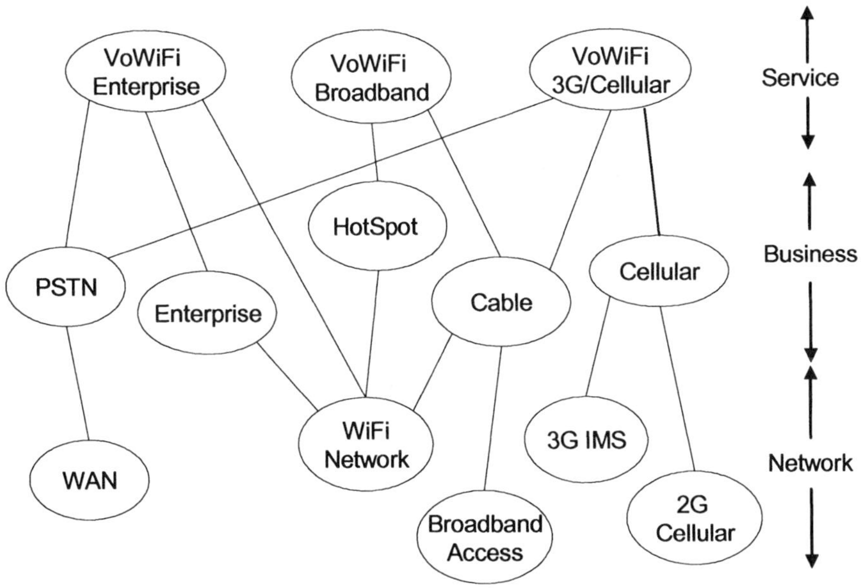

Figure 7.1 Voice over WiFi high-level service model.

7.2.1 Service Perspective

The general service category is VoWiFi. Depending on the business arrangement, the service can be VoWiFi enterprise, VoWiFi broadband, or VoWiFi/3G. It is based on carrying voice over multiple media with termination at a WiFi network. The cellular network is circuit switched (near-term) or 3G UMTS-based (long-term). To ensure interoperability, other terminations such as PSTN are included. Therefore, scenarios such as using WiFi on a cordless phone inside a residential home with broadband access to VoIP service is also covered in the VoWiFi broadband scenario.

As mentioned above, we have focused on the service scenario where one of the termination points is the VoWiFi network. With this restriction, we have eliminated pure wireline VoIP such as VoIP PBX or a digital PBX terminating at a cellular network. However, when we illustrate the idea of how to construct a service model for VoWiFi, it should be clear that the basic building blocks and modeling methodology can easily be applied to cover other variations of voice services and offerings.

7.2.2 Business Perspective

One aspect of the service model that was not elaborated in the last chapter is how the business relationship can be captured in a way that reflects the business need for inter-carrier operations. The business arrangement is more complex in the packet voice service scenario (peer to peer) than in client-server services, such as a GPRS data service. This is because in the former case, multiple providers are involved, including complicated wholesale and retail relationships. For example, a VoWiFi/3G dual-mode service may involve as many as six (or even more) different operators and respective networks, arranged in the following combination:

1. Hotspot provider A for the WiFi access of the originating side;
2. Cable network provider for broadband access of the originating side;
3. 3G visiting network provider X as a transit network;
4. 3G home network provider Y providing service control;
5. DSL provider for broadband access at the terminating side;
6. Hotspot provider B for WiFi access of the terminating side.

On the contrary, a GPRS service is peer to content and is generally offered by a single cellular provider who plays the role of both a retailer and a wholesaler. To model the VoWiFi/3G service as shown in Figure 7.1, the arc connecting the business components to the next lower layers is now interpreted more generally as a “has” or “is implemented by relationship” rather than the dependency relationship as described in the last chapter. For thc cxample given in Figure 7.1, we have focused on the following three business models:

- Enterprise VoWiFi;
- Hotspot VoWiFi via a broadband network;
- VoWiFi and cellular integration.

Further description of these business models is given next. It should be noted that these models represent some of the most common arrangements but are by no means exhaustive.

7.2.2.1 Enterprise VoWiFi

This is basically an extension of the enterprise VoIP or voice PBX scenario. The providers are usually enterprise IT organizations that are constantly under pressure to reduce the cost of providing phone services for enterprise employees. In general, this business model is an extension of the VoIP enterprise model in which the fixed wired phone termination is replaced by a WiFi connection. WiFi adds the tetherless aspect to VoIP PBX without incurring the higher cost of cellular phone service. Mobility within the enterprise is offered as an additional incentive. Key management issues are related to voice quality, security, capacity limitation, power consumption, handoff, AP layout, coverage, and availability of handsets. Since this business model is an extension of the VoIP PBX model, it is also necessary to address the ability to connect to the PSTN, as well as to broadband networks.

7.2.2.2 Hotspot VoWiFi over Broadband

This arrangement refers to a scenario where WiFi technology is used to provide features similar to a cordless phone in a WiFi hotspot venue. The VoWiFi hotspot is an add-on feature for traditional WiFi data users. The technology used for supporting this business arrangement is very similar to that of the enterprise scenario, except that the management boundary and the operations procedure can be vastly different. For example, for service assurance, a remote diagnostic capability initiated from a centralized location is extremely important, as dispatching maintenance personnel to the hotspot area to solve every service problem is not economically desirable.

7.2.2.3 VoWiFi and 3G Integration

Some call this the dual-mode service. This is probably the most demanding service from a technical viewpoint. Although this is not yet a proven business proposition from the revenue-generation perspective, it is nonetheless strategically important to many cellular providers. Some may see this as a "tag-on" feature that gives them differentiation. Others believe that they need to provide WiFi access to maintain control over the growth of the cellular business. Most agree, however, that WiFi access may provide a conduit to attract more applications developers thus customers using their next generation multimedia network (e.g., the 3GPP IMS).

In this service scenario, the user uses a dual-mode handset. The two modes of operations are circuit-switched cellular phone service and VoWiFi/3G VoIP telephony. The user switches to a WiFi voice call when he or she is inside a WiFi network. The call becomes a regular cellular call once the user departs from the WiFi network and enters the cellular network (by default, everywhere where WiFi is not available). When the user is using the WiFi network, there can be three types of terminations. First, the call is terminated by the circuit-switched cellular network (1G or 2G). In this case, a voice gateway converts the packet voice traffic into traditional cellular voice traffic. In the second case, both ends of the call are on WiFi networks, but the transit network is the 3G IMS network. In the third case, the WiFi voice call is terminated by a 3G UMTS network via the IMS core. This is not yet generally available as UMTS is still in the introduction stage, where UMTS voice is mostly circuit-based. In both the second and third cases, the end-to-end call is packet voice. In all the three cases, call setup may be implemented by a protocol such as SIP over the IP network. There are still a lot of issues to overcome before this service can become commercially practical. For example, issues including customer acceptance, billing models, seamless handoff, roaming support, and NAT/firewall all need to be resolved. Nonetheless, this service may be the best example for the illustration of how a complicated business and technological problem can be systematically analyzed with a service model. This is the reason that the next chapter is completely devoted to studying and analyzing the service assurance and operations aspect of the dual-mode VoWiFi/3G service.

7.2.3 Network Perspective

This layer of the service model deals with various modules of technology. The idea is to define network technology components in such a way that flexible business models and services can be supported without a large change of the modular structure. To construct the network model, we first need to understand

and capture the basic building blocks of VoWiFi. We attempt to modularize these building blocks so that various business arrangements can be readily supported.

As the network service model is the vehicle for supporting the assurance operations function, it is imperative that many of the operations aspects be considered at the early stage. One way to approach the design is first to list issues that operations personnel would be concerned about, such as:

- What voice quality is perceived by end users?
- Do customers have trouble getting authenticated?
- Do customers have trouble initiating calls?
- How do we locate network problems that cause degradation of service quality?
- How do we detect and deal with traffic-load and network-resource-allocation problems?
- What are the priorities of various service-affecting problems? How many users are affected?
- How do we write SLAs with peering partners?

We will see in the following how we can structure the network service model, aiming at solving these problems. First, we define the basic building blocks of the network.

7.2.3.1 Basic VoWiFi/3G Integration Building Blocks

We begin by decomposing the VoWiFi service into three basic components: *voice data transfer*, *voice call setup*, and the *AAA* functions. A high-level diagram of control signaling and data flow of these three basic functions is shown in Figure 7.2.

Voice data transfer: This deals with the transfer of voice stream payload data. It also includes voice coding and decoding functions, transcoding (conversion of different compression schemes), packetization, and transport and routing of voice packets.

Registration: SIP registration provides mobility support at the user level. As SIP user A moves around, it registers with the home network registrar where its IP location is recorded. When another user B wants to call A, B first goes to the SIP registrar; from there, A's IP location is found. During registration, the user is authenticated based on an authentication procedure. For an IMS user, the authentication is IMS Subscriber Identity Module (ISIM)–based. For general SIP authentication, an HTTP digest mechanism is defined in RFC 2617 [1]. Registration may also include downloading of user profile and filters that are specific to the user.

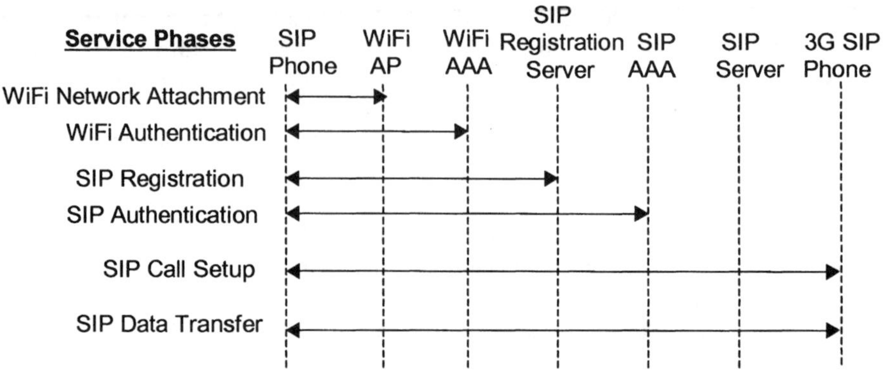

Figure 7.2 Basic signaling and data flows.

Authentication: This refers to the end user's showing his or her identity to the network provider. There are two levels of authentication and authorization functions. The first level is at the WiFi layer, where an end user is authenticated for WiFi access. After the WiFi authentication, the second layer of authentication occurs at the SIP level.

Authorization: Authorization refers to the service provider's checking the legitimacy of the user and the use of the service. The provider needs to know whether the user is a subscriber or an intruder. The provider also needs to know whether the user has paid the bill and therefore is authorized to use the service.

Accounting: Accounting allows collection of usage information for billing and other purposes. The provider assigns a charging identifier to the user and collects usage data for charging purposes. Accounting information may need to be collected at various places, for example, at the SIP proxy, where a SIP session is monitored, and also at the IP level, where digital usage is counted. A correlation identifier will need to be used to correlate the accounting measurements at different levels.

Voice call setup: This deals with the setup and teardown procedures and associated signaling of a SIP call. It may also include value-added services such as voice features or intelligent network (IN) functions like prepaid or 800 number services. Related network components include signaling devices, feature servers, signaling gateways such as SIP to ISDN User Part (ISUP) conversion, signaling transport, and signaling switching networks. We will see that the different ways of supporting these functions very often distinguish the nature and the performance of the service. By further decomposing these elements into supporting components, we can create the basic building blocks of the service model.

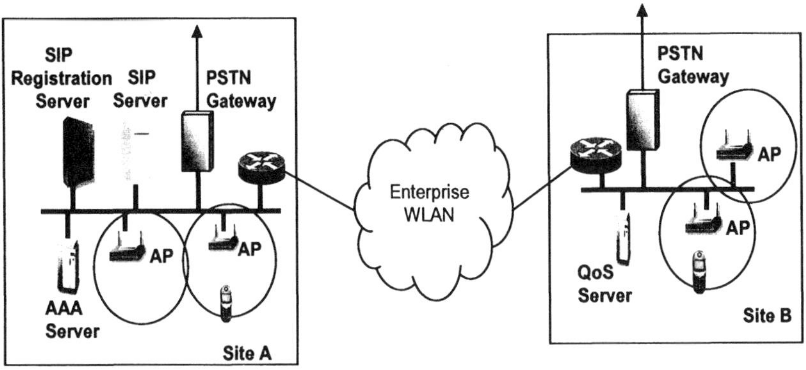

Figure 7.3 Simple VoWiFi enterprise architecture.

7.2.3.2 Voice-Enabled Enterprise WiFi Network Model

In this scenario, we are dealing with a VoWiFi enterprise network. As described in Section 6.4.6, the enterprise VoWiFi model evolves from the IP-PBX model, with the addition of SIP call-control functionalities. WiFi access is an evolution to replace wired access. The key components of a simple two-site VoWiFi enterprise network are shown in Figure 7.3. The key components include

- WiFi registration via AAA server;
- SIP registration;
- SIP call and feature controller;
- Data transfer;
- Gateway to PSTN.

In the service model illustrated in Figure 7.4, the WiFi and SIP registration functions are grouped together as the registration and AAA components. Since most WiFi authentication and authorization is implemented using RADIUS, the RADIUS server is a subcomponent. The RADIUS server authenticates and authorizes WiFi users and also keeps track of the accounting information. These registration and AAA functions are necessary for all WiFi users and are independent of the SIP registration, since not all WiFi users are necessarily VoWiFi users. After performing the WiFi registration, authentication, and authorization functions, the WiFi user has established a digital pipe to access data services. To initiate a SIP call, the SIP UA must next acquire permission to use the SIP service. This process includes another registration and authentication between the SIP phone (UA) and the SIP registration server. Therefore, the SIP registration server is another separate component under the registration and AAA component. Since all SIP messages are transported and routed via an IP network and the WiFi network, the IP network component and the WiFi network are also subcomponents of the registration and AAA component. After the user is authenticated and authorized to use the WiFi network and the SIP service, the SIP UA can now initiate a SIP call via the SIP call and feature server component. This component is responsible for all the SIP signaling and call-control functions and depends on the SIP proxy server and SIP redirect server, and the media gateway controller which is responsible for converting signaling messages to and from the PSTN (SIP-ISUP). In addition, all the voice call features, such as voice mail, PBX features, and announcement, are supported by the application server component. Since the same network is used to carry all the control messages, the IP network, the WiFi network, and the SIP phone components are all common subcomponents.

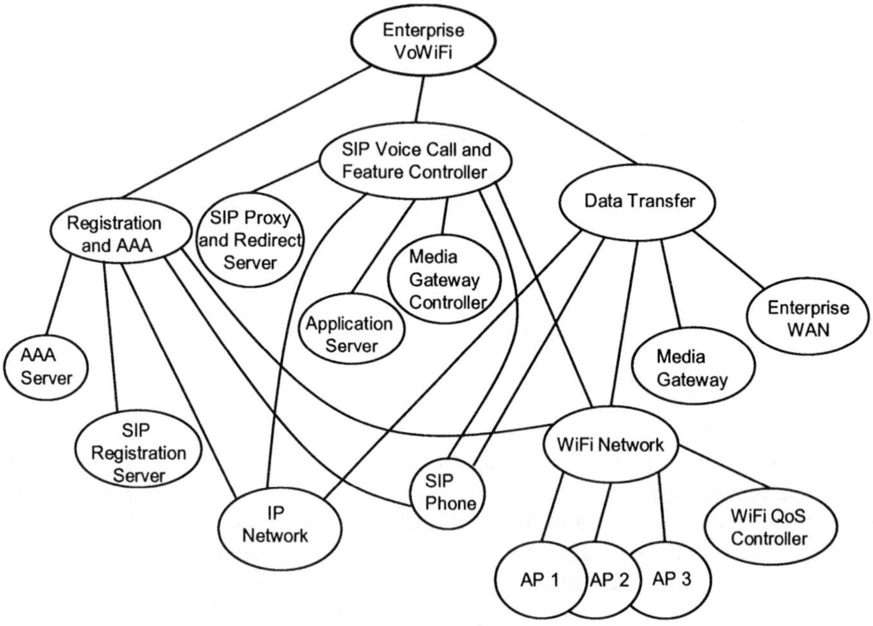

Figure 7.4 Enterprise VoWiFi service model.

The third key service component is the data transfer component, which depends on the WiFi network, the IP network, the enterprise WAN that connects multiple sites, and the media gateway, which handles format conversion (i.e., transcoding) of compressed voice packets. The WiFi network component is further decomposed into APs and a WiFi QoS controller. The AP component provides performance data regarding the AP and the corresponding 802.11 basic service set. The WiFi QoS controller can substantially improve the number of simultaneous voice sessions via admission control or priority queuing mechanisms. It is therefore important to model this function separately from other WiFi or IP network components as shown in Figure 7.4. In the future, the WiFi QoS function will likely be absorbed into APs or routers.

The service model of Figure 7.4 is structured in such a way that the following set of assurance problems can be easily identified and resolved:

- *Inability to make calls:* Likely cause is WiFi or SIP registration failure.
- *Revenue assurance:* Includes losing revenue as a result of users unable to get onto the network or fraud (people stealing services). Related components are the registration and AAA functions.
- *High call setup failure:* This points to the voice setup/teardown components.
- *High call drop rate:* Probably related to the SIP signaling component or network performance.
- *Long latency in setting up calls:* Collecting statistics from the SIP signaling component will give insight into this problem.
- *Low voice quality:* Related to the data transfer component.
- *Voice PBX feature problem:* Application server.

It should be noted that any of the above potential assurance problems can be very complicated with confusing symptoms. At this point, we are just structuring the service model and pointing out at a high level what components are likely places for problem isolation or diagnosis. Detailed root-cause analysis will be much involved and is the key topic in Chapter 8.

7.2.3.3 VoWiFi Hotspot Scenario

Supporting VoWiFi in public hotspots such as airports, train stations, or coffee shops is an attractive added feature for business customers. Yet, another business application of this scenario is that wireline phone operators are contemplating using the VoWiFi technology to try to win back some of the lost telephony customers. In this arrangement, the VoWiFi network is intended to be offered to the residential market. Whether it is a VoWiFi hotspot or a residential wired replacement, the architecture and service model relatcd to the network implementation are very similar and will be described here as the same scenario. This scenario, however, has a few distinguishing features compared to the enterprise scenario described above. The difference lies in the fact that in the hotspot scenario, the scale of the network is much larger, and the AAA structure is also different since a third party or vendor may be involved in settlement and billing. Imagine a deployment scenario as illustrated in Figure 7.5, where there are 500 WiFi hotspots. The WiFi sites are connected to a centralized control site via a broadband network such as cable or DSL networks. Moreover, since this model evolves from the existing WiFi hotspot model, there is likely to be a clearinghouse handling the accounting, billing, and settlement issues. Therefore, at least four administrations are involved: WiFi venue owner (e.g., coffee shop), WiFi network operator, broadband provider, and the clearinghouse.

A clear distinction between the VoWiFi architecture in Figure 7.5 and the enterprise architecture of Figure 7.3 is that the WiFi and SIP AAA functions are localized and managed by the same network operator in the enterprise scenario, whereas they are managed by a different administration entity in the hotspot scenario. In the hotspot environment, the need to support roaming among different WiFi operator networks favors the clearinghouse model. In that model, the AAA function is centralized, and accounting information is also handled by a centralized clearinghouse architecture.

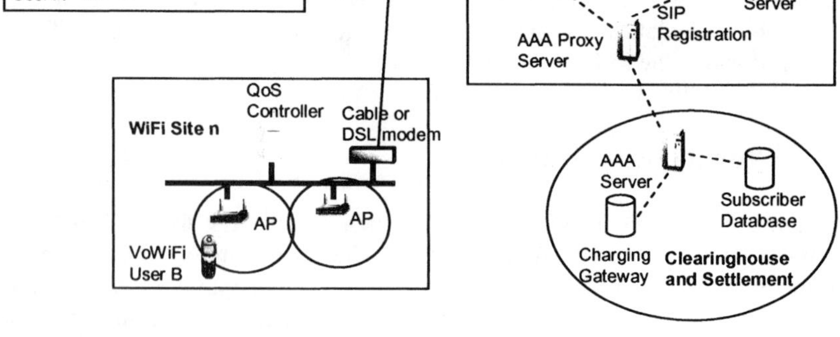

Figure 7.5 VoWiFi hotspot architecture.

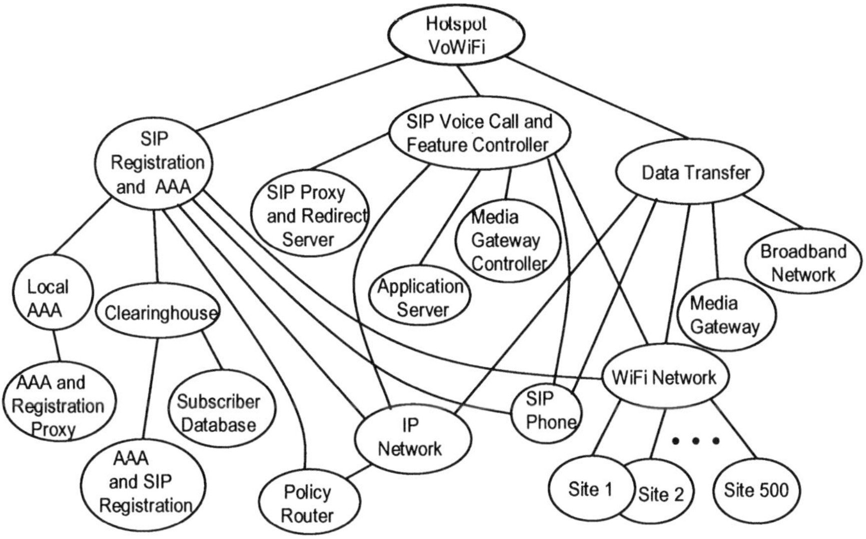

Figure 7.6 Service model for hotspot VoWiFi scenario.

The service model for the hotspot VoWiFi arrangement is shown in Figure 7.6. As we can see, the main difference compared to the enterprise service model is that in the hotspot model, the AAA component is now decomposed into the local AAA function, which can be just an AAA proxy function, and the centralized AAA function that is part of the clearinghouse component. In addition, a policy router, which is located at the edge of all the WiFi networks, acts as a "throttle" point for accessing the WiFi service. If the authentication result is negative, the policy router will deny access for that particular user. The policy router is thus considered as part of the AAA component. It should be noted that there are variations of how AAA is handled in practice.

For example, in the case that the VoWiFi service operator is different from the WiFi hotspot provider, the WiFi provider will need to perform its own AAA function, in addition to passing the same information to the clearinghouse.

7.2.3.4 VoWiFi and 3G Service Model

The 3G IMS architecture is based on the concept of separating the home network and the visited network. When a mobile user is roaming in a foreign (visited) network, all the signaling and control information will be routed back to the home network. The purpose of this design philosophy is that all the service features offered by the home network are now available to the roaming user. Extending this fundamental design philosophy to the integration of WiFi network with 3G

IMS suggests that the WiFi network can be treated as a visited network. This means all the signaling and control information will first be routed back to the home network.

Following the design philosophy of the IMS, the WiFi network can be operated by a separate operator or by the same home network operator. For a voice call that involves termination in both a WiFi and a 3G IMS network, multiple operators, as shown in Figure 7.7, are involved. Referring to the figure, Caller A is located in a visited network operated by Operator V, who provides a WiFi access network and service. Operator V has a roaming agreement with the home network provider of Caller A, which is operated by Operator X. The responsibility of Operator V with respect to the roaming agreement with Operator X includes:

- Relaying AAA information between Operator V's WiFi network and the AAA server in the home network of Caller A;
- Reporting charging information to Operator V's charging function and relating charging information back to Caller A's home network's charging gateway;
- Relaying SIP signaling message between Operators V's and X's SIP servers;
- Implementing filtering policy with respect to voice streaming data;
- Enforcing SLAs derived from the roaming agreement between Operators V and X;
- Enforce the routing tunnel between Operators V and X.

On the terminating side of the voice call, a similar role is played by Operators Y and U. The interface between Operators X and Y, however, will involve SIP signaling and data transfer but does not need to exchange AAA information.

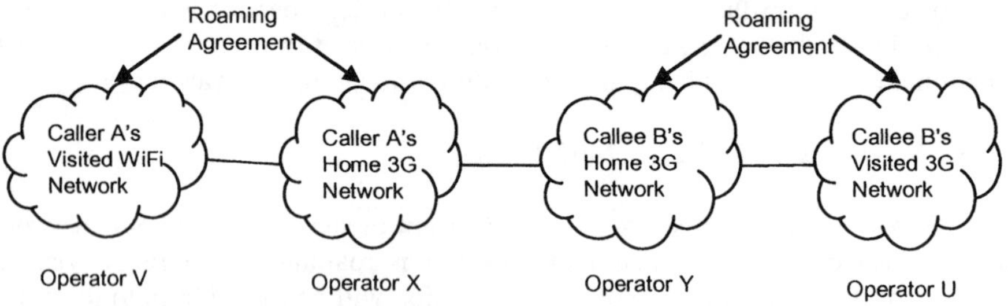

Figure 7.7 End-to-end roaming operators for VoWiFi/3G integrated networks.

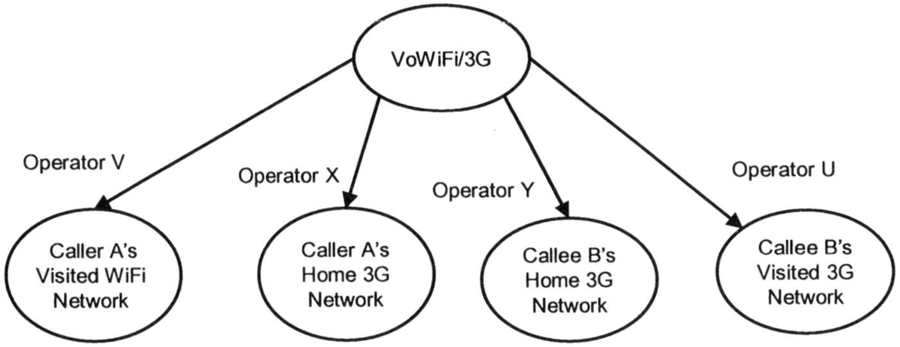

Figure 7.8 High-level service model.

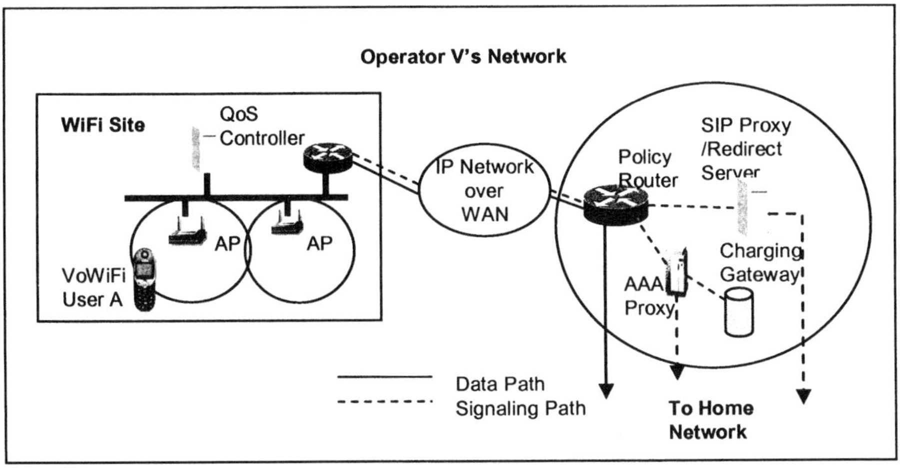

Figure 7.9 Caller A's visited WiFi network infrastructure.

To create a service model of the end-to-end VoWiFi/3G integrated networks, we start with a high-level model based on the operator's network involved as shown in Figure 7.8. For each of the operator's components, we can further define a subcomponent service model as detailed in the following sections.

Operator V: Caller A's Visited WiFi Network

Caller A's visited WiFi network is operated by Operator V, as illustrated in Figure 7.9. The service model (shown in Figure 7.10) for Operator V's WiFi access is composed of the following subcomponents:

- AAA functions;
- Charging gateway;
- DNS and ENUM servers;
- SIP proxy or redirect server;
- Voice data transfer via policy router;
- SIP UA.

As mentioned above, all SIP signaling and control functions will be routed between the visited network and the home network. The AAA function performed in the visited network (Operator V) is a proxy AAA function implemented by an AAA proxy server. The authentication and authorization will be performed in the AAA server in the home network. However, Operator V will likely also keep track of the accounting information via the charging gateway. The actual charging and settlement may be achieved via a clearinghouse outside the operators (not shown in diagram), similar to the hotspot scenario described before. 3GPP Release 6 defines two types of charging functions: offline charging where the service is not impacted by the charging information, and the online charging system (OCS), where charging information can affect the service rendered in real time. Both the WiFi AAA and the IMS SIP level AAA can assume either type of charging methods. More detail about WLAN charging from the IMS perspective can be found in [2, 3].

The SIP proxy or redirect server implements the P-CSCF function of IMS and is described in Chapter 6. It relays the SIP signaling messages to the home SIP server. Since registration and SIP-level authentication and authorization happens in the home network, the SIP proxy performs simple relaying of SIP messages only.

To find the home network, the SIP proxy uses the DNS server to locate the domain address of the entry point (I-CSCF) of the home network. In the case that the callee SIP address is a non-URL address (phone_number@domain_name), a telephone number to SIP-URL conversion takes place in the ENUM server. DNS and ENUM are often implemented in the same server.

The key components in the data path include the IP network, the WiFi network, and the SIP phone. The IP network transports the RTP voice packets from the WiFi network to the home network. A key subcomponent is the policy router (called wireless data gateway in 3GPP), which is controlled by the AAA function so that the actual voice stream may be blocked (e.g., prepaid credit is empty) or mixed with other voice streams as a result of policy execution (e.g., an announcement to add more prepaid).

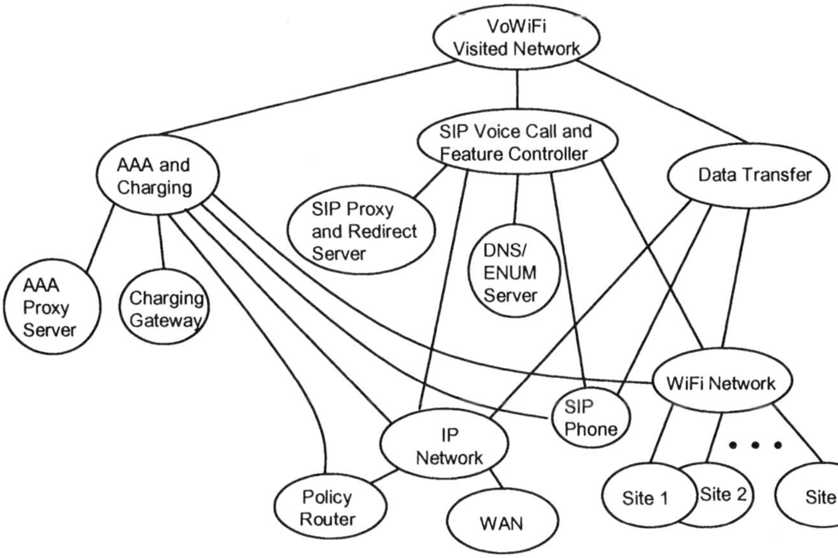

Figure 7.10 Service model for VoWiFi visiting network (Operator V).

The WAN transport provides the lower-layer transport function. Examples of WAN transport technologies include frame relay, IP VPN, or ATM services. WAN may be provided by Operator V, or obtained as leased service from another provider. In the latter case, an SLA between Operator V and the WAN provider is used to guarantee the quality of the WAN service.

The WiFi network consists of multiple 802.11 (a, b, g) APs basic service set (BSS) connected in an extended service set (ESS). To support VoWiFi, 802.11e, which allows priority access to WiFi radio resources, is assumed. In some implementation scenarios, a QoS controller is used to provide voice admission control and differentiated QoS support by giving higher queuing priority to voice packets compared to data packets. Admission control keeps track of the number of voice calls in a WiFi network. Differentiated queuing treatment is necessary to give queuing priority to voice calls to minimize latency.

The SIP phone is likely to be a dual-mode phone that supports both VoWiFi and a traditional cellular phone service. It contains a WiFi client adaptor (e.g., personal computer memory card international association (PCMCIA) card, SIM card (for GSM), SIP UA, and vocoder (compress and decompress voice signal), in addition to all the 2G cellular signaling and control functions. It is a sophisticated device with hardware and software complexity approaching that of a personal computer. The SIP phone is a key part of the service model, as many assurance issues are related to the end devices.

There are variations in the architecture and the corresponding service model that are not elaborated here. For example, in the WiFi domain, the WiFi operator may have another WiFi AAA server for the authentication and authorization of WiFi users. In such case, there may be another WiFi proxy agent before the AAA server.

Operator X: Caller A's Home Network

The architecture of caller A's home network is illustrated in Figure 7.11. This is based on the 3GPP IMS network, with focus on the VoWiFi/3G integration. The key components are:

- SIP proxy servers (I-CSCF, S-CSCF);
- The feature application server;
- Home subscriber server (HSS);
- Home AAA server;
- Packet data gateway (PDG);
- Charging gateway;
- Media gateway control function (MGCF);
- Media gateway (MGW);
- Signaling gateway (SGW).

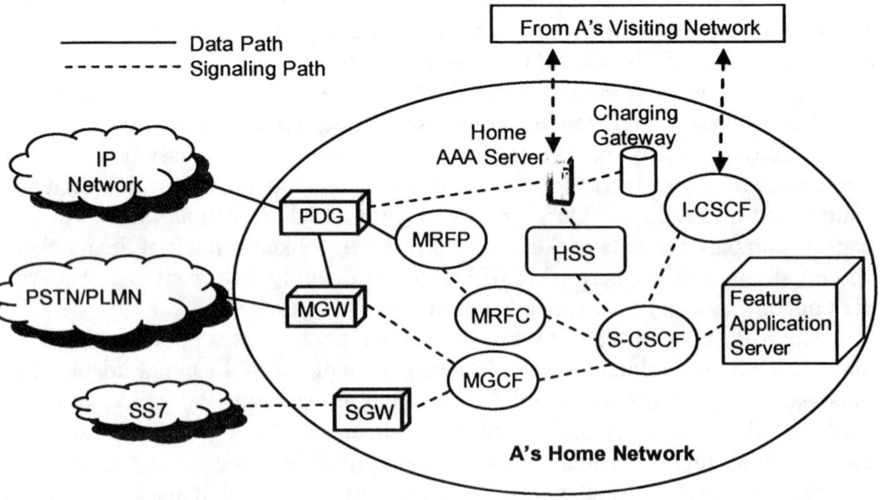

Figure 7.11 Caller A's home network (Operator X).

Continuing the scenario that User A makes a VoWiFi/3G call from Operator V's network, intended for User B located at B's visiting network, three types of message and signal come from Operator V's network. First, the AAA message from the AAA proxy server will be processed by the home AAA server, which looks up the HSS where the subscriber's IMS authentication vector [4], the user profile, and the service authorization information are stored. The home AAA server determines the response to the authentication and authorization request. It also keeps track of the change of location status for the end user.

Second, the SIP message forwarded by the P-CSCF in Operator V's network is processed by the I-CSCF, S-SCSF, and feature application server for both SIP registration, as well as call signaling and message processing. Third, the voice payload packet is routed by the packet data gateway (PDG), which supports functions including a user IP address allocation, association of a foreign IP address with a home IP address, and collection of charging information. Depending on the type of interworking, various gateway functions are performed. For voice intended for termination at the PSTN or 2G PLMN, the media gateway (MGW) may provide voice coding conversion function. The media gateway control function (MGCF) and signaling gateway (SGW) are needed to convert between SIP signaling and PSTN ISUP signaling. The MGW is also a convenient location where a performance monitoring agent, using protocols such as real time control protocol (RTCP) is located.

If the voice packet is intended to be terminated at another VoWiFi network, SIP signaling can be used end to end, and an SS7 signaling gateway will not be involved. In addition, the multimedia resource function controller (MRFC) controls the multimedia resource function processor (MRFP), which supports multiparty call-bridge function, message announcement, voice recognition, and transcoding of audio signals. It may also include conversion of voice coding format such as from adaptive multirate (AMR) to G.711 format.

The description of the message and signaling flows and the functional components of the home network architecture provide a logical separation of key components of the service model. As illustrated in Figure 7.12, the service model follows the same three components of AAA and charging, SIP call processing and features, and media control and transport. Compared to the service model described earlier for the enterprise or the hotspot models, this home network service model has to deal with many gateway functions, including the signaling gateway function between SIP signaling and ISUP signaling for call processing, as well as the transcoding gateway function for the signal conversion between packet voice and PSTN or 2G PLMN voice networks. As we will see in the next chapter, the gateways are critical locations where many assurance functions are enforced. In addition, a lot of the voice features, such as call forwarding, voice mail, and user preference, are implemented in the home network feature application server. Another important service component that needs to be monitored carefully is the charging gateway function, which can be either an offline or online charging system and is supported in the IMS charging architecture. The charging gateway receives charging information from various sources, including CSCF, application server (AS), MRFC, GGSN, SGSN, and WLAN network elements [2]. It performs functions including session charging, bearer charging, and event charging, rating, and account balance management. The charging gateway function subcomponent shown in Figure 7.12 is vastly simplified to make the figure simpler. Readers are referred to [3] for a complete description of the 3GPP IMS charging architecture.

Operator Y: Callee B's Home Network

For calls initiated in the WiFi network and terminated in the PSTN or a cellular 2G PLMN, the service model boundary is already covered in the last two sections. For end-to-end packet voice calls, such as when Callee B is located in another WiFi network or Callee B is an IMS/UMTS user, the Callee B home network and the visited network also need to be modeled to manage the end-to-end voice service. Since Callee B's home network and its service model are very similar to that of Caller A's home network, it will not be repeated here.

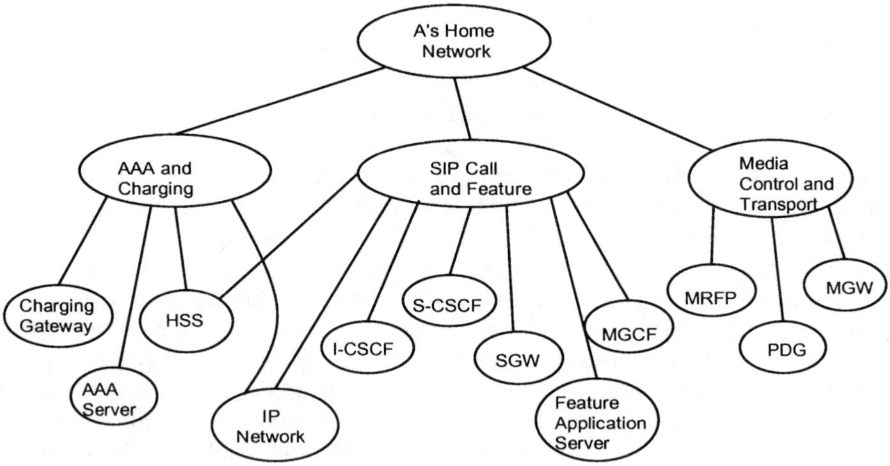

Figure 7.12 Service model of Caller A's home network.

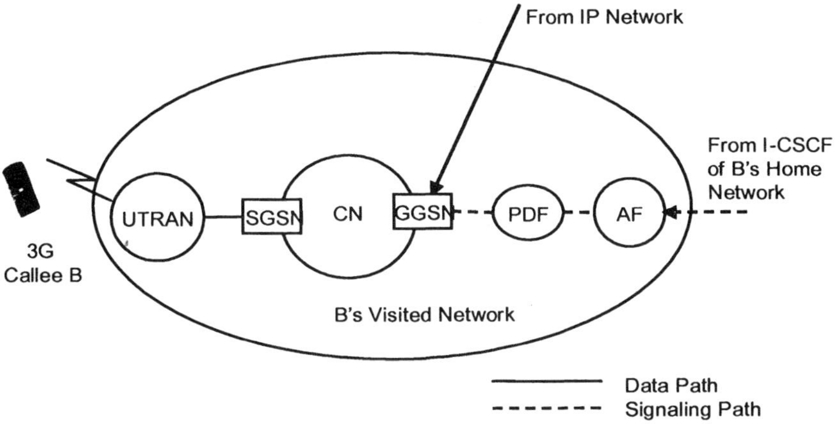

Figure 7.13 Callee B's visited 3G network.

Operator U: Callee B's Visited Network

The terminating side of the SIP call is at Callee B's visiting 3G network, operated by Operator U. As shown in Figure 7.13, the main components of this network are the application function (AF), the policy decision function (PDF), the UMTS CN, UTRAN, and the 3G SIP phone (SIP UA). With respect to the VoWiFi/3G service, the AF is a P-CSCF. In 3GPP Release 5, the PDF is contained inside the P-CSCF. In Release 6, an explicit interface G_q is defined between the AF and the PDF. The reason for opening up this interface is in case such an interface can be used for access networks other than mobile networks, which may have different requirements with respect to controlling the access network resource. In Release 6, the G_q interface is defined as a DIAMETER interface.

The PDF is a key entity that controls the reservation of QoS on a per-SIP-session basis. QoS is allocated during an SIP session description protocol (SDP) offer/answer exchange between the SIP UAs. When the P-CSCF receives an SDP offer, it sends a DIAMETER message to the PDF and asks for permission to support the media and the corresponding QoS. The PDF checks the local policy and if the request is authorized, it creates a token that is sent back to the P-CSCF, which is then forwarded back to the SIP UA. The SIP UA next requests the necessary UTRAN resources, which are identified by the token. This reservation is then sent back to the PDF via the Go interface [4] using the Common Open Policy Service (COPS) [5] protocol, located between the GGSN and the PDF. After checking that the requested resource is indeed what was originally authorized, the PDF grants final approval and opens the GGSN gating function. In this manner, the QoS can be assigned and kept track of on a per-SIP-session basis. The PDF also provides a charging mechanism, as each session can be assigned a charging identifier corresponding to the resource granted at the GGSN.

The service model of B's visited network is shown in Figure 7.14. Notice how the components of PDP setup is decomposed into subcomponents, including HLR, SS7, and UMTS components. Description of these components can be found in Chapter 3.

The GPRS CN model is discussed in Chapter 5. Here we show the UMTS model with UMTS terrestrial radio access network (UTRAN) and the 3G CN components 3G SGSN and GGSN. Figure 7.15 shows the high-level components of UMTS and the associated logical connections. The detailed description of the UTRAN resource can be found in [6]. The corresponding UMTS service model of Figure 7.15 is shown in Figure 7.16, where the same modeling steps described in Chapter 5 are applied to the UTRAN and the UMTS core network.

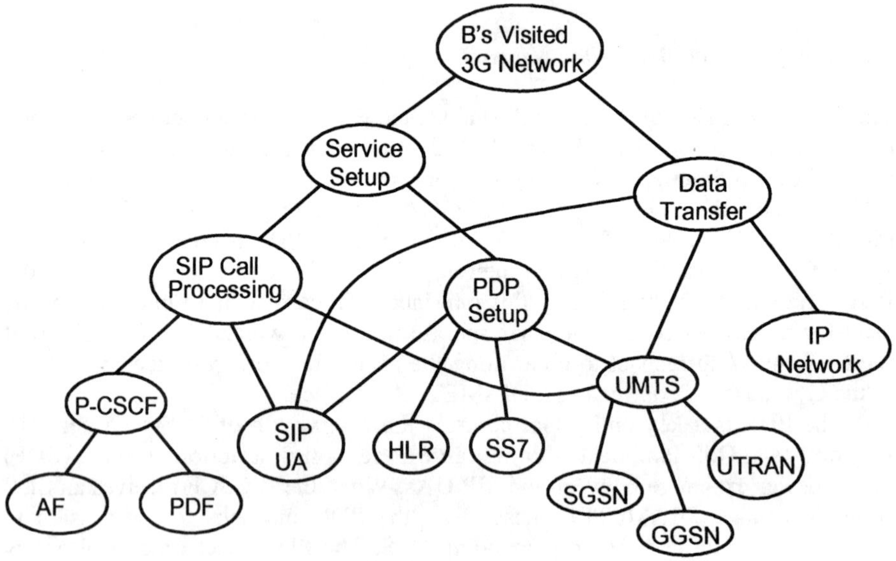

Figure 7.14 Service model of Callee B's visited network (Operator U).

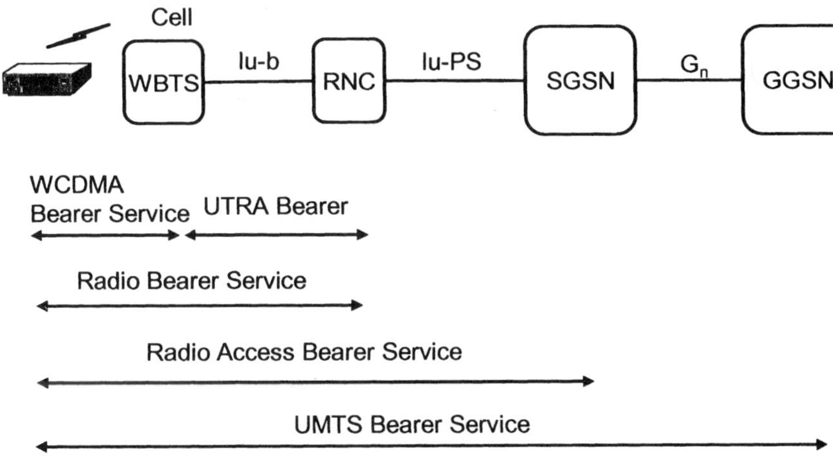

Figure 7.15 Network model for UMTS.

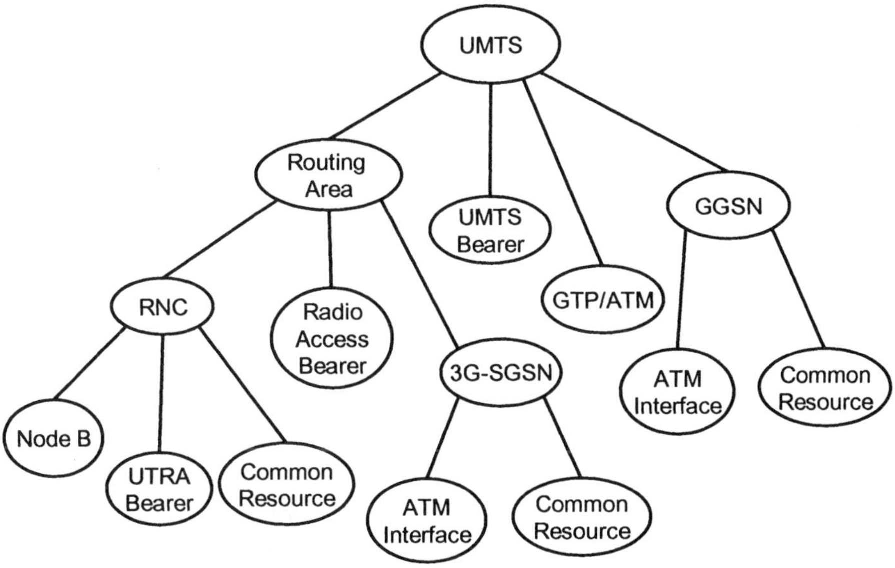

Figure 7.16 UMTS service model.

7.3 QUALITY-MONITORING PERSPECTIVE

The next step in defining the service model is to define the relevant performance data, how it is collected with respect to the service model, and finally how it can be used to solve assurance problems. In this section, we first discuss voice quality metrics, then focus on the data-collection issue.

7.3.1 General Voice Performance Metrics

Managing a VoWiFi and 3G integrated network, in certain ways, is similar to managing a traditional voice service, while in other aspects, it may be completely different. At a high level, the quality of the voice service, whether it is the expectation of end users or the metrics for measuring the quality, will be very similar to traditional telephony. A user experiencing latency larger than half a second will not care whether the underlying technology is a state-of-the-art voice over WiFi/3G service or traditional POTS; he or she will still be unsatisfied since the delay makes conversation difficult. Similarly, users will only accept a shorter time to dial and receive a ring tone rather than longer. Furthermore, people also expect that the phone should always be available to make and receive calls. A malfunctioning phone may only be tolerated if there are occasionally large disasters such as massive power blackouts. Service disruptions caused by human errors or network failures without built-in recovery or tolerance are considered unacceptable. These "metrics," or service quality perceptions, have developed and evolved since there were first telephone services and are therefore independent of networks or new technologies.

While human perception of telephone service quality is unchanged, the network and end devices and the associated technology of delivering the new phone service are vastly different from the traditional circuit-switched or 2G cellular technology. As an example, the advent of digital technology allows the voice signal first to be digitized and compressed into a digitally coded signal, and represented as a sequence of ones and zeros. From the network perspective, these digital signals can be conveniently transported and routed in a packet format. This provides tremendous flexibility and efficiency as it is now economical to transport hundreds of gigabits of digital data across optical fibers reliably and relatively inexpensively. In addition, the ability to mix voice, data, and multimedia signals provides a very flexible and powerful means for introducing many novel and interesting services, of which we are just now scratching the surface. However, the new way of representing and transporting voice calls also brings with it a new set of problems with respect to assuring that the new paradigm can and will deliver equivalent or better quality compared to the traditional telephone network.

In the following, we first review the metrics and methodology for quantifying voice quality.

7.3.1.1 Voice Quality Measures

No matter how a provider advertises its new technology and network, it is very easy for a telephony user to evaluate the quality of a telephone call. However, it is a lot harder to define objectively a set of measurable metrics regarding voice quality, as one would need to model accurately the human auditory system, as well as the psychological and sociological implications of a voice conversion. While these are topics beyond the scope of this book, there are many research results in the literature [7], which can be used as the basis for a practical voice-quality model suitable for the discussion of voice-quality assurance. Figure 7.17 captures the essence of the key metrics for voice quality. The first key metric is mean opinion score (MOS). MOS is a traditional way of quantifying the quality of a digitally coded and decoded voice signal. It is a calibrated, subjective measurement process in which a group of people is instructed to listen to a number of speech phrases before and after degradations then asked to rank the quality on a scale of 1 (worst) to 5 (best). When all the scores are averaged, the result is called the MOS.

For example, toll quality as defined by ITU-T Recommendation P.800 [8] means that an MOS of 4.0 is attained. Although MOS may not sound very scientific or completely reproducible, it is nonetheless surprisingly useful and serves as a reliable way to measure and compare the quality of various voice and network technologies.

The second key voice quality metric is dialing quality. Dialing quality is generally related to the setup delay time, which depends on many factors, including signaling delay, processing time by the switches and feature servers such as number translation in 800 services, or IN services such as prepaid. Dialing quality has been improved substantially over the history of the telephone.

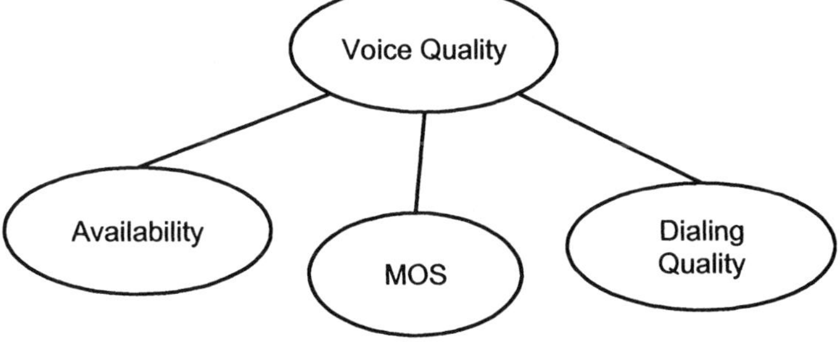

Figure 7.17 Voice quality model.

In PSTN, the key signaling network is the signaling system 7 (SS7) network. The SS7 network is a store-and-forward packet-based network over 64 Kbps transmission links. The cellular network has also adopted the SS7 signaling network as the primary mechanism for carrying call-related messages. In addition, the short message service (SMS), which is the most popular packet service worldwide, is also carried over the same SS7 network.

Regardless of technology, the end user has learned to expect dialing delay to be categorized by the type of call. Figure 7.18 shows that the overall dialing quality depends on the type of call including local calls, toll calls, international calls, and the new SIP calls. In general, the user would expect an average of 3 seconds for basic local calls and an average of 8 seconds for toll calls, and can tolerate an average of 11 seconds for international calls [9]. It is important that the advent of new technology such as the SIP-based networks provide at least an equivalence of dialing quality compared to existing technologies. Since SIP is carried over the IP transmission lines, which can generally have a higher transmission capacity than the 64 Kbps SS7 links, it is expected to have an advantage over traditional networks regarding dialing quality. More on the assurance and diagnosis regarding setup delay will be discussed in Chapter 8.

The third key voice quality metric is availability. This is actually the most critical quality measure, as one cannot discuss other quality metrics when the service is not available. Availability may appear to be less important to PSTN users as they have come to expect that the network is almost always available. The same may not be true in cellular, IP networks, or cable networks.

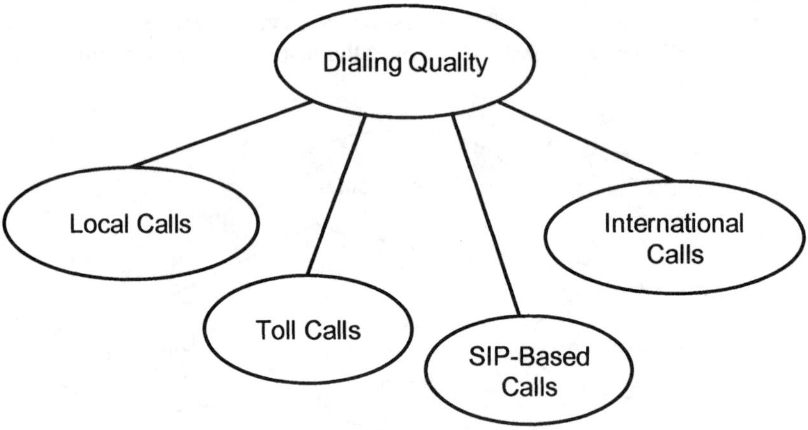

Figure 7.18 Dialing quality expectation.

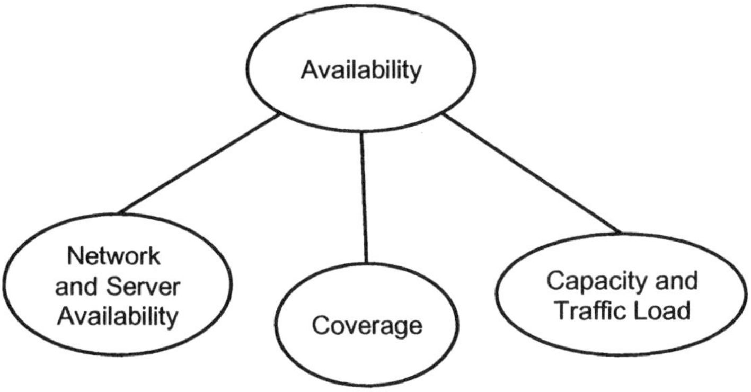

Figure 7.19 The availability metric.

Figure 7.19 shows that availability depends on multiple metrics, including the availability of the network and associated servers (e.g., HLR, SIP proxy server). Coverage is related to cellular or WiFi networks, where blindspots (signal strength is weak) will render the service unavailable. Another critical measurement is call failure as a result of handoff of calls between cell sites, routing areas, or between cellular and WiFi locations. It is important that the assurance system has the ability to identify the type of problems that impact the availability of the overall service. Finally, the network capacity and the offered traffic load can also cause severe blocking of call setup or even termination of an ongoing call and must be monitored separately for root-cause identification or for network and traffic planning purposes.

Availability of the voice service is directly tied to the availability of the network. More on the use of the service model to monitor service and network availability will be discussed later in this chapter.

7.3.1.2 Practical Measurement of Mean Opinion Score

As mentioned earlier, while MOS is a useful and commonly agreed-on method to describe voice quality, it is a subjective measurement and cannot be readily used to monitor voice quality for assurance operations. A more useful metric for voice quality assurance is one that tells what the end user perceives and also provides a good indication of where the possible problem is when the quality degrades. To derive such a measure, it is useful first to understand the underlying factors that affect the quality of voice. Figure 7.20 shows the underlying factors that impact voice quality as perceived by the end user. One can relate the voice quality as measured in MOS, which is affected by the following key factors:

- Codec-related performance;
- Network impairments.

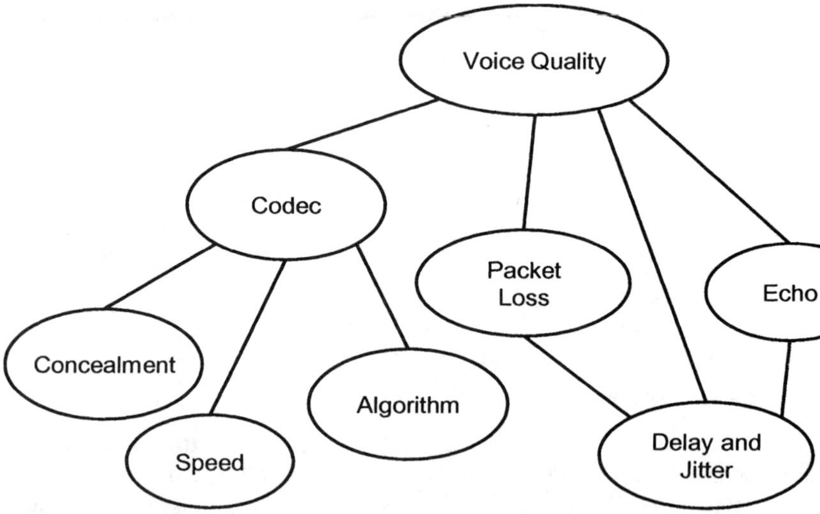

Figure 7.20 Underlying factors impacting the quality of voice.

There is a large number of coding and decoding algorithms, which vary in the ways they compress the digitized voice signal. The quality of compression depends also on the coded bit rate and the sophistication of the error concealment mechanism, in addition to the coding algorithm itself and the amount of back-to-back coding and decoding in an end-to-end call. Codec quality is a matter of system and design choice, in which OSS has little influence directly.

The second factor that impacts the voice quality, network impairments, is related to network quality and tightly coupled with assurance operations practices. Network impairments include packetization delay, packet-loss ratio, network-induced delay and jitter, codec playout buffer delay, echo compensation, and the residual echo level. The relationship of these impairments among themselves and their impact on voice quality are illustrated in Figure 7.21. Note that delay and jitter not only affect voice quality directly, but they also indirectly affect the packet-loss rate and echo level. When a playout buffer suffers from depletion or overflow, it has to repeat or drop voice data since the playout data stream has to sustain a constant data rate. In either case, there is data interruption that can be very disturbing. The user may hear sudden interruption of speech, depending on the sophistication of the error concealment deployed, coupled with an annoying clipping sound effect (a more detailed description is given in Chapter 8). Similarly, the level of delay and jitter affects the effectiveness of echo cancellation, and residual echoes, become extremely annoying if larger than 100 ms. The need for the voice- modeling capability is readily recognized in the research and standards arena. The following section describes the ITU standardized E-model, which captures many of the criteria described above.

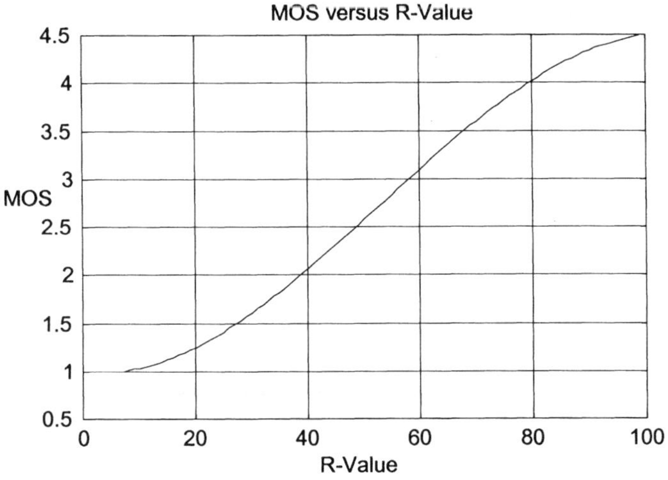

Figure 7.21 MOS to R-value mapping.

7.3.1.3 The E-Model

The E-model is recommended by ITU G.107 [10]. In the E-model, the evaluated speech quality is summarized by a single number called the R-value. The R-value ranges from 0 (worst quality) to 100 (best quality). The relationship between the R-value and MOS is given by the following equation and shown in Figure 7.21.

$$\text{MOS} = 1 + 0.035R + 7 \times 10^{-6} R(R - 60)(100 - R) \qquad (7.1)$$

The E-model incorporates impairments composed of two key parameters:

Ie effect of equipment (e.g., codec)
Id end-to-end delays

Various impairments, including codec, network delay (one way and round-trip), jitter, echoes, data loss due to network or other means, are captured in *Ie* and *Id.* Calculation of *Ie* and *Id* is rather involved, but the overall R-value is expressed as

$$R = Ro - Id - Ie \qquad (7.2)$$

The default value of *Ro* is 93.34 when there is no degradation due to network or equipment. The inherent degradation of (100 − *Ro*) is a result of converting an actual spoken conversation to an electrical signal and back to an audio waveform. As shown in Figure 7.21, *Ro* corresponds to an MOS of 4.5.

7.3.1.4 Codec Algorithms and Performance

A speech coder and decoder (codec) is used to convert an analog voice signal into a digital representation and transported via a digital medium. There are generally two types of codecs:

- *Waveform-based codec* samples the voice waveform and codes each sample with 8 bits of quantization codes. If the sampling rate is 8 kHz, the resulting coded rate is 64 Kbps. There is no knowledge of how the waveform is generated and compression rate is usually low.
- *Linear prediction coding (LPC)* takes advantage of how the human voice is created. It uses a voiced/unvoiced model of human speech and uses a digital linear filter to model the human vocal tract. The filter is excited by a computer-generated signal representing the voiced or unvoiced signal. The filter and the excitation are then represented in a compact set of parameters, which are to be transmitted to the decoder. The decoder can then reconstruct the speech based on the received parameters. LPC vocoders are usually much more efficient with respect to the required bandwidth, but are usually more susceptible to network-induced errors.

Based on these two coding types, there are dozens of algorithms available. Compressed capacity ranges from the little compression of pulse code modulation (PCM, 64 Kbps) to high compression, such as algebraic code excited linear prediction (ACELP, 5.3 Kbps). Usually, high compression has lower quality and is more susceptible to network impairments such as packet loss. As shown in Table 7.1, many popular codecs give toll quality under little network degradation. Table 7.1 shows the relationship between various codec algorithms and corresponding parameters. Referring to Table 7.1, the PCM technology is a technique based on scalar quantification of the voice stream. The analog voice signal is directly coded in binary format. Quantification may be uniform or nonuniform, depending on the application. The PCM method was first defined in ITU standard G.711. It is based on the modulation of coded pulses, and uses 64 Kbps. After nonlinear compression is applied, the amplitude of samples is quantified over 8 bits.

Table 7.1

Codec Algorithms and Performance

Codec Algorithm	*Speed (Kbps)*	*Coding Delay (ms)*	*MOS*	*IP Bandwidth (Kbps)*	*Deployments Example*	*E-Model R-Value*
G.711 (PCM)	64	0.125	4.4	80	Yahoo Messenger	90–100
G.721 (ADPCM)	32	0.125	4.2	45	Telephony	80–90
G.728 (LD-CELP)	16	0.625	4.2	32	Low-delay version	80–90
G.729 (CS-ACELP)	8	15.0	4.2	24	Cellular, PCS, VoIP	80–90
G.723.1 (ACELP) (MP-MLQ)	5.3 6.3	37.5	4.2	16.27	IS-136 VoIP	60–70
AMR	4.75–12.2	20.0	3.0–4.0	10–25	GSM WCDMA PoC, VoIP	60–90

ITU standard G.721 defined a 32 Kbps coding method called adaptive differential pulse code modulation (ADPCM). This is also a form of waveform coding, but rather than measuring the sampling amplitude, as in PCM, this method quantifies the difference between the amplitude and a predetermined value, which is predicted based on an adaptive digital filter. This allows removal of significant redundancy information carried in PCM codecs and is thus more efficient.

ITU standard G.728 is called a low-delay code excited linear prediction (LD-CELP) coder. The algorithm takes five voice samples (at 8 kHz) and assigns a codebook vector of 10 bits for each of the five samples. Hence, the delay is 0.625 ms, and the coding rate is 16 Kbps. LD-CELP has a MOS of about 4.2 and is comparable to ADPCM quality.

ITU standard G.729 uses conjugate structure algebraic code excited linear prediction (CS-ACELP) and codes 80 voice samples (10 ms) as a frame. It uses 5 ms of look-ahead samples and thus incurs a total delay of 15 ms. It transmits 80 bits of LPC parameters and codebook indexes in 10 ms and achieves a coding rate of 8 Kbps. Similar to G.728, it also has a MOS of 4.2.

ITU standard G.723.1 is based on two algorithms. The algebraic code excited linear prediction (ACELP), an LPC voice compression algorithm, has a coding rate of 5.3 Kbps while the multipulse maximum likelihood quantization (MP-MLQ) has a coded rate of 6.3 Kbps. It is ideal for multiplexing applications, can handle dual-tone multifrequency (DTMF) codes, and provides a low-cost solution to maintaining voice quality in high-traffic networks. The ACELP comfort noise version also offers bad or lost packet interpolation and reduced bandwidth during silence. The quality of ACELP voice has been extensively tested, with results indicating that it is equal to or better than the industry-standard 32-Kbps ADPCM. ACELP has a MOS of approximately 4.2, which is in the toll-quality range.

The adaptive multirate (AMR) codec is defined in GSM specification 06.90. It has eight modes of operation. At the highest rate of 12.2 Kbps, it is the same as the GSM enhanced full rate codec using regular pulsed excited long-term predication (RPE-LTP) algorithm. When used at the lowest rate of 4.75 Kbps, it is equivalent to the IS-641 codec used in TDMA cellular systems. Its MOS ranges from 3.0 to 4.0 depending on the rate.

7.3.2 Service- and Network-Level Monitoring

After defining the voice-quality metrics in the last section, we now focus on how to collect necessary data from both the end-user device and the network. Ideally, the performance monitoring system should not interfere with the actual service. It should also satisfy the requirements that the collected performance parameters reflect the user's experience of the service. In addition, there should be enough details in the collected data that further statistical analysis is possible. The analysis aims at answering the following questions:

- What is overall the voice quality perceived by end users?
- Does the quality fluctuate depending on time of day or location?
- How do we locate network problems that cause degradation of service quality?
- If multiple providers are involved in the service delivery, how is the responsibility of guaranteeing end-to-end quality achieved?
- How do we detect and deal with traffic-load and network-resource-allocation problems?
- What are the priorities of various service-affecting problems? How many users are affected?

In this section, we focus on the monitoring structure and related protocols. Because a monitoring structure is usually tightly related to the key protocols and the corresponding communications layers, we start with a description of the VoIP-related protocols. The monitoring structure and further analysis are given in the subsequent sections.

7.3.2.1 Protocols for Transport of Voice Packets

The two most popular protocols in the Internet are User Data Protocol (UDP) and Transmission Control Protocol (TCP). UDP provides simple encapsulation, source and destination port addresses, and simple checksum functions. There is no acknowledge, lost-packet detection, or other sophisticated control functions in UDP. UDP is usually used for sending short control or management data such as Internet Control Message Protocol (ICMP) packets (which PING is based on) or Dynamic Host Configuration Protocol (DHCP) requests (for requesting IP addresses). TCP, on the other hand, is a reliable protocol. In addition to the UDP header functions, it also provides functions including data session setup and teardown, end-to-end data-unit acknowledge, and a rather complex congestion window-based flow-control mechanism. At the time of this writing, TCP is still the dominant protocol on the Internet, constituting over 70% of all Internet traffic.

Although TCP is the more sophisticated protocol compared to UDP, it is not suitable for carrying VoIP packets. The major problem is that because of its built-in acknowledgement and flow-control functions, the incurred end-to-end delay is too long to be acceptable for transporting voice. UDP meets the delay requirement for voice transport, but is not sufficient alone to satisfy requirements such as source timing information, synchronization of various media sources, or detection of lost packets. To fill these gaps, the RTP is used over UDP for the transport of voice (or other multimedia) packets.

7.3.2.2 Real-Time Protocol

The Real-Time Protocol (RTP) [11] has two parts: the RTP Data Transfer Protocol, and the RTP Control Protocol (RTCP) [11]. RTP is used to carry the payload of the multimedia source. Its payload format is shown in Figure 7.22, which includes the following fields:

- *V—Version number:* Set to 2 for current RTP version.
- *P—Padding indicator:* If set to 1, this indicates that there is padding. The pad length is identified by the last octet of the RTP packet. Pad length can be 1 or more octets.

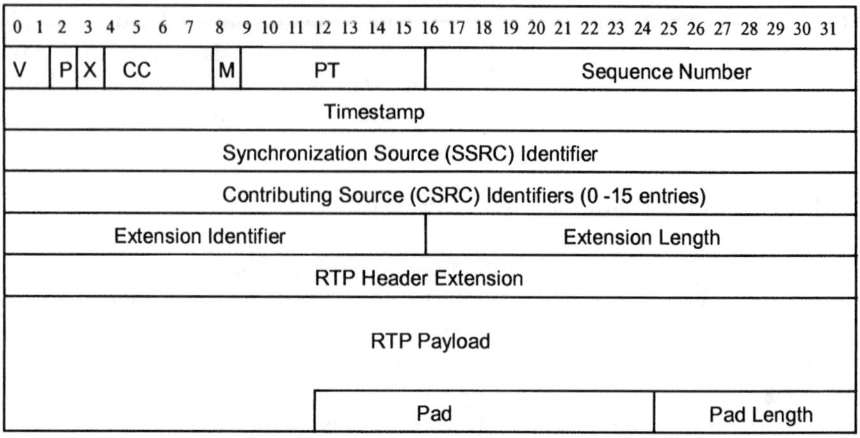

Figure 7.22 RTP payload format.

- *X—Extension:* If set to 1, the fixed header is followed by a header extension.
- *CC—CSRC Count:* Number of contributing source identifiers.
- *M—Marker:* Interpretation is defined by the profile. Example of use includes marking the beginning and ending of a talk spurt.
- *PT—Payload type:* Indicates format of the RTP payload.
- *Sequence number:* A random number for the first RTP packet that increments by 1 in subsequent RTP packets.
- *Timestamp:* Indicates the instant when the first sample in the payload was generated. Timestamp can be used to compute jitter as well as reconstruction of the sampling frequency of the source.
- *Synchronization source (SSRC):* Identifies the source and is globally unique within an RTP session. Packets from the same SSRC form part of the sequence and the timing space. Examples of an SSRC are packets from a VoWiFi SIP phone, or the video stream from a video camera. In the case that the RTP stream comes from the output of a mixer, the SSRC identifies the mixer and not the original media source.
- *Contributing source (CSRC):* A source of a stream of RTP packets that has contributed to the combined stream produced by an RTP mixer. An example is an audio conference where the original SSRCs of RTP voice streams from the conference participants are carried as CSRCs of the mixer output.
- *Extension identifier:* Defined by specific profile.
- *Extension length:* The length of the extended RTP header.

- *RTP header extension:* Defined by a specific profile.

Real-Time Control Protocol (RTCP)

RTCP is a companion protocol that comes with the RTP. Its purpose is to provide end-to-end feedback information about the data quality to all participating sessions. With a distribution mechanism like IP multicast, it also supports use as a network management tool for a network service provider that is not otherwise involved in the session to receive the feedback information and act as a third-party monitor to diagnose network problems. The RTCP therefore provides a fundamental monitoring mechanism for the VoWIFI/3G application.

As with RTP, RTCP also rides on top of UDP and uses the same data distribution mechanism. The underlying protocol provides multiplexing of the data (RTP) and control (RTCP) packets, for example, by using separate port numbers with UDP. RTCP provides the following five types of control information:

- Sender report (SR): for transmission and reception statistics from participants that are active senders;
- Receiver report (RR): for reception statistics from participants that are not active senders;
- Source description (SDES): including canonical name (CNAME);
- BYE: which indicates end of participation;
- Application-specific functions.

Sender Report: For the VoWIFI/3G application, the most important type is the sender report. Its format is shown in Figure 7.23. It has three sections: header, sender information, and report blocks. The header and the sender information are common to all sender reports, and report block depends on the number of sources.

The header section contains the following fields:

- *V—Version:* Set to 2 for current RTCP version.
- *P—Padding:* The last octet indicates the number of padding bytes.
- *RC—Reception report count:* Number of reception blocks carried in this packet
- *PT—Payload type:* Set to 200 for SR packet.
- *L—Length:* The length of this RTSP packet minus 1 (so that zero is a valid length) including header and padding.

The next section contains the sender information as follows:

- *SSRC:* Synchronization source identifier of the originator of this SR packet.
- *Network Time Protocol (NTP) timestamp:* Indicates the time elapsed since 00:00 January 1, 1900, when this report was sent. It may be used in combination with timestamps returned in reception reports from other receivers to measure round-trip propagation to those receivers.
- *RTP timestamp:* Corresponds to the same time as the NTP timestamp, but in the same units and with the same random offset as the RTP timestamps in data packets. This correspondence may be used for intra- and inter-media synchronization for sources whose NTP timestamps are synchronized, and may be used by media-independent receivers to estimate the nominal RTP clock frequency.
- *Sender's packet count:* Total number of RTP packets sent since beginning of session to the time the SR is sent.
- *Sender's octet count:* Total number of octets sent since the beginning of session to the time the SR is sent.

The third section contains zero or more reception report blocks, depending on the number of other sources heard by this sender since the last report. Each reception report block conveys statistics on the reception of RTP packets from a single synchronization source:

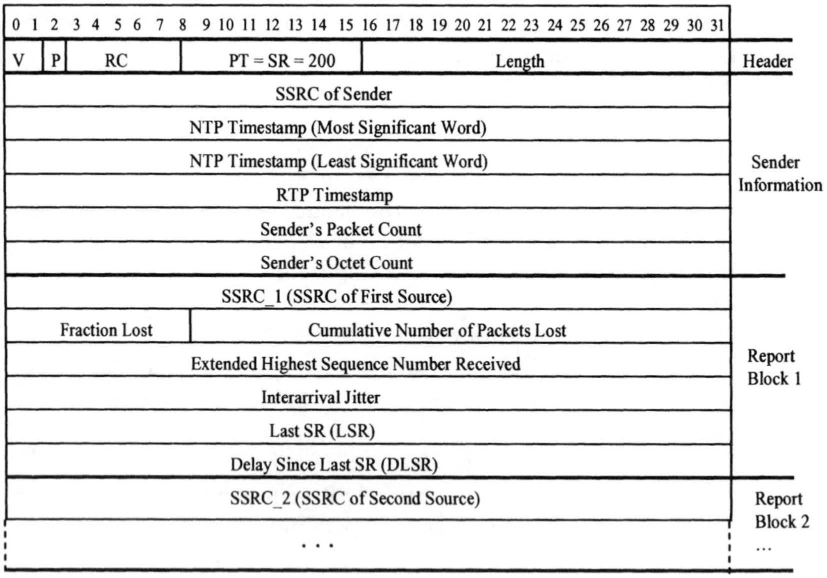

Figure 7.23 RTCP sender report.

- *SSRC_n:* The SSRC identifier of the source to which the information in this reception report block pertains.
- *Fraction lost:* The fraction of RTP data packets from source SSRC_n lost since the previous SR or RR packet was sent.
- *Cumulative number of packets lost:* The total number of RTP data packets from source SSRC_n that have been lost since the beginning of reception. This count is computed from the expected received packets minus the actual received packets, including duplicated and out-of-order packets. Therefore, a negative count is possible.
- *Extended highest sequence number:* The sequence number in the last RTP packet from SSRC_n.
- *Interarrival jitter:* An estimate of the statistical variance of the RTP data packet interarrival time, measured in timestamp units, and expressed as an unsigned integer. The interarrival jitter J is defined to be the mean deviation (smoothed absolute value) of the difference D in packet spacing at the receiver compared to the sender for a pair of packets. This is equivalent to the difference in the relative transit time for the two packets; the relative transit time is the difference between a packet's RTP timestamp and the receiver's clock at the time of arrival, measured in the same units.
- *Last SR:* The middle 32 bits out of 64 in the NTP timestamp received as part of the most recent RTCP SR packet from source SSRC_n. If no SR has been received yet, the field is set to zero.
- *Delay since last SR (DLSR):* The delay, expressed in units of 1/65,536 seconds, between receiving the last SR packet from source SSRC_n and sending this reception report block. If no SR packet has been received yet from SSRC_n, the DLSR field is set to zero.

Receiver report (RR): The RR is issued by session participants who receive RTP packets but do not send their own RTP packets. One application of RR is for the participants in a multicast scenario. The format of an RR packet is the same as that of the SR packet except that the packet-type field contains the constant 201 and the NTP and RTP timestamps, and the sender's packet and octet counts are omitted. The remaining fields have the same meaning as for the SR packet.

SDES: The SDES provides information about the RTP source and the identification of the session participants. Its format is shown in Figure 7.24. The first four octets have the same field as that of the SR (Figure 7.23), except that the payload type has a constant of 202. After the header, the SDES has chunks of information identifying an SSRC or CSRC, followed by session participant information such as name, e-mail, or phone number. All these SDES items are optional except for the CNAME, which has the format of user@host.

0 1	2	3 4 5 6 7	8 9 10 11 12 13 14 15	16 17 18 19 20 21 22 23 24 25 26 27 28 29 30 31
V = 2	P	RC	PT = SDES = 202	Length
SSRC/CSRC_1				
SDES Items (CNAME, ...)				
SSRC/CSRC_2				
SDES Items (CNAME, ...)				

Figure 7.24 SDES format.

RTCP BYE packet: The BYE packet is used to signify the termination of the RTP session. It contains an optional field regarding the reason for the termination.

Application-defined RTCP packet: The application-defined packet is intended for experimental use as new applications and new features are developed, without requiring packet-type value registration.

RTCP Encryption

Either RTP or RTCP packets can be encrypted to achieve confidentiality. For RTCP, a 32-bit random number is prepended to the RTCP packet before encryption to deter known plaintext attack. For RTP, such precaution is not necessary since the time stamp and sequence number provide the randomness.

For RTCP, it is allowed to split a compound RTCP packet into two lower-layer packets, one to be encrypted and one to be sent in the clear. For example, SDES information might be encrypted while RRs are sent in the clear to accommodate third-party monitors not privy to the encryption key.

7.3.2.3 RTCP Extension for VoIP

RFC 3611 [12] defines an extension to RTCP to include seven extended report blocks, including block types for reporting upon received packet losses and duplicates, packet reception times, receiver reference time information, receiver inter report delays, detailed reception statistics, and VoIP metrics. In this section we focus on the VoIP RTCP extension as it is directly relevant to the management of VoWIFI/3G applications. Refer to RFC 3611 for a description of other extensions. The general RTCP XR format is shown in Figure 7.25. It has the same general format as other RTCP packets. The header is the same as other RTCP headers except that the payload type is 207 for XR reports. The first SSRC (in the second 32-bit word) is the sender of the RTCP XR report. For VoIP metrics, the report block type (BT) is 7. The block length is 8 (four-octet words).

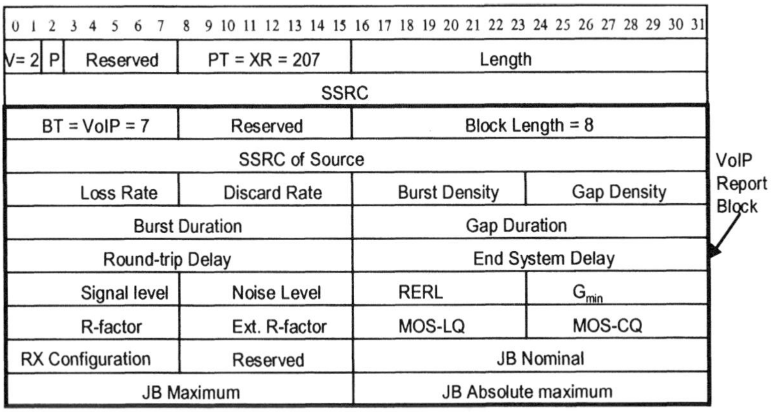

Figure 7.25 RTCP XR VoIP format.

The SSRC inside the VoIP XR block refers to the SSRC of the RTP source (i.e., the voice caller in our VoWiFi/3G application). The metrics defined in the VoIP XR packets include the following:

- *Packet loss rate:* Packets are lost in the network.
- *Packet-discard rate:* Packets are discarded due to being late or because the receiving buffer experiences overflow or underflow.
- *Burst metrics:* These include burst density, gap density, burst duration, and gap duration. Burst is the period during which the packet discard rate is high. On the contrary, a gap is the period where the discard-rate is low. A burst does not contain G_{min} consecutive undiscarded packets. RFC 3611 recommends a value of 16 for G_{min}. By choosing G_{min} to be 16, the resulting gap will correspond to good signal quality.
- *Delay metrics:* These include round-trip delay and end-system delay.
- *Signal metrics:* These include signal level, noise level, and residual-echo return loss.
- *Call quality metrics:* R-factor [12], Ext R-factor MOS for listening quality (MOS-LQ), and MOS for conversational quality (MOS-CQ) [12].
- *Configuration parameters* [12]: These are the gap threshold (G_{min}) and the receiver configuration byte, which includes loss concealment, jitter buffer adaptive, jitter buffer rate, jitter buffer nominal, and jitter buffer absolute maximum.

7.3.3 Quality Monitoring Using RTCP in VoWiFi/3G

The RTCP XR [12] for VoIP defines the attributes on a per-SIP-session basis that reflects the quality of the call. However, RTCP itself is designed for collecting performance information without explicitly designating how the RTCP data is to be collected by a network or service management system. Currently, at least two methods of collecting of RTCP XR VoIP data have been proposed in IETF. These methods address similar data-collection needs of a management system but may be applied in different ways. The first method is via a traditional SNMP MIB, as defined in an Internet draft [13]. The second method of conveying RTCP VoIP metrics is via a new approach based on SIP event package [14]. Although both methods carry similar VoIP quality information, the way to collect the data is very different and subsequently has different implications on whom the data is collected for. The mechanism of data collection and the pros and cons are described below.

7.3.3.1 The RTCP XR SNMP MIB Approach

The SNMP MIB is defined with respect to a SIP session. The objects defined in the RTCP XR VoIP MIB can be collected by any SIP host system. This MIB defines the following five tables:

- Voice termination point;
- Parameters associated with voice coders;
- Call records with call identifying and quality information;
- Extended call records with additional metrics;
- Termination point groups with one entry per logical group.

The SNMP MIB objects reside in an SIP endpoint or in a monitoring agent that collects performance data for a group of SIP sessions. In addition, traps can be defined to send alerts from either hosts or monitoring agents to network or service management systems when voice quality falls below a certain threshold. When the notification mechanism is enabled, SNMP traps can be sent to management stations with respect to a single session or a group of sessions.

In the VoWiFi/3G scenario, each handset can potentially implement an SNMP MIB that can be periodically polled by a monitoring or device management system. This will be a very useful feature for monitoring the performance of all voice calls. However, there are a few concerns with using the MIB in this way. First, having an SNMP MIB in every handset increases the cost substantially. Second, sending SNMP data may cause further resource issues in radio networks where capacity is scarce. Finally, there may be a security issue if the SNMP agent is located inside a firewall since external SNMP clients will not generally be allowed to access SNMP data inside the firewall. Because of these concerns, the RTCP XR MIB will be more suitable for managing objects inside the network rather than for managing application-level objects from the SIP endpoints.

7.3.3.2 The RTCP XR SIP Event Package Approach

The IETF draft RFC on the SIP service quality reporting system defines another way to report SIP session quality to data-collection system. It proposes to use the SIP event package that is based on the SIP PUBLISH method. In this scheme, RTCP XR VoIP metrics are collected and published by the SIP UA at the end of a SIP session as illustrated in Figure 7.26.

The event package is called *svcqual*, which is made up of a list of metrics derived from the RTCP XR on VoIP metrics. As defined in [14], the PUBLISH method can also be used to report events when certain metrics cross a threshold during the SIP session. This facilitates real-time reporting of performance problems on a per-session basis, even before the session is over. This in-session alerting capability is not usually supported in SNMP. Overall, the SIP method is more suitable as an application layer reporting system compared to SNMP, which is traditional for network management rather than service management.

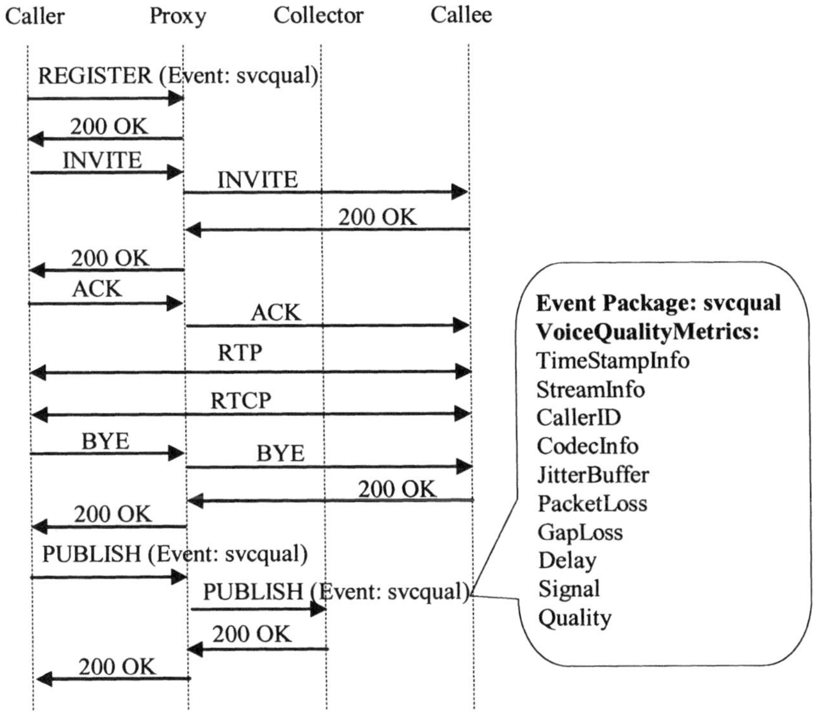

Figure 7.26 Reporting voice quality metrics using SIP PUBLISH.

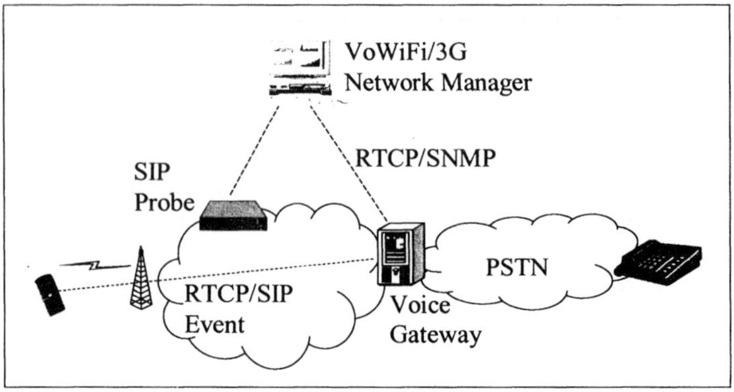

Figure 7.27 Use of SIP event and SNMP.

In the VoWiFi/3G scenario, one may consider using a combination of both SIP event reporting as well as the SNMP approach, as shown in Figure 7.27. A SIP event notification has advantages when used between each SIP UA and a data-collection system. After collecting relevant voice-quality metrics, the data-collection system can organize the performance data in the SNMP MIB, which is to be used by the network and service management systems, where the frequency of reporting is less critical.

7.3.4 Critical Monitoring Points

When a service problem is detected either via customer feedback or service-level monitoring reporting such as RTCP, the source of the problem is usually unknown. Network operators will get a lot of insight if they can monitor and observe different critical points in the network, which correspond to various protocol layers of the end-to-end service. The monitoring should include both the data path and the signaling path. Monitoring can be in the form of:

- A time series display of a particular attribute (e.g., throughput, loss);
- A statistic summary of a KPI (e.g., number of failed calls, total number of PDPContext setup attempts);
- Alerts for KPI threshold crossing;
- Alarms from equipments (e.g., SNMP trap from a router interface);
- Active testing results (e.g., IP Ping, emulated SIP UA).

These monitoring results provide valuable information for the analysis of the root cause of a problem. They also supply performance data for future traffic analysis and network planning purposes. By mapping the monitored data into the

VoWiFi/3G service model, it is possible to organize the observed data and relate the critical information to solving a business problem such as SLA violation, supporting a help desk regarding customer complaints, or performing a revenue assurance function such as checking the call detail record. The amount of data that needs to be collected depends on the applications and resources available since monitoring also requires equipment, management, and technical resources to manage the collected data.

By taking a top-down approach with guidelines derived from the service model, it is possible to design a monitoring infrastructure to meet near-term needs and be expandable when new services are deployed. Figure 7.28 shows the critical monitoring points of a VoWiFi/3G network. The monitoring structure follows closely the architectures of UMTS, IMS, and 3GPP-WLAN interworking, and illustrates the logical paths for a SIP voice call from a UMTS UTRAN to a WiFi location via a Multiprotocol Label Switching (MPLS) network. Although the service is a dual-mode voice service, the voice path shown in Figure 7.28 is focused on the end-to-end SIP call (omitting the circuit-switched part).

To have proper observation of critical sections of the end-to-end voice path, we have identified in Figure 7.28 the following key monitoring points, which are labeled, and correspond to the following numbered list. For each logical monitoring point, we describe the purpose and provide a summary of the functions.

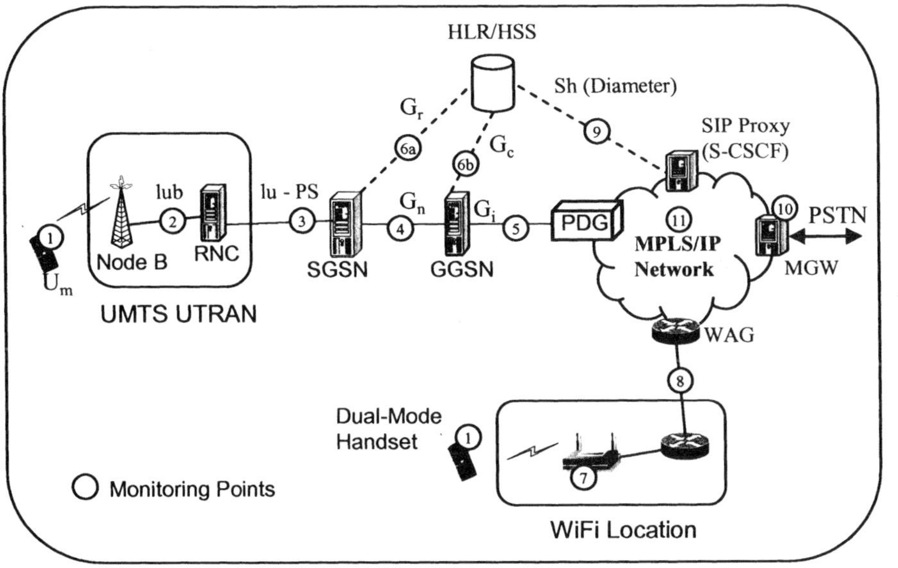

Figure 7.28 Network monitoring points.

1. *SIP UA monitoring:* This is based on RTCP as discussed in the last section. The SIP UA publishes the quality report after the termination of the SIP call. (The collector agent is not shown in Figure 7.28.)
2. *The Iub interface:* This interface is based on ATM Adaptation Layer 2 (AAL2) for the user plane and AAL5 for the network control plane. It supports the establishment and release of a radio link between NodeB and the radio network controller (RNC). It also supports the management of cell management and reporting of measurement information such as power measurements and general error situations.
3. *Iu-PS interface:* This interface is based on the ATM Adaptation Layer 5 (AAL5) protocol and carries messages related to PDPContext setup and teardown, mobility (attach, detach, routing area update), UMTS subscriber identification module (USIM) and ISIM (IMS SIM) authentication messages, and ciphering, as well as the traffic and buffering parameters of the mobile station. The Iu-PS is a critical interface for trouble sectionalization for the following reasons. First, it carries messages and signaling information close to the RF parameters and contains enough detail about the general operations related to the radio domain. It is not as detailed as the Iub interface but captures enough detailed parameters necessary for trouble analysis. It carries all the PDP session signaling messages and mobility messages, capturing two critical functions that are very often the source of performance problems such as call drops during handoffs or high blocking ratios. Moreover, monitoring the Iu-PS interface is much more scalable than it is for the Iub interface since in a typical cellular network, there are two orders of magnitude more Iub interfaces than Iu-PS interfaces.
4. *The G_n interface:* This is an IP over ATM interface and carries the GPRS Tunneling Protocol (GTP) that connects an SGSN to a GGSN. The G_n carries traffic information and provides statistics on a per-access-point-name (APN) basis. From an assurance viewpoint, this interface is less problematic and very often is monitored for collecting traffic information rather than real-time assurance functions.
5. *The G_i interface:* This is also a critical interface as the KPIs give the overall performance and traffic statistics of the entire mobile network. As an example, if one wants to learn the traffic mix of the entire network with respect to application protocols such as RTP, HTTP, TCP, UDP, SMTP, and WAP, the G_i is the ideal place to monitor. This interface is also scalable as there are usually no more than a dozen of them, even in a tier 1 provider's network.
6. *The G_r and G_c interfaces:* These are SS7/MAP interfaces and carry the signaling messages to and from the HLR/HSS for authentication and authorization of the UMTS call. Monitoring G_r is more important as the G_c is an optional interface. Monitoring G_r may provide key knowledge about why a call is not accepted or why it suffers from a large setup time.

7. *WiFi access point monitoring:* This provides information about the performance of the WiFi AP and can provide indicators for detection of wireless interference, signal-to-noise ratios, traffic parameters, including the number of simultaneous users, and the total capacity seen by the AP.
8. *WiFi location monitoring:* This interface captures all the traffic and messages originating and terminating from the WiFi site. In addition to monitoring the IP-level performance parameters, it also captures AAA messages and can be used to check and detect authentication failure causes and statistics.
9. *The Sh interface:* Between the S-CSCF and the HSS. This interface is DIAMETER as defined in 3GPP IMS. The Sh interface supports SIP registering (including authentication and authorization of use of the service).
10. *PSTN gateway monitoring:* This is a key interface for debugging performance problems, including voice transcoding and SIP message problems, and is also a critical point for monitoring RTCP reports.
11. *IP network monitoring:* Monitoring the IP (including MPLS) network is important. However, many cellular operators are beginning to implement their own private IP networks based on MPLS, and at the early stage of 3G deployment, the core IP network is unlikely to be the performance bottleneck.

7.3.4.1 Monitoring with Respect to the Service Model

The logical monitoring points described above give a good summary of how the various layer of the voice call should be monitored. The importance of these monitoring points and their relationship to the service components of the service model are illustrated in Figure 7.29.

The monitored points provide performance data, feeding into the computation of the relevant KPIs corresponding to the service components. Depending on the nature of the assurance problem, the mapping in Figure 7.29 provides guidance on drilling down to relevant monitors for detailed analysis. As an example of problem sectionalization, suppose an end-to-end call experiences severe voice clipping. The diagnostic process may involve examining the performance of the end-to-end path, section by section. Referring to Figure 7.29, the data-transfer sections to be examined include:

- WiFi AP and QoS server;
- Home UTRAN Iu-PS, G_n, and G_i interfaces;
- UTRAN Iub interface;
- MPLS network;
- SIP UA via RTCP.

Once a particular section is identified, more detailed analysis and drill-down diagnosis can be performed.

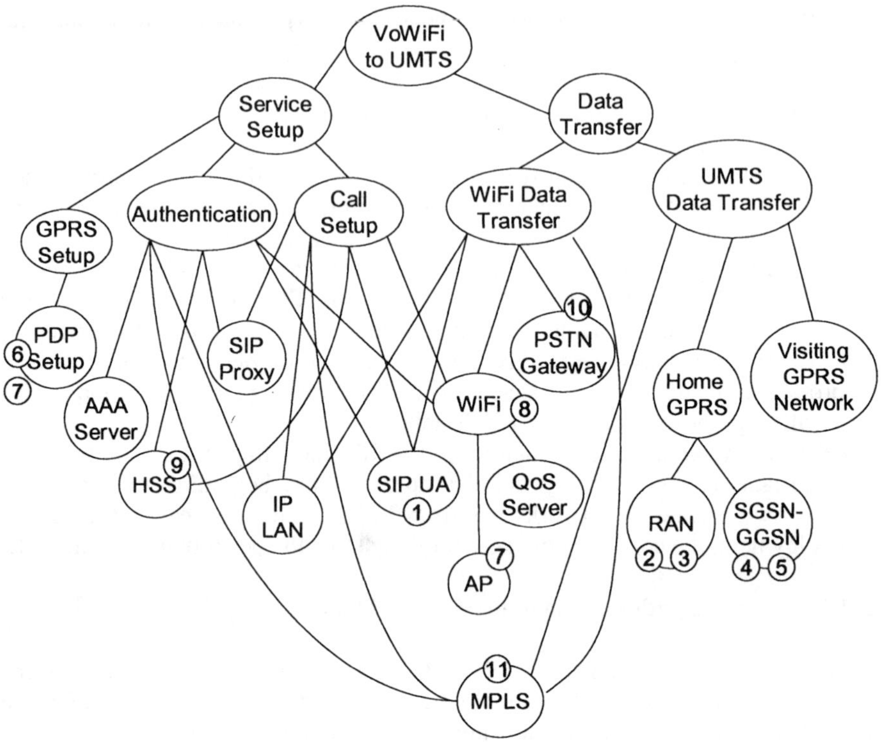

Figure 7.29 Monitoring points and KPIs with respect to the service model.

The monitoring points given in Figure 7.29 are not meant to be exhaustive. For example, we have not shown any monitoring of SIP-level performance or AAA server performance. Depending on the specific need of the provider, the monitors may be implemented in phases. Generally, for problem diagnosis purposes, G_i (5), Iu-PS (4), and the AP (7) monitoring are significant. For service-level problem detection and diagnosis, the PSTN gateway (10) and the RTCP reports (1) are critical. The service model thus provides a high-level view of how the monitoring points can be prioritized.

KQI and KPI

Many metrics and parameters related to voice quality and network performance can be monitored. However, only a few of these metrics are most important from the user perspective. These are the same metrics that providers should focus on with respect to monitoring, analysis, and subsequent optimization. These are also the metrics that provide differentiation of the service or provider and therefore directly affect the generation of revenue and reduction of churning. At the user level, these metrics are independent of the underlying network. However, to ensure that these metrics meet a certain QoS depends very much on how the voice service is implemented and the type of network carrying the voice. Our goal in this section is to identify a critical set of metrics (KPI/KQIs) that corresponds with the components of the previously defined VoIP/3G service model. Using the voice service model defined earlier and in conjunction with the set of KPI/KQIs, we will be equipped with the right tools to tackle assurance operations in Chapter 8.

7.3.4.2 Basic KPIs and KQIs

The basic KPIs and KQIs are shown in Table 7.2. Although these are not exhaustive, they represent a common set of attributes that provide good information about the performance of the service and the underlying network. Many of the techniques for statistical analysis described in Chapter 5 can be applied to subsets of these KQIs and KPIs. When associated with the proper service model, the alarmed KQIs and KPIs and the resulting CSI alerts provide a starting point for trouble localization. As shown in Table 7.2, the illustrated KPIs and KQIs belong to the following categories:

- User voice quality;
- IP network;
- Servers or gateways;
- UMTS network;
- WiFi network.

The objective is to illustrate how the user-level KQIs are related to the KPIs of the service components including servers, gateways, and the network. Readers are referred to a more detailed description in [15], where the KPIs of UMTS are described in detail.

The next section shows an example of how KQIs and KPIs are related to the VoWiFi service model. Further root-cause analysis and assurance operations are discussed in Chapter 8.

Table 7.2

General KPI and KQI

Level	*KPI/KQI*	*Definition*	*Unit*	*Comment*
User	MOS	Mean opinion score	1–5	1 – best, 5 – worst
	R-factor	From E-model	0–100	100 – best
	Postdialing delay	Delay from last digit to receiving ringback	ms or sec	Setup delay; typical few seconds
	Block call	Unsuccessful call due to network resource	%	Typical 1–2 %
User	Drop call	Completed call lost	%	Typical 1–2 %
	Echo	Returned signal bounced from callee to caller	%	Sensitive to latency
	Handoff failure	Drop call due to handing off from one network to another	%	Specific to mobile networks Typical 1–2 %
	Availability	% of time of end-to-end service	%	Typical 99.9%
IP network	Delay	Latency	ms	Typical for IP is 1–100 ms
	Jitter	Delay variation	ms	Peak to peak or average
	Packet-loss ratio	Ratio of packet loss to total number of intended transport packets	%	Typical range: 0.1–20%
	Utilization	Ratio of traffic to capacity	%	0–100%
	Throughput	Bandwidth of an IP connection	Kbps	
Server or gateway	Availability	Percentage of time that server is operational	%	Typical: 99.999%
	Transit latency	Latency in transit time	ms	e.g., gateway

Table 7.2 (continued)

Level	*KPI/KQI*	*Definition*	*Unit*	*Comment*
Server or gateway	Availability	Percentage of time that server is operational	%	Typical: 99.999%
	Transit latency	Latency in transit time	ms	e.g., gateway
Server or gateway	Response time	Latency in processing	ms	e.g., server
	Accuracy	Percentage of error messages or total messages	%	Can be applied to transactions
	CPU loading	Percentage of CPU resource used	%	Abnormally high (e.g., 80%) indicates problems
	Traffic load	Number of messages or transitions per unit time	Unit/hour	1,000 transactions per hour
	Total capacity	Total traffic supported	Unit/hour	# of messages or transitions
UMTS	Attachment latency	PDPContext attachment	ms	Average, peak
	Attachment failure	Ratio of PDPContext failure to total number of PDPContext attempts	%	Average, peak
	DNS latency	From GGSM to DNS server	ms	Average, peak
	DNS failure	DNS conversion failure	%	Average, peak
	Number of PDPContext activation	Total number of PDPContext attempts	Integer	Average, peak
	Number of failed PDPContext activation	Total number of failed PDPContext attempts	Integer	Average, peak
	User data volume	Measured on a per PDPContext basis	Kbps	Average, peak
	Traffic mix	Mix of traffic type observed at G_i	—	TCP, UDP, WAP, SMTP, FTP, RTP, HTTP
	Circuit switched usage	Minutes of voice calls	Erlang	On a per-cell basis
	Packet loss	From U_m to G_i	%	Average, peak

Table 7.2 (continued)

Level	*KPI/KQI*	*Definition*	*Unit*	*Comment*
UMTS	Delay	From U_m to G_i	ms	Average, peak
	Number of failed PDPContext activation	Total number of failed PDPContext attempts	Integer	Average, peak
	User data volume	Measured on a per PDPContext basis	Kbps	Average, peak
	Traffic mix	Mix of traffic type observed at G_i	—	TCP, UDP, WAP, SMTP, FTP, RTP, HTTP
	Circuit-switched usage	Minutes of voice calls	Erlang	On a per cell basis
	Packet loss	From U_m to G_i	%	Average, peak
	Delay	From U_m to G_i	ms	Average, peak
	DNS latency	From GGSM to DNS server	ms	Average, peak
	DNS failure	DNS conversion failure	%	Average, peak
	Number of PDPContext activation	Total number of PDPContext attempts	Integer	Average, peak
	Number of failed PDPContext activation	Total number of failed PDPContext attempts	Integer	Average, peak
	User data volume	Measured on a per-PDPContext basis	Kbps	Average, peak
	Traffic mix	Mix of traffic type observed at G_i	—	TCP, UDP, WAP, SMTP, FTP, RTP, HTTP
	Circuit-switched usage	Minutes of voice calls	Erlang	On a per-cell basis
	Packet loss	From U_m to G_i	%	Average, peak
	Delay	From U_m to G_i	ms	Average, peak
WiFi network	Capacity	Total capacity allowed	Mbps	802.11b – 11Mb/s
	Operating rate	Actual capacity	Mbps	1, 2, 5, 11 Mb/s
	Number of voice call user	Number of simultaneous voice calls	Integer	Typical < 10 for 802.11b

Table 7.2 (continued)

Level	*KPI/KQI*	*Definition*	*Unit*	*Comment*
WiFi network	Authentication failure	Count of AAA failure	Integer	Useful for revenue assurance
	Packet-error rate	Error rate of packet transfer	%	After correction
	Frame-count sequence error	CRC-error ratio	%	Indication of interference level
	Handoff latency	Delay time when traversing from one AP to an adjacent AP	ms	Tens of milliseconds
	Retry-count ratio	Resend of WiFi frames	%	Indication of lack of coverage or interference

7.3.4.3 Example of Use of KPIs and KQIs in the Service Model

Defining KPIs and KQIs with respect to the service model defined previously provides the basic metrics necessary for supporting assurance operation functions. By applying the statistical model to the KPIs and KQIs, operations such as assurance, impact analysis, SLA planning, and monitoring can be defined in a systematic manner. These operations procedures will be investigated further in Chapter 8. An example of how the KPIs are related to the service model is shown in Figure 7.30, which shows the KPIs and KQIs with respect to the VoWiFi service as seen from Operator V (Caller A's visited network) scenario. KPIs for other operators (X, Y, and U) can be a similar extension.

7.4 SUMMARY

The focus of this chapter has been on two fundamental steps toward creating a service model for VoWiFi services. First, we defined a layered model capturing the service, business, and network aspects. This layering approach makes the resulting service model modular and addresses the four key aspects of administrative boundaries, end-user experience, technology domain, and monitoring for assurance. We then illustrated the service model framework with three variations of VoWiFi service, including enterprise, hotspot, and dual-mode VoWiFi/3G integration. While these examples are not exhaustive, they serve as a good indication of how service model components can be combined to satisfy the desirable features necessary for assurance operations.

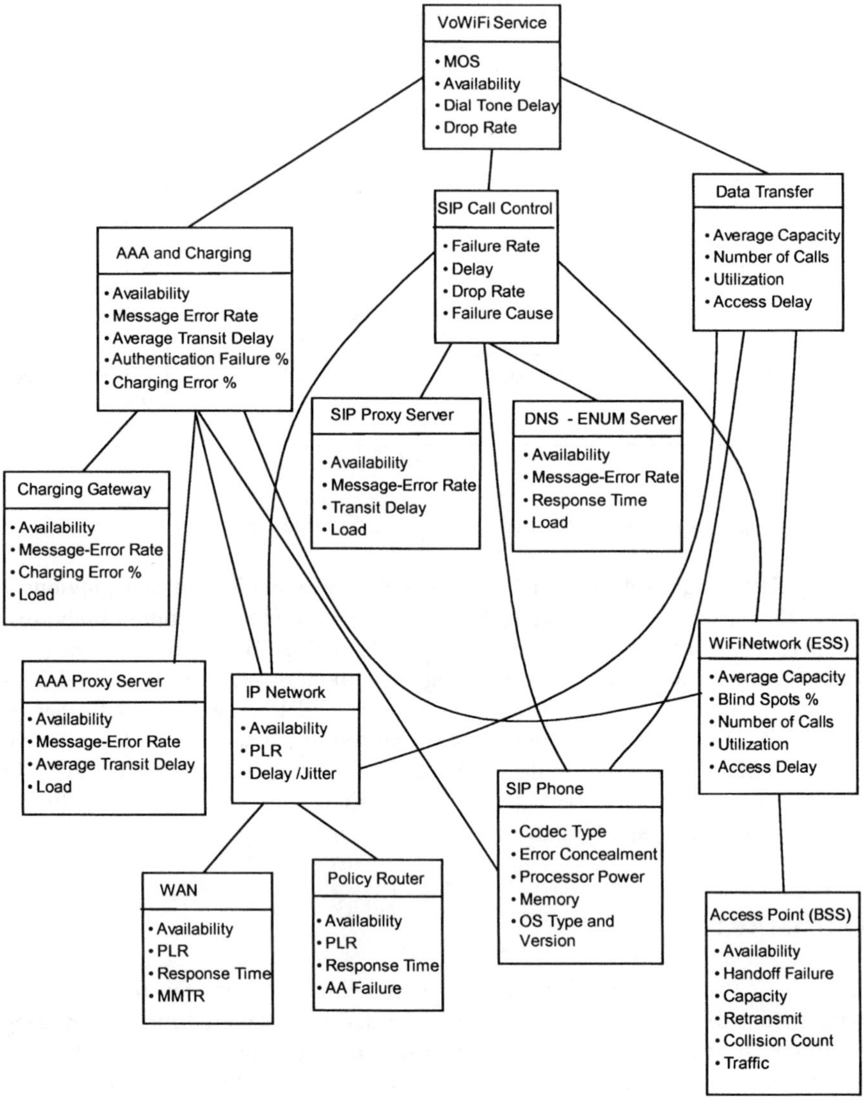

Figure 7.30 KPIs and KQIs in service model for Operator V.

We also focused on the monitoring structure, which is a key building block of the service model intended for assurance operations. We reviewed the key protocols and data-collection architecture and defined the fundamental service-quality metrics and necessary KQIs and KPIs. We provided an example of how these KQIs and KPIs are used in one of the subcomponents in the VoWiFi/3G integration scenario.

The next chapter will build on what we have established and focus on using the service model to address the assurance operation of the dual-mode VoWiFi/3G service. The focus there will be from a mobile carrier viewpoint where, in large-scale operations, efficiency can be a key differentiator for the business.

References

[1] Frankel, J., et al., "HTTP Authentication: Basic and Digest Access Authentication," RFC 2617, June 1999.

[2] 3GPP TS 32.252, "3rd Generation Partnership Project; Technical Specification Group Services and System Aspects; Telecommunication Management; Charging Management; Wireless Local Area Network (WLAN) Charging," (Release 6), March 2005.

[3] 3GPP TS 32.815, "3rd Generation Partnership Project; Technical Specification Group Services and System Aspects; Telecommunication Management; Charging Management; IP Multimedia Subsystem (IMS) Charging," (Release 6), September 2003.

[4] 3GPP TS 29.207, "3rd Generation Partnership Project; Technical Specification Group Core Network; Policy Control over Go interface," (Release 6), April 2005.

[5] Durhamet, E., et al., "The COPS Common Open Policy Service Protocol," RFC 2748, January 2000.

[6] 3GPP TS 25.410, "3rd Generation Partnership Project; Technical Specification Group Radio Access Network, UTRAN Iu Interface, General Aspects and Principles," (Release 6), December 2004.

[7] Papamichalis, E., *Practical Approaches to Speech Coding*, Upper Saddle River, NJ: Prentice Hall, 1987.

[8] ITU-T Recommendation P.800, "Methods for Subjective Determination of Transmission Quality," August 1996.

[9] ITU-T Recommendation E.721, "Network Grade of Service Parameters and Target Values for Circuit-Switched Services in the Evolving ISDN," Telecommunication Standardization Sector of ITU, Geneva, Switzerland, May 1999.

[10] ITU Recommendation G.107, "A Computational Model for Use in Transmission Planning," February 2003.

[11] Schulzrinne, H., et al. "RTP: A Transport Protocol for Real-Time Applications," RFC 1889, January 1996.

[12] Friedman, E., et al., "RTP Control Protocol Extended Reports (RTCP XR)," RFC 3611, November 2003.

[13] A. Clark, and A. Pendleton, "Proposed RTP Control Protocol Extended Reports (RTCP XR) VoIP Metrics Management Information Base," February 2005, Internet Draft, draft-ietf-avt-rtcp-xr-mib-01.txt.

[14] Johnston, A., A. Clark, and A. Pendleton, "SIP Service Quality Reporting Event," October 2004, Internet Draft, draft-johnston-sipping-rtcp-summary-06.

[15] 3GPP TS 32.403: "3rd Generation Partnership Project; Technical Specification Group Services and System Aspects; Telecommunication Management: Performance Management; Performance Measurements – UMTS and Combined UMTS/GSM," (Release 6), March 2005.

Selected Bibliography

3GPP TS 33.203, "3rd Generation Partnership Project; Technical Specification Group Services and System Aspects; 3G Security; Access Security for IP Based Services," (Release 6), March 2005.

3GPP TS 33.102, "3rd Generation Partnership Project; Technical Specification Group Services and System Aspects; 3G Security; Security Architecture," December 2004.

3GPP TS 23.234, "3rd Generation Partnership Project; Technical Specification Group Services and System Aspects; 3GPP System to Wireless Local Area Network (WLAN) Interworking," (Release 6), March 2005.

3GPP TS 25.410, "3rd Generation Partnership Project; Technical Specification Group Radio Access Network; UTRAN lu interface: general aspects and principles," (Release 6), December 2004.

3GPP TS 21.101, "3rd Generation Partnership Project; Technical Specification Group Services and Systems Aspects; Technical Specifications and Technical Reports for a UTRAN-Based 3GPP System," (Release 6), March 2005.

Bannister, J., P. Mather, and S. Coope, *Convergence Technologies for 3G Networks*, New York: John Wiley, 2004.

Jayant, N., and P. Noll, *Digital Coding of Waveforms, Principles, and Applications to Speech and Video*, Upper Saddle River, NJ: Prentice Hall 1984.

Chapter 8

VoWiFi/3G Service Assurance Operations

8.1 INTRODUCTION

In Chapter 7, we described the service model intended for the integrated VoWiFi/3G application. The design of this service model makes use of many of the modeling methodologies discussed in Chapter 5, and it also reflects a lot of the network aspects of the integrated VoWiFi/3G technology detailed in Chapter 6. From a modeling perspective, we have shown how the general service model concept can theoretically be applied to provide operators with a unified approach to solving the complicated service assurance problem.

In this chapter, we take two additional steps to illustrate how the VoWiFi/3G service model is applied in a practical operations environment. First, we look at practical scenarios involving the most commonly encountered service-quality problems of an integrated VoWiFi/3G network described from a user perspective. Some of these voice-quality problems are typical quality problems documented in the literature. For example, while the echo problem is well known and generally well controlled in the PSTN, it reappears differently in a VoWiFi/3G network. The nature of the echo problem changes as a result of the artifacts introduced by the IP packet network. Consequently, many known theories and solutions behind echo modeling and analysis from the PSTN need to be modified to match new network scenarios. Other service problems are manifestations of the problems in the new WiFi or cellular radio environment. Examples include RF interference and handoff-related performance problems. These are relatively new problems that the industry is still grappling to understand. No matter what the causes, having a fundamental understanding of the nature of these problems is a first step toward solving them. Moreover, we believe that having deeper knowledge of these problems will provide insights into how to design a proper framework for the targeted assurance process.

With a good understanding of the nature of the voice-quality problems at both the service and network levels, we will, in the second step, describe how a set of operations best practices can be designed and applied to support the assurance of end-to-end voice quality. We point out that the current state of the art may not realize many of the desirable features of the flow-through assurance process since operations procedures are traditionally put into practice long after the deployment stage. We argue that having a long lag time between service deployment and optimized assurance operations is undesirable, as a smooth operations process is no longer a luxury but a necessary feature that is critical for helping reduce customer churn and assure revenue. In the rest of this chapter, we describe the features of the targeted assurance architecture and explain the role of the service model in supporting the best practices of assurance operation. The discussion builds on the general assurance flows described in Chapter 4 and heavily uses the service model concept with a focus on how it is applied to the VoWiFi/3G assurance problem.

8.2 SERVICE ASSURANCE FOR VOWIFI/3G

Traditionally, the assurance process deals with operations problems such as occasional equipment and network failures; network and facility equipment monitoring; software upgrades and installation; special handling of focused traffic congestion problems; and customer-call-center support. The operations process flow implemented in an operator's NOC is required to deal with large-scale operations needs. For example, a customer call center typically records on the order of 100,000 customer service calls per month. From a network management viewpoint, the number of alarms and alerts generated by a typical tier 1 (over 10 million subscribers) mobile operator's network is generally in excess of 100,000 per day.

8.2.1 Scalable Assurance Operations

For the case of a large-scale deployment of VoWiFi integrated with a 3G network, Table 8.1 gives estimated data that provides an indication of the scale of operations required for supporting a typical tier 1 or tier 2 operator's network. Inferring from Table 8.1, it is clear that NOC operations need to be streamlined with well-defined processes. Moreover, when a new technology such as VoWiFi is deployed, there are additional operational concerns as summarized next:

Table 8.1

Typical Size of VoWiFi/3G Operators

Operator Type	*Number of Cell Sites*	*Number of RNC and BSC*	*Number of HLRs*	*Number of WiFi Sites*	*Number of VoWiFi Subscribers*	*Number of SIP Proxy*	*HelpDesk Calls per Day*	*Number of NOCs*
Tier 1	15,000	300	50–100	2,000	1,000,000	70–100	100,000	5–10
Tier 2	1,500	30	5–10	300	100,000	5–10	10,000	1–2

- New equipment and software usually takes years before software problems are clear. These problems may take a while to detect, and they may behave differently depending on the load of the network. The NOC maintenance engineers will need to expend extra effort to deal with complaints due to the new technology. Vendors providing the new equipment and software will need to get involved in solving complicated network problems.
- Many operational parameters need to be fine-tuned, and there is little expertise with respect to setting these new parameters.
- The industry as a whole is not yet comfortable dealing with the new requirements of packet voice.
- Coordination among peering operators becomes part of the operations assurance flow. This needs to be properly defined and incorporated in the overall process flow.
- It is desirable to implement the new operations process flow as an extension of the existing flow so that minimal operational disruption or retraining is needed.

On the positive side, due to the experience gained in the deployment of packet-voice services over the last 5 years and the commitment of many large providers to deploying new technologies, many operators have gained enough experience and confidence to be ready to offer a VoWiFi/3G service commercially. However, we will see in the remainder of this chapter that there are still many operations issues to be solved before high-quality, flowthrough assurance operations can be achieved. In the following, we first review key performance and assurance practices in the PSTN, then point out a number of operations-related challenges when deploying VoWiFi/3G technology. It is important to understand these challenges since the ultimate targeted assurance OSS will need to deal with them in a scalable fashion.

8.2.2 PSTN Assurance

The PSTN service has been around for over a century, and its performance metrics are well understood. From the service quality perspective, these metrics include the following:

- Availability (dial tone);
- The likelihood of completing a call;
- The likelihood of an established call being dropped;
- The expected time to establish a call;
- The quality of the voice conversation perceived by the end user;
- Proper functioning of various call features.

End-user perceptions of a voice call's quality are independent of the underlying technology. In fact, many of these metrics are identical in the VoWiFi/3G service. In the PSTN, these quality metrics are influenced by factors including:

- Traffic forecasting and proper network dimensioning;
- Voice codec selection;
- Signaling network design and engineering;
- Design of high-availability network;
- End-to-end delay budgeting;
- Echo compensation and control.

Service assurance operations for the traditional PSTN voice network have mostly focused on monitoring and assuring the availability of the network and monitoring the traffic load and network capacity to ensure end-to-end QoS. Assuring network availability includes monitoring and resolution of failures of network elements and transmission facilities and configuration problems such as loop and switch port assignment errors or physical wiring problems. In the traffic engineering area, operations involve proper monitoring of usage statistics and blocking probability and proper dimensioning of network capacity (both bearer and signaling). PSTN engineering usually talks about "five 9s," meaning its availability is 99.999%, which corresponds to approximately 5 minutes of downtime per year. From an end-user perspective, it is indeed true that when one makes a call in the PSTN, there is very little expectation of problems related to network failure or dropping of calls.

8.2.3 New Challenges in Assurance of VoWiFi/3G Service

In the process of migrating from the PSTN to the new VoWiFi/3G network, providers are faced with many challenges. First of all, the IP network introduces packet loss and unpredictable delay. The Internet was initially designed to transport data efficiently. While there have been many efforts in the last decade to make the Internet QoS capable, it is fair to say that the success has been limited. Today, the most reliable method of achieving QoS is still via bandwidth over-provisioning. While this works in some special cases, there are many issues that make over-provisioning impractical as a long-term solution. First, as described in Chapter 4, the trend in the service value chain favors a multiprovider and rather complex retailer-wholesaler relationship. In such a model, it is simply impractical to assume that all the providers and business partners can and will overprovision their respective networks effectively. The SLA among peering partners will become impractical, since it is not easy to define quantitatively the degree of capacity over-provisioning. Second, in many access scenarios where QoS is most needed, there is simply no spare capacity for over-provisioning. A typical example is radio spectrum, including cellular radio and the WiFi ISM band. Keep in mind that radio spectrum is usually the most expensive network asset, and therefore, spectral capacity needs to be used efficiently to ensure proper return on investment (ROI). Finally, when IP packets carrying delay-sensitive and best-effort services are mixed together, an over-provisioning strategy would require an over three- to four-fold increase of capacity. If such additional capacity is not available, users will still encounter service-quality problems.

There is no doubt that the industry as a whole will continue to make the Internet more reliable and more suitable for carrying QoS-sensitive service, but the road to a QoS-enabled Internet may take another 5 to 10 years. In the mean-time, operators cannot afford to wait to begin to offer the new VoWiFi/3G service and make the best use of the newest QoS technologies whenever possible. Before the Internet can be considered a five-9s network, the challenge of IP packet loss, burst loss, packet jitter and latency in excess of tens, sometimes hundreds, of milliseconds, and unpredictable bandwidth needs must be taken into consideration in assurance operations planning.

The second challenge is related to the WLAN technology. The introduction of WLAN technology and widespread WiFi deployment both in commercial hotspots and residential areas is a worldwide phenomenon. While the extension of WiFi to support VoIP is a natural and logical next step, its deployment offers nontrivial challenges. The QoS issues, which have not yet been solved in IP networks, offer another level of complexity when it comes to WiFi networks. The 802.11 MAC layer uses a Carrier Sense Multiple Access/Collision Avoidance (CSMA/CA) protocol [1]. Before an AP or a wireless station transmits data, it listens to the radio link first. If someone else is using the RF channel, it backs off for a random period before trying again. After sending a frame, the wireless terminal will wait for the acknowledgement (ACK) signal. If the ACK is not received after a certain time, the wireless terminal will retransmit assuming that the frame has been lost. The 802.11b CSMA/CA mechanism does not distinguish among different types of traffic. Thus, a bandwidth-hungry application such as file transfer over TCP can dominate the usable RF spectrum for a significant duration and cause performance problems to smaller, delay-sensitive voice packets. This situation is being addressed in the follow-on standard in 802.11e, which is targeted to provide a framework for assuring WiFi QoS.

The 802.11e draft specification defines an enhanced distributed control function (EDCF) that allows a WiFi access point to provide up to eight virtual channels to every end station with simple traffic differentiation capability. Each of these channels has associated QoS parameters in order to ensure the highest-priority channel is transmitted first. This differentiated QoS capability will be very useful for allowing voice packets to gain easier access to the RF spectrum than other less-delay-sensitive data services. However, although 802.11e is actively working on providing much improvement with respect to the support of QoS, it will take some time before such improvement is implemented by the next generation vendor equipment. Moreover, the assurance system must now be ready to deal with a mix of QoS-ready and QoS-unaware APs, making provisioning as well as assurance functions more complicated.

Besides IP-packet-level QoS, another necessary component in supporting VoIP in the WiFi is call admission control. Call admission control ensures that the number of simultaneous calls does not exceed a certain number, depending on the conditions of the WiFi network. It has been shown [2] that a typical 802.11b network may only be able to support approximately 10 simultaneous voice calls using G.711 coding with no silence compression. Beyond this number of simultaneous calls, the WiFi network will be overloaded to the point that the quality of all existing calls may be jeopardized. There is currently no standard on admission control, although there are proprietary solutions offered by vendors. However, the assurance system should be aware of whether admission control is implemented, and make sure that admission control is functioning properly if used.

Another challenging requirement for deploying a VoWifi/3G service is the support of seamless handoffs, both between WiFi access points, and between the cellular and WiFi networks. Handoffs between access points, even within the same extended service set (ESS) [i.e., APs with the same service set identification (SSID)], will incur delay as a result of reassociation of the mobile station with the new AP. This delay can easily be on the order of half a second or longer. When the handoff is across ESSs, the problem is even harder since a new IP address now has to be acquired for the mobile station. The handoff problem between 3G and WiFi networks will be even more complicated. It involves detection of the proximity of the handset from the WiFi access point, signaling between the handset and the network, location update, reauthentication, and the actual transfer of the voice payload between the cellular network and the WiFi/3G network. As we shall see later in this chapter, setting up a call may take seconds. A seamless handoff must take into account all of the above setup delays, yet maintain the call state in such a way that the user never loses connection. Although the components of VoIP, WiFi, cellular, and 3G networks are independently available, it will still be some time before full integration and smooth operations can be achieved.

Each of the brief descriptions above is an oversimplification of the actual complexity. The issues are complicated partly because there are many new network and service requirements as a result of the need to integrate different technologies. Also, some of these issues represent the lack of actual deployment experience. There is no doubt that the industry will solve many of the technical problems over time. However, before the deployment becomes mature and the operations of the entire VoWiFi/3G integrated systems are optimized, the assurance operations system will be even more important. It is therefore highly desirable that the assurance process be put in place at an early stage of service deployment and not as an afterthought.

8.2.4 Desirable Features of Service Assurance

Proper instrumentation for collecting critical performance data is a key aspect of an advanced service assurance infrastructure. A well-designed service model that directly ties user experience and network aspects is also desirable. Putting all these features and intelligence in a coordinated process flow is essential for supporting large-scale and effective operations, which can be the key factor in a successful rollout of VoWiFi/3G service. Specifically, the following set of principles constitutes the design guidelines for an advanced assurance operations system.

- Measure the overall quality of VoWiFi/3G from the end-user perspective. This measurement should be made over a large population of users in order to collect enough data for statistical analysis.
- Detect the problem before the customers do! A well-defined performance alarm and warning system is necessary to alert the operator when something is not working properly.
- Be prepared to work with peering providers using well-defined SLAs, and include the handling of partnering agreements in the assurance operations flow.
- Have a model of how things work, who is being impacted, and a quantitative measurement of the impact of the problem.
- Be prepared to deal with nondeterministic problems, and have a proper statistical model to predict the likelihood of future failures.

In the rest of this chapter, we will explore in greater depth how these principles can be applied and implemented in a practical manner.

8.3 VOWIFI/3G PROBLEM DESCRIPTIONS

This section describes a number of known problems in VoWiFi/3G networks. It provides a categorization of some of the problems and also describes the likely scenarios in which these problems will be observed. However, since the problem space is large and complex, a small change in the network state and conditions may totally change the symptoms of the problem. Consequently, understanding the conditions that lead to the problems provides a framework for tracking the problem space but does not necessarily give a complete solution. More about resolving these problems will be discussed in Section 8.5.

8.3.1 The Echo Problem

Echo is a familiar phenomenon in which the speaker constantly hears her or his own voice after a noticeable delay. Echo is more annoying if it is loud or if the echo delay is longer than 50 ms. When the delay of the echo is less than 25 ms, it is usually not noticeable. Echo is different from a side tone where a portion of the voice signal is intentionally coupled back to the earpiece. Side tone is generated so that the speaker hears her or his own voice and, therefore, infers that the receiver at the other end of the phone call also hears the conversation.

In the PSTN, echoes are generated whenever there are impedance mismatches in the 2-wire to 4-wire hybrid circuits. A hybrid circuit, or simply a hybrid, is a device which terminates a 4-wire (two wires for transmit and two for receive) connection into a 2-wire (both transmit and receive) connection. A hybrid may be located in the local central office or a Private Branch eXchange (PBX) system.

There is also a hybrid in the telephone set where a 2-wire signal from the loop is converted back into a 4-wire signal. As shown in Figure 8.1, the outgoing signal from phone A is leaked back to the return path at the hybrids in local switch A, at the switch in B's local switch, and at phone B. The echo produced at A's local switch is negligible because the delay will be small. However, the echo from the remote hybrids may be noticeable, especially for long-distance calls, where the delay can be over 50 ms. To eliminate or reduce the effect of echoes, echo cancellers are placed in local switch B for echoes affecting phone A. The echo canceller illustrated in Figure 8.2 models the delay characteristics of the local loop of B, where an estimate of the echo, $d_e(n)$, is subtracted from the actual echo, $d(n)$. Echo cancellers can be effectively implemented digitally using digital signal processing, where the delay characteristics of the echo return path are modeled by an adaptive digital delay filter [3, 4]. Usually, due to the short distance between the phone and the location of the echo canceller, a delay filter of less than 20 ms will be sufficient. The echo problem in the PSTN (for local telephone calls) is not very severe since the end-to-end delay rarely exceeds a few tens of milliseconds. For packet voice networks interworking with PSTN, however, the problem is more severe because the IP packet network can potentially incur a delay exceeding 50 ms or even hundreds of milliseconds. In addition, the packet voice codec and the buffer at the VoIP gateway also create delay in excess of a few tens of milliseconds. As shown in Figure 8.3, the echo returning to phone A is likely to be unacceptable.

When there are echo problems, the VoIP gateway is a likely focal point for debugging. Possible root causes include:

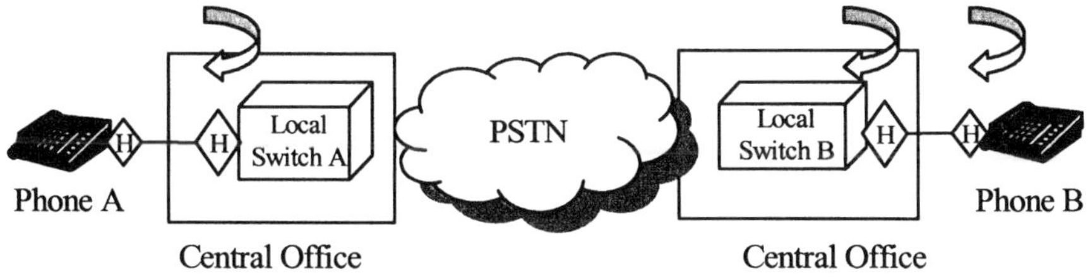

Figure 8.1 *Echoes in PSTN.*

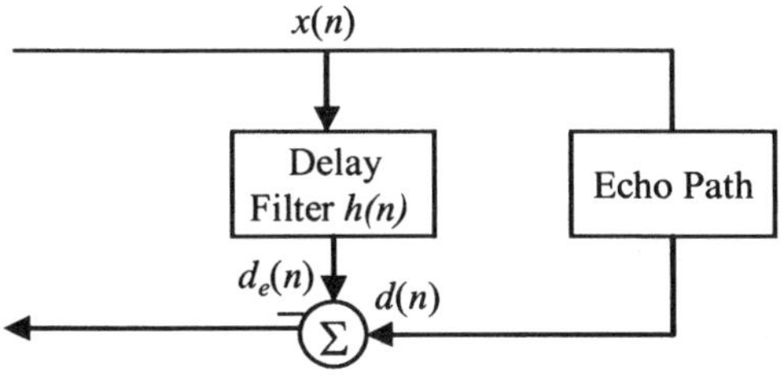

Figure 8.2 *Echo canceller.*

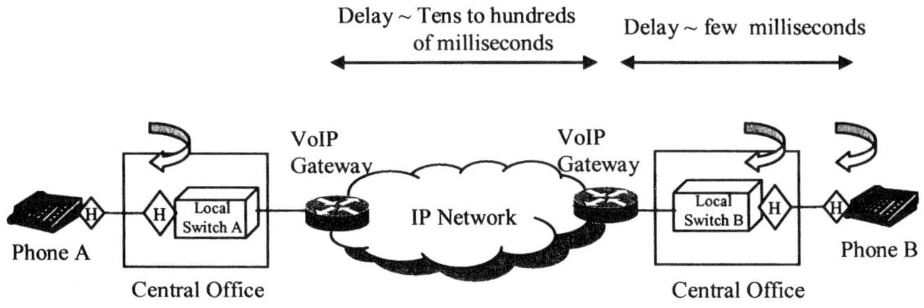

Figure 8.3 Echo in VoIP networks with PSTN termination.

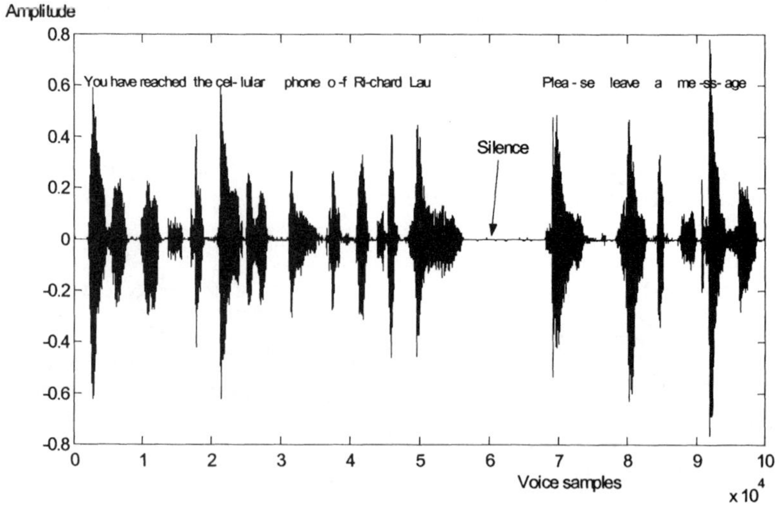

Figure 8.4 Digitized samples of a voice stream.

- The echo is too loud (i.e., the echo return loss (ERL) is not large enough due to insufficient echo cancellation).
- The echo delay may be too long such that the echo canceller's filter does not have a long enough response to cover the duration of the delay.
- The echo canceller is not configured to be on, or it is not working properly.
- The analog phone is on speaker-phone mode, which produces too much acoustic feedback. Headsets can be used to reduce the acoustic feedback significantly.

8.3.2 Clipping of Voice Sound

VoIP users sometimes complain that the voice conversation clips off badly, as if a large section of the conversation were being lost. Clipping of a voice conversation is very unusual in the PSTN, but it happens occasionally in a packet voice network. As will be shown in the following, this problem is likely related to the jitter incurred in packet networks.

8.3.2.1 Voice Packetization

We first describe how the voice signal is converted into IP packets. An illustration of a digitized voice stream is shown in Figure 8.4. There are 100,000 voice samples shown in the figure. With a sampling rate of 8,000 samples per second, the voice segment in Figure 8.4 corresponds to 12.5 seconds of speech. Each of the clusters (~100–500 ms) of voice samples is called a phoneme (a vowel or a consonant). A voice coder usually takes a fixed duration of samples (e.g., 20 ms) and encodes the voice samples via coding algorithms, such as waveform coding or linear prediction coding. The coded bits are then put into the RTP-encapsulated IP packet.

As an example, a voice codec may produce a 16-Kbps coded stream, which corresponds to a coding rate of 2 bits/sample if an 8-kHz sampling rate is used. A 20-ms voice frame then corresponds to 320 encoded bits, or 40 octets. These 40 octets of encoded voice samples are then placed into one IP packet to be transported. Although the packetization efficiency is low due to the large RTP/UDP/IP overhead, putting more encoded bits into an IP packet would increase packetization delay. Thus, a typical IP voice packet is relatively small compared to a typical data packet (~1,000 octets).

8.3.2.2 A Latency Model for an IP Telephony Network

Once the voice signal is packetized, we follow the path of the voice packet and identify the critical elements that are responsible for incurring significant delays. A simplified model describing the allocation of end-to-end latency for an IP telephony network is shown in Figure 8.5. In this model, the voice signal is first sampled and quantized into a 64-Kbps stream of data and then coded into a variable packet stream of $U(n)$ packets/second. Note that a constant-rate voice coder as described in the last subsection (coded rate of 16 Kbps) is a special case of this variable-size packet model. The transmitter buffer is needed in case its output is a shared medium (e.g., Ethernet) with an output packet rate of $V(n)$ packets/second, or if the voice codec is a variable rate as in the case of silence suppression.

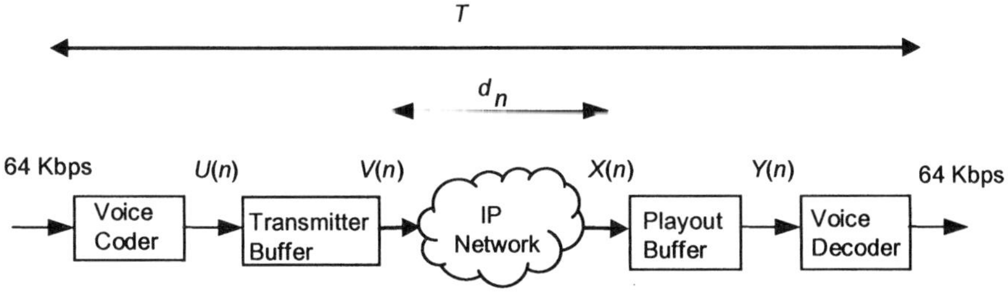

Figure 8.5 A latency model for an IP telephony network.

A voice packet entering the IP network experiences a delay of d_n ms, where d_n is a random process that represents the delay through the IP network. Note that jitter of the IP network is given by the variation of d_n as a function of n. The receiver buffer, also called a playout buffer, is responsible for buffering the incoming packets [$X(n)$ packets/second] and plays out at a rate of $Y(n)$ packets/second demanded by the voice decoder. The decoder then decodes the data to produce a 64 Kbps output voice-data stream. In case silence suppression is used, the decoder will insert silence back into the output decoded voice stream.

An interesting design parameter of such a voice model is T, which denotes the end-to-end delay from the input to the output. When the system is initiated, the transmitter buffer and the playout buffer will be allowed to start at a certain fill level in such a way that the chance of overflow and underflow of the buffers is minimized. Once the system is in operation, the dynamics of the system will dictate the buffer-fill level in such a way that the end-to-end delay T is a constant. This does not mean that the T for each telephone call is the same, but once the system starts, T does not change. As shown in the following section, the transmitter and playout buffers are probably the most important elements to monitor since they directly impact the quality of the voice call.

8.3.2.3 Sound Clipping

The dynamics of the buffer occupancy, the latency of the IP network, and the different rates of the encoded and decoded bit streams are governed by a set of rather complex equations (outside the scope of this book). Although we will not go into detailed mathematical analysis of such a latency model, we would like to point out that the transmitter and receiver buffers are key elements with respect to sound-clipping artifacts. The sizes of these buffers are critical in determining:

- The end-to-end latency (from mouth piece of the caller to the ear piece of the receiver) that affects the amount of echo cancellation required;
- The ability to absorb the jitter resulting from IP latency and voice codec;
- The chance of buffer overflow or underflow.

While these factors all influence the overall quality, the buffer overflow or underflow directly leads to the clipping of the voice sound. Unlike isolated IP packet loss, which can usually be concealed, buffer over or underflow causes loss of a large number of voice samples, which is unrecoverable. This manifests itself as an abrupt clipping of a voice sound.

Thus, sound clipping is the loss of a large piece of data beyond repairing by concealment. The design of the buffers is tightly related to the codec, channel rate, network conditions, including delay, loss, and jitter, and desirable MOS. A large buffer reduces the chance of underflow or overflow, but will incur larger end-to-end latency. Reducing the buffer size will increase the likelihood of losing a large piece of voice data. Therefore, buffers should be configured to match the characteristics of the underlying end-to-end network. When problems related to sound clipping are detected or reported, the first step in debugging is thus to detect whether there are abnormal buffer overflows or underflows. The condition of the end-to-end voice packet path can then be traced to find out the root cause. More about the tracing and sectionalization of troubleshooting will be discussed in Section 8.4.2.3.

8.3.3 Dropping Calls

For a VoIP service based on SIP, the signaling path is logically independent of the RTP voice payload path. However, when the IP network experiences degradation, such as excessive packet loss or delay, both the signaling and the RTP payload packets can be affected. In some situations, the impact due to the degradation in the signaling path may be more severe than that of the payload path. For example, a large burst of loss packets may manifest itself as a noticeable clipping of the voice conversation if the lost data is beyond reconstruction via error concealment. Although annoying, the call will still be active. However, when the large burst of packet loss occurs in the SIP signaling path, it is possible that the call will be terminated, an effect that is usually more annoying to the user than a temporary voice clipping.

In an SIP session, after the SIP call is set up, there is no separate SIP message defined for informing the SIP UAs at both ends or SIP proxies that the session is still active. If the BYE message is not transmitted or lost in the network, a stateful proxy or the SIP UAs at the endpoint may not be aware that the session is already closed and will keep the call state active, potentially indefinitely. To avoid this situation, IETF defined a mechanism [5] for conveying a "keep alive" state via the expiration timer and session refresh fields. When a session is initiated, an expiration timer value is agreed upon among all SIP UAs and proxies. A typical value of this timer is 30 minutes. If more time is needed for the session, SIP will convey the "keep alive" state of the session via either a re-INVITE or UPDATE message, which is called a session refresh request. During initial setup of the session, one of the UAs is designated to be responsible for generating the session refresh request. It is recommended that session refresh requests be sent after half of the session expiration time has elapsed. If the session continues beyond the expiration time without a refresh, both UAs are expected to send a BYE message to terminate the session just before the expiration. The SIP proxy, after session expiration, will terminate the session and remove all resources allocated to the session. The proxy, unlike the UA, will not generate a BYE message after expiration.

However, the following situations are possible: An SIP session is configured to be expired in 30 minutes but lasts longer than the expiration time. After 15 minutes, the designated UA generates an UPDATE message. However, the UPDATE message does not get through the IP network due to a high burst of lost IP packets. Without receiving an acknowledgement, the UA's retry timer times out and sends out another UPDATE message. Suppose this message is also lost. According to the mechanism defined in RFC 3261 [6], the UA will continue to send out retries but double the retry timer every time it expires. RFC 3261 also recommends that the retry timer to be 500 ms, and that the expiration timers for the retries be 0.5, 1, 2, 4, 8, and 16 seconds. The times when retried UPDATES are sent are 0.5, 1.5, 3.5, 7.5, 15.5, and 31.5 seconds. After 32 seconds, the UA will give up and declare the destination unreachable. Therefore, if all the seven UPDATE messages are lost consecutively, the SIP session continues until it passes the 30-minute expiration time, and then the session will be terminated. From a user perspective, the call is suddenly dropped for no apparent reason.

Normally, losing seven UPDATE messages consecutively is very unlikely. For a packet-loss rate of 10%, the chance of termination due to premature expiration is 10^{-7}, which is small enough. However, when there is a burst of packet loss, a large packet-loss rate that lasts for a minute is not unlikely (much larger than 10^{-7}). This situation is even more likely for a wireless environment, especially when handoffs of call sessions between radio cells are involved.

8.3.3.1 Misconfigured Timers

Another scenario in which SIP calls can be dropped is when timers are misconfigured. Imagine that the expiration timer is misconfigured to be 30 seconds instead of 30 minutes. At 15 seconds, the designated UA sends out a re-INVITE message. If this message is lost, there will be no chance to perform a retry. The reason is that the rule for sending out a BYE due to session expiration is that the designated UA sends out a BYE x seconds before the expiration time, where x is the smaller of one-third of the session expiration time or 10 seconds. In this scenario, the designated UA will send out a BYE at $t = 20$ seconds, allowing only 5 seconds (from the 15-second mark) for the refresh process to complete, during which only three retries can be attempted. Obviously, the chance of losing three retry packets is not that small, especially in a hostile environment.

8.3.3.2 What Can Be Done?

In some hostile environments (wireless is an example) where a burst of packet loss is likely, the timers may need to be readjusted so that the above situation can be avoided or occur less frequently. For example, the duration of retries may be increased from 32 seconds to 64 seconds, so that the chance of the retries going through the degraded network increases. Another possibility is to increase the expiration duration from 30 to 60 minutes, which will drastically reduce the number of calls that require refresh. However, changing the timer setting needs to be carefully weighed against other considerations since it may affect other factors such as increasing call setup time. There are many more scenarios where misconfiguration of various SIP timers, in combination with degraded network conditions—such as a large burst of packet loss, excessive network delay, slow processing time of the proxies or UAs, or software bugs—can cause calls to be dropped. If a problem is identified to be SIP related, a SIP signaling analyzer that displayed SIP messages and their associated timestamps would be a very useful tool for debugging.

8.3.4 The NAT/Firewall Problem

The use of network address translation (NAT) to deal with the IP version 4 (IPv4) address shortage problem creates a fundamental conflict in security as the end-to-end IP model is broken. Recognizing that the IPv4 to IP version 6 (IPv6) migration will take a long time, the industry has worked very hard to try to solve this conflict. This section describes how the NAT/firewall interacts with the SIP infrastructure and summarizes some currently proposed solutions. This is an evolving research area where there seems to be no completely satisfactory solution. Nonetheless, it is important to understand the nature of this problem as any solution will likely have an important impact on the design of the assurance system.

8.3.4.1 Problem Description

NAT is designed to deal with the shortage of IPv4 addresses. It allows the use of private IP addresses within the private domain, thus reserves scarce public IP addresses. NAT is usually implemented in software within a router, which demarcates the boundary of a public and a private IP domain. It uses a network address translator to convert private IP addresses and port numbers into a public IP address or addresses with new port numbers. The NAT maintains a table that keeps track of the mapping between the private addresses or ports and the public IP addresses or ports. Since the mapping from private to public addresses is usually many to one, public IP addresses are conserved.

With NAT, when an external IP host tries to communicate with an IP device behind the NAT router, the NAT will convert the incoming public IP destination address into a private IP address before forwarding the packet to the local private LAN, assuming that the incoming IP address is allowed. However, mapping between the private and public address in the NAT is usually created by the outgoing IP traffic. Therefore, unsolicited external IP clients will normally be unaware of this mapping. In the case where NAT does not recognize an incoming IP packet destination address, it will simply discard the packet.

For VoIP using SIP, there are two potential problems [7]. Referring to Figure 8.6, the first problem is the traversal of SIP signaling messages through the NAT. Suppose an SIP user agent client (UAC) (i.e., the caller within a private domain) initiates an SIP INVITE message over UDP. This message is transported to the SIP user agent server (UAS) (i.e., the callee, through the NAT and one or more SIP proxies). The SIP UAS nominally responds and sends a SIP message back to the source IP address and port number of the caller. In some scenarios, the first proxy after the NAT may be able to use the public IP address of the NAT [$A'(p)$ in Figure 8.6]. However, since the source port address, $r_1(i)$, is still incorrect, the SIP reply message will be blocked by the NAT router.

The second SIP NAT traversal problem is related to RTP going through the NAT and is also illustrated in Figure 8.6. In SIP, the signaling path and the media path (RTP) are in general transported on separate port numbers. SIP uses session description protocol (SDP) inside the SIP message to signal to the SIP user agent server to set up the RTP transport in an SIP offer/answer exchange.

When the advertised IP address and port numbers are those of the private domain, the incoming RTP packets will again be stopped by the NAT. In this scenario, the caller and the callee will hear the ringing and see the caller ID but will not be able to hear each other.

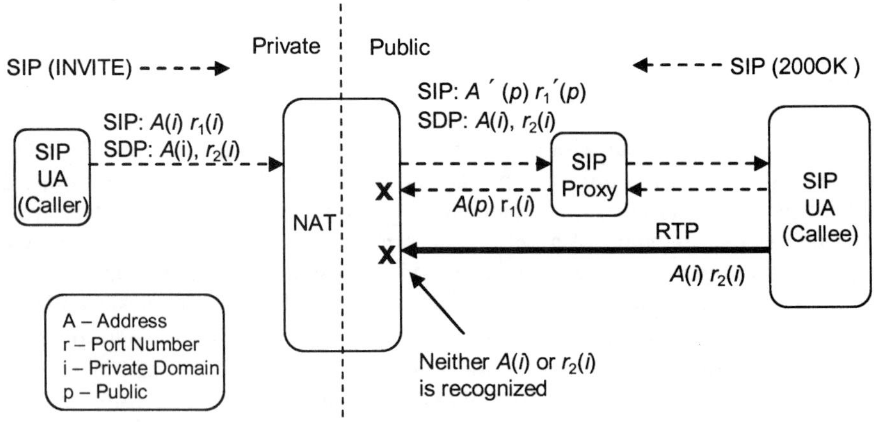

Figure 8.6 SIP-NAT traversal problem description.

The firewall issue is similar to NAT blocking. A firewall allows the protection of an IP domain (e.g., the residential home network) by blocking unsolicited or untrusted access. The basic method is to examine the IP source and destination addresses, port number, and application type of an incoming IP packet. If unauthorized, the IP packet is simply dropped. In VoIP, an external incoming call will generally be unsolicited and untrusted, therefore will likely be blocked by the firewall.

8.3.4.2 NAT/Firewall Solutions

The NAT/firewall is a common problem in VoIP networks. However, the solution to the problem is complex, costly, and potentially problematic. Various solutions have been proposed and studied by the IETF and offered by the vendor community. These include:

- Simple traversal of UDP through NAT (STUN) and traversal using relay NAT (TURN);
- Session border controller (SBC);
- Application level gateway (ALG);
- Universal plug-and-play (uPnP).

We first briefly describe these approaches. Each of these solutions has its advantages and disadvantages. It is currently unclear which will become the dominant scheme for mass deployment. It is also likely that a combination of different approaches may be needed to solve this problem. Our purpose here is to illustrate the complexity of the proposed solutions and point out the possible implications with respect to service assurance.

STUN and TURN

In this approach, a Simple Traversal of UDP through Network Address Translation (STUN [8]) server is placed in the public network. The STUN server collects information about the NAT mapping. An STUN client can first discover an NAT's existence and type. When an SIP UA wants to make a call to outside the NAT, it first establishes a dialog with the STUN server and obtains the routable IP address for that particular SIP session. Once this information is available, the SIP client knows to place the "correct" SIP URL into the SIP message. In addition to requiring a new STUN agent in the SIP application, STUN has limitations in that it only supports UDP and does not work for symmetrical NAT, which uses different NAT mappings for different destination IP addresses. Since the destination IP address for the SIP callee is different from that of the

STUN server, the address learned by the STUN procedure would not be valid. The solution for symmetrical NAT traversal is called Traversal Using Relay NAT (TURN [9]). TURN is similar to STUN, except that the TURN server is located in the path of the SIP message and relays the SIP signaling to the destination. Because all SIP messages will pass through the TURN server, it can be used for symmetrical NAT traversal.

Session Border Controller

Both STUN and TURN require the SIP client to be modified. Another solution that has received a lot of attention in the industry involves an intermediate device, the session border controller (SBC) [10], placed in the network to correct the SIP address problem. The SBC performs two functions. First, it acts as a session signaling gateway and functions as a signaling proxy [e.g., a back-to-back user agent (B2BUA)]. The SIP signaling message from the caller is terminated at the SBC, and a new SIP UA is created on behalf of the caller. This new SIP UA of the SBC becomes the contact point for all the returned SIP messages. The assignment of the proxied SIP UA is achieved during the initial SIP registration process as shown in Figure 8.7. After registration, when the caller sends out the SIP INVITE message, the SBC signaling gateway performs the following functions. It changes the "Via" header so that it appears that the INVITE message is sent by the SBC. It also changes the return address and port number of the RTP path in the SDP message so that the media stream will be sent to the SBC media proxy gateway as shown in Figure 8.7. In addition, the SBC signaling gateway (SIP B2BUA) informs the SBC media proxy (inside the SBC) of the newly assigned IP addresses and ports for the return signals and RTP media via a signaling protocol such as H.248. In the return path, the SIP signaling will be sent back and terminated at the SBC signaling gateway. The SBC next creates a new SIP message and sends it back to the caller via the pinhole of the NAT. For the media path, the RTP is processed at the media gateway and sent back to the caller via symmetrical RTP. For symmetrical RTP to work, it is assumed that the caller will first generate the RTP stream and send to the callee. The media gateway will use the same port number in the return RTP stream and is thus guaranteed to be able to go through the NAT.

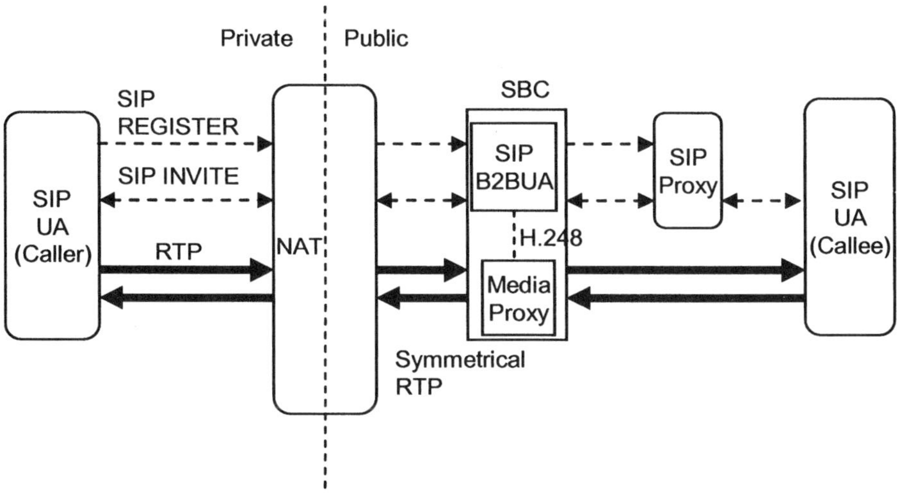

Figure 8.7 Session border controller for NAT traversal.

Application Level Gateway

An SIP application level gateway (ALG) requires a new NAT device, which is capable of understanding the SIP-level message and performing the necessary conversion of the relevant address parameters. Conceptually, this approach will solve the problems introduced by the NAT since the NAT has all the address-mapping information. Similar to the SBC, the ALG does not require a new SIP client. However, the ALG requires an upgrade to the NAT device. Moreover, there are also concerns about security as the SIP ALG breaks the end-to-end security model. Either the SIP message cannot be encrypted, which is a severe limitation, or a more complicated trust model is required between the two providers, which are linked by the NAT. Because of all of these concerns, ALG is not expected to be a popular solution.

Universal Plug and Play (uPnP) [11]

In this approach, an SIP client will discover the public address information from the NAT. When this information is known, the SIP UA will work as if the SIP call were initiated by the NAT so that the SIP URL carried inside the SIP messages will be correct and routable. In this approach, both the SIP client and the NAT have to be supported by uPnP software. Migration to this approach will take time as most existing SIP software and NAT equipment do not currently support this capability.

8.3.4.3 Operations Considerations

Given that the NAT is an unavoidable migration step and that there are many varieties of NAT traversal solutions, it is likely that many assurance issues of SIP telephony will be related to the NAT and the firewall. NAT/firewall problems can manifest as failed connectivity, one-way connection, or allowing of only the outgoing calls but not accepting incoming calls. Due to the variety of equipment and solutions, the overall problem may become intermittent. From an assurance operations perspective, it is therefore important that the associated VoWiFi/3G service model and operations procedures take into account the implications of this type of problem. It is also desirable that providers select an NAT traversal solution that requires minimal maintenance and configuration effort as it will clearly be very demanding to have operations personnel deal with these types of problems.

8.3.5 Long Call-Setup Delay

In Section 7.3.1.1, we cited that a dialing delay of 3 seconds is expected for local telephone calls. In this section, we look at the potential problem of long call-setup delays in an SIP environment. Needless to say, the 3 seconds of dialing delay is a starting benchmark. As described in Chapter 5, the SIP call setup is composed of several phases. When the user first initiates an SIP call at a new site, such as a new WiFi hotspot, the SIP UA is first registered and authenticated. This requires that the SIP UA sends a registration message to an SIP registration and authentication server, via one or more SIP proxy agents. Once an SIP UA is registered, it needs to be authenticated before it is authorized to initiate a call. This authentication and authorization procedure is independent of the access network and is also in addition to any equipment authentication procedure that is required for the access network. For example, UMTS SIM card authentication is independent of SIP-level authentication. After the SIP UA is authenticated, it can initiate a call by sending the INVITE message. As shown in Figure 8.8, a key attribute for the setup time is post dial delay (PDD), which is the latency between sending the INVITE message and the receipt of the RINGING message. The PDD is a subset of the total setup time, which is the latency between INVITE and the response that the call is accepted. The overall setup time of SIP calls depends on multiple factors, including registration, authentication, and setup delays (PDD and total call-setup delay). The message flow of a simple SIP call setup is shown in Figure 8.8. Long setup delay can result from various reasons, including:

- Heavy traffic load in SIP proxy servers or SIP redirect servers;

- Heavy traffic in or failure of DNS/ENUM servers;
- Congestion or failure in the IP network, which delays SIP messages;
- Loss of SIP messages due to network congestion and retransmission delay;
- Misconfiguration of SIP timers.

Diagnosing excessively long setup delays requires measuring the latency of SIP messages and correlating the latency with KPIs of network failures or congestion. An effective way to collect useful data for analysis and debugging setup latency problems is the use of active SIP emulation agents, which emulate both SIP UAC and SIP UAS and record all the timestamps of the emulated call. The emulating UA can also periodically generate dummy SIP calls to collect statistics related to one-way latency of registration, authentication, PDD, and total setup delays. This data can then be correlated with KPIs collected from relevant servers and the network.

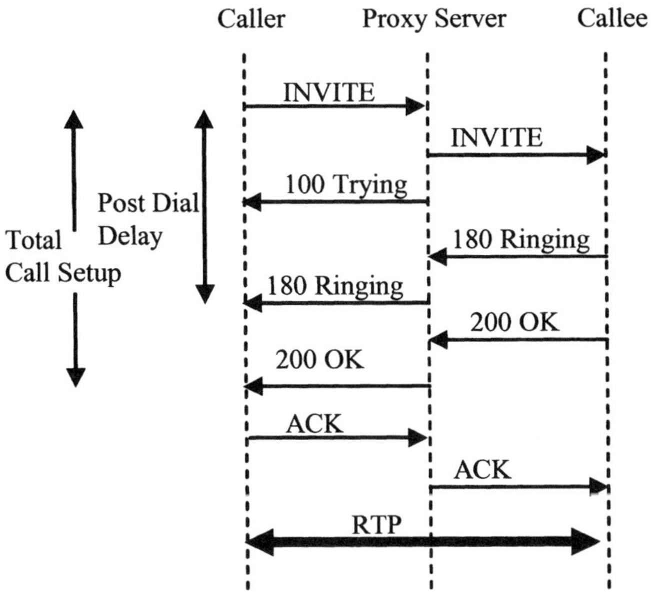

Figure 8.8 Post dial delay and call-setup time.

8.3.6 One-Way Voice-Quality Problem

One-way voice problems occur when one of the RTP paths is somehow blocked or suffers from severe performance degradation. The problem can be deterministic, in which one party cannot hear any conversation at all, or it may be intermittent such that there is a temporary one-way loss of conversation and, after a period of silence, the conversation becomes normal again. The artifact may occur intermittently and can happen multiple times within one SIP call, making it very annoying to the user.

One-way voice problems may be due to a number of different causes. We use Figure 8.9 to show a few scenarios in which this can happen. In Figure 8.9, the SIP proxy server is independent of the payload path and is located in a different IP subnet from those of the SIP endpoints.

Scenario 1: When an SIP call is set up, caller A can hear the ring tone, and callee B can see the calling number of A via caller ID. However, after callee B picks up the phone and begins to talk, caller B cannot hear the conversation, while callee A is able to hear B. This happens because the RTP packets from A to B are unable to reach callee B. As illustrated in Figure 8.9, one possible cause can be an IP routing problem along the IP path from A to B. Since the IP paths for the two directions of the RPT packets are independent, and there is no problem in the B→ A path, caller A is able to hear caller B's voice. Also, keep in mind that SIP signaling goes through a different IP route; therefore, the call setup is not affected.

Scenario 2: Another cause (see case 2 in Figure 8.9) of the one-way problem can be the NAT/firewall problem as described in Section 8.3.5. The NAT/firewall at the A location simply blocks the RTP path from callee B to caller A, but the RTP path from A to B does not have a problem because the NAT is able to recognize the IP source inside the NAT/firewall and allows the outgoing RTP packets through.

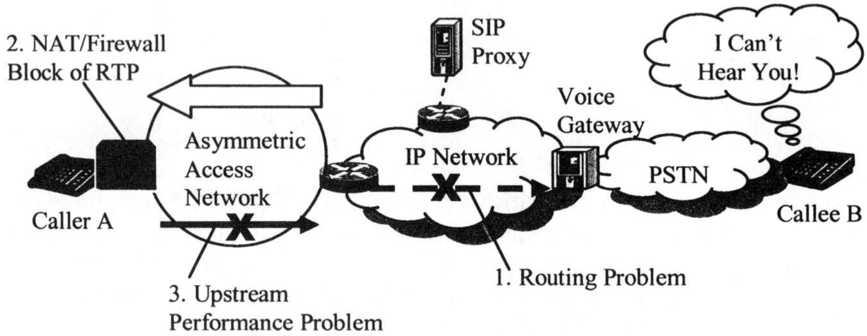

Figure 8.9 One-way voice problem.

Scenario 3: When the one-way voice problem occurs intermittently with temporary loss of conversation, the root cause can be quite different from that of the deterministic case. Case 3 in Figure 8.9 shows that Caller A's access network is asymmetrical such that the upstream path has smaller capacity than the downstream one, as in the case of cable or DSL networks. During the SIP call, it happens that the upstream traffic suffers from a large burst of IP packet loss due to congestion in the upstream direction for a duration of, say, 15 seconds. During this time, callee B experiences a large degradation in voice quality, such as choppy voice or even total silence.

The intermittent problem can also result when the IP core network experiences large delay or delay fluctuations in one of the paths, which results in a large number of IP packets being dropped, either at the routers in the core network or at edge devices such as access points, edge routers, or end devices. If the delay fluctuation lasts for too long, it can also cause the call to drop.

Another possible source of one-way intermittent problems can be related to the timing synchronization between the two ends of the SIP call. In practice, the timing sources for the transmitter codec and the receiver codec are not perfectly synchronized. Assume that the clocks at the two ends of the conversation are off by 100 parts per million. For a speech signal sampled at 8 kHz, the clocks at the two ends will be off by 0.8 sample per second. For a 15-minute conversation, this corresponds to an additional (or depletion of) 720 samples, or 720 extra bytes of data for a 64-Kbps speech, to be buffered at the receiver. This can cause an overflow situation and therefore may clip sound as described earlier. Moreover, as the conversation goes on, as in a long conference call, the synchronization can cause buffer overflow to happen periodically. This may occur at both ends of the call at different occurrence frequencies, or the artifact may be observed at one side only, because the implementation of the buffer in the codecs and the amount of jitter incurred by the network may not be symmetrical.

What Can Be Done?

To debug an one-way problem, it is important that the conditions that cause the problem can be observed in each direction independently. For example, an IP Ping test of end to end delay will not be sufficient to pinpoint the problem since ping measures round-trip delay and loss. One-way delay, one-way jitter, and one-way packet-loss measurements are critical for debugging one-way quality problems. It may also be helpful if there are asymmetrical parts in the network where traffic conditions can be monitored, again each direction independently. Moreover, checking the configuration of NATs/firewalls and associated gateways, such as those discussed in Section 8.3.4.2, is essential.

8.3.7 WiFi Cochannel Interference

Since IEEE 802.11 uses the industrial, scientific, medical (ISM) unlicensed band, RF interference can result from various signal sources, including microwave ovens, Bluetooth devices, or 2.4-GHz cordless phones. If these sources are close enough, they can completely dominate the RF spectrum and adversely affect WiFi performance or even render the network unusable. Another common source of interference is WiFi internal interference. This is the interference created by another nearby WiFi network that overlaps with the footprint of the existing WiFi network. Sometimes the overlap is intentional for the purposes of covering a large area or to provide more capacity. However, if interference is observed in a residential scenario, it is more likely that WiFi networks from neighboring houses happen to be using an overlapping channel. Since there are only three nonoverlapping channels in 802.11b (channels 1, 6, and 11), the chance of interference in a densely populated residential area is high. This scenario is shown in Figure 8.10.

With cochannel interference, the available bandwidth may be drastically reduced, and the degradation can be intermittent. The degradation experienced by users of the VoWiFi service may be most noticeable when the SIP phone is used in conjunction with other simultaneous applications such as an FTP file download.

Interference is not difficult to detect. Symptoms and performance indicators include sudden reduction in capacity, slow access to IP services, high packet retransmission rate, and high AP frame check sequence (FCS) error. However, in the residential VoWiFi service scenario, the end customer is not expected to perform diagnosis. It will be desirable to have remote access point management capability so that problems can be identified without sending a maintenance person to the household. More about this capability will be discussed in Section 8.6.2.3.

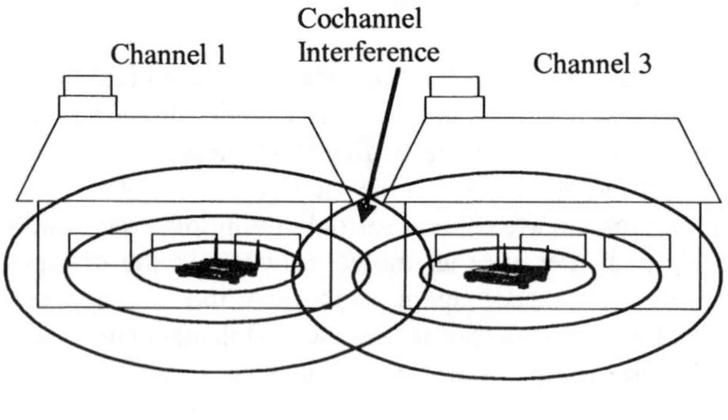

Figure 8.10 WiFi cochannel interference.

8.4 ASSURANCE METHODOLOGY

When a problem is detected, it is usually noticed and presented as a symptom. A symptom is an undesirable state or characteristic of a system. Symptoms may be due to a faulty component (hardware or software), a protocol design flaw, an unplanned event, or an unforeseen condition such as traffic congestion (e.g., IP route flap) or RF interference. Basically, symptoms are warning signs about the operation of a system. Depending on the severity and the impact of the underlying problem, immediate actions may be warranted. Unfortunately, the relationship of symptoms to the underlying root cause of the problem is usually so obscure that the observed symptom gives few hints about the root cause.

As a result, troubleshooting a communications problem is usually the responsibility of a few experts who know the details of operational aspects and even the design of the system. However, as the overall system gets larger and more complex, as in the case of VoWiFi/3G service, where multiple providers and networks are involved, it becomes clear that a more formal assurance process is required to deal with the scale and complexity of service problems. In the following, we focus on the description of such a process and the corresponding assurance methodology.

8.4.1 Assurance Process

In general, there are two categories of network, or service-related problems: hard faults, which are deterministic, undesirable, and permanent changes to the state of a communications system, and soft faults, which correspond to unwanted, nondeterministic, and intermittent states that cannot be easily recreated.

Whereas hard faults are usually a result of network equipment malfunctioning, soft faults may be due to many different kinds of problems, including misconfiguration of network- or service-related parameters, software, traffic overloading of network or equipment capacity or degradation of physical conditions such as radio fading or interference. Hard faults are usually easier to detect since the faulty state does not change with time. Moreover, operations personnel are more familiar with many of the symptoms resulting from network-related hard faults, such as facility failure or failure of an interface card. Soft faults, on the other hand, may occur intermittently or be triggered by usage or hard-fault events. The ensemble of soft-fault states is much larger than that for hard-fault events; therefore, they are harder to detect and replicate. With respect to VoWiFi/3G service, soft faults are more important as we have seen that many of the problems described in Section 8.3 belong to this class. Regardless whether a problem is due to a hard or soft fault, the goal of root-cause analysis is to discover the cause of the problem, remove the faulty component or undesirable state, and reset the system to the normal operational state.

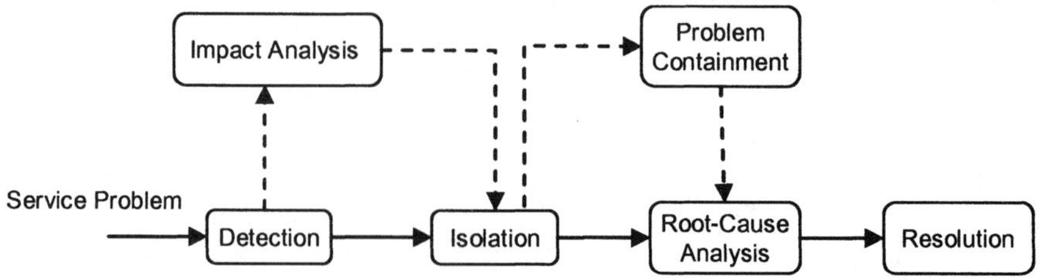

Figure 8.11 High-level assurance process flow.

8.4.2 Assurance Flows

The overall flow of the assurance process is illustrated in Figure 8.11. The process takes the service problem as input, performs diagnosis, and finally resolves the problem after the root cause of the problem is identified. The assurance process generally involves the following components:

- Detection of the problem;
- Impact analysis and prioritization;
- Problem isolation and localization;
- Problem containment;
- Root-cause analysis;
- Resolution.

We will describe each of these components in more detail in the following sections.

8.4.2.1 Problem Detection

Problems are usually detected as either service-level degradations or network behavior symptoms indicated as alarms or alerts. Symptoms are manifestations of an undesirable state or characteristic of a system.

The first step of a root-cause analysis procedure is the detection of the problem in the form of identified symptoms. In modern networks, a service layer is normally supported by many underlying protocol layers and is implemented across many technology boundaries that are distributed across geographically disperse networks. When a problem occurs, many symptoms may be observed corresponding to various parts of the protocol layers in different parts of the end-to-end network. This makes the task of root-cause analysis difficult, especially for soft problems where temporal effects and statistical variations make the symptoms appear intermittently, further complicating the analysis. It is therefore critical that sufficient service and network-level monitoring, as was described in Section 7.3, be placed in critical points in the network so that key performance data is collected to give early warning signs of a service problem.

8.4.2.2 Impact Analysis

As shown in Figure 8.11, after detection of the symptoms, it is desirable to perform an impact analysis before further diagnosis is continued. As explained in Section 5.7.4, impact analysis is important when a large number of symptoms, in the form of alarms or threshold-crossing alerts, are detected. Results of impact analysis can be used as a basis for prioritizing the order of treatment, as resources are limited. Impact analysis is usually achieved based on high-level information such as business and revenue impact. Impact analysis ensures that problems with severe business impact are given immediate attention. Impact analysis also should be completed within a short period so that the normal debugging process can start quickly.

8.4.2.3 Problem Isolation

The next step in the overall assurance process is to isolate or localize the problem [12]. Isolating the problem has two benefits. First, isolated and localized problems are easier to handle than distributed ones. It is also easier to access relevant performance data for analysis, and subject matter experts are generally knowledgeable about specific domains. Second, by localizing the problem, it is usually possible to contain the problem or temporarily bypass it before the root cause is found. This may be significant in certain scenarios where it takes a long time to find the root cause.

Problem localization can be achieved via various methods. One example is the correlation of observed symptoms. By identifying the set of symptoms, such as threshold crossing of KPIs, and performing correlation tests on these symptoms (see Section 5.6.2), one can identify KPIs that are highly correlated with the observed service-level problem. Identifying the set of KPIs for localization analysis is greatly facilitated by a service model in which dependence of KPIs is explicitly available as a parent-child relationship. Once the highly correlated and troubled KPIs are identified, the service model inherently suggests how the identified KPIs may have any localized properties, such as their all coming from the same network component or all being related to the radio access network. Another way to perform problem localization is to engage in systematic problem sectionalization as described in Section 7.3.4, where each network section can be independently tested and analyzed.

In addition to these benefits, problem localization can be considered an initial step toward root-cause analysis. Subsequently, the root-cause-analysis procedure is required to search a smaller and bounded domain for the cause of the problem.

8.4.2.4 Root-Cause Analysis

The final step of problem diagnosis is to try to explain the observed phenomenon. The explanation needs to be presented in such a way that the root cause, which can be viewed as a symptom that does not further depend on other symptoms, is identified. Although there are many methodologies discussed in the literature on this topic, their effective application, especially to large and complicated networks, is still unsatisfactory. In most root-cause-analysis applications, a specific solution has to be applied on a case-by-case basis. Root-cause analysis applied to a VoWiFi/3G integrated service is certainly an active research area. However, it should be noted that, in general, the more accurately the system is modeled, the more effective is the subsequent root-cause-analysis algorithm.

The detection, localization, and correlation procedures are usually necessary steps for collecting information before making one or more hypotheses about the root cause. To create a hypothesis, one usually needs to have prior knowledge about the symptoms associated with possible root causes. This knowledge may come from experiences or from a detailed model of the system.

Table 8.2 shows the possible root causes for different types of commonly encountered problems. For each row, the "x" identifies the set of possible root causes for the observed service problem. Notice that a single symptom may point to different possible root causes (see the possible root causes of the call-dropping symptom in Table 8.2). To further reduce the number of possible root causes, one may apply a method, called signature analysis. A signature is a set of characteristics of a root cause, illustrated as the marked ("x") symptoms in the columns in Table 8.2.

For example, suppose the root cause is that the SIP proxy server is down. This usually means that no SIP messages can be processed. Therefore, if there is no backup server, it is likely that many users will not be able to make a call at all. If there is a backup server or load sharing, calls can still be established, but users may experience larger than normal call-setup delays. In this case, the signature set may include the following symptoms:

- Server alarms are observed.
- Many existing calls are dropped.
- Many users cannot complete call establishment.
- If there is server load sharing, the load of the backup server is heavy, and users experience long delays for call setup.

Table 8.2

Correlation Between Perceived Problem and Hypothesis

Root Cause / *Problem*	*High IP Delay*	*IP Packet Loss*	*Buffer Problem*	*SIP Timer Problem*	*Loss of WiFi Connect*	*SIP Server Down*	*IP Routing Problem*	*NAT Problem*
Echoes	x	x						
Call blocking	x	x			x	x		x
Bad voice quality		x	x		x			
Call dropping	x	x		x	x	x		
Long setup Delay	x	x		x		x	x	
Sound clipping	x	x	x					
Asymmetric voice quality				x			x	x
Handoff failures				x	x			
SIP server alarms						x		

If all these symptoms are observed, one may conclude that there is a high probability that a server being down is in fact the root cause of the call-dropping problem. However, if a slightly different set of symptoms is observed, such that there are no server alarms but handoff failures and bad voice quality are reported, then the root cause is likely to be the loss of a WiFi connection.

Conceptually, the relationship between the observed problem, possible root causes, signature sets, and symptoms is illustrated in Figure 8.12. It should be noted that since signature sets may (usually) overlap each other, and not all symptoms of a signature set are always triggered, signature analysis is best described by a statistic model. In general, if accurate data reflects the statistics of the occurrence of the symptoms and signature set with respect to root causes, one can compute the most likely root cause based on the observed symptoms.

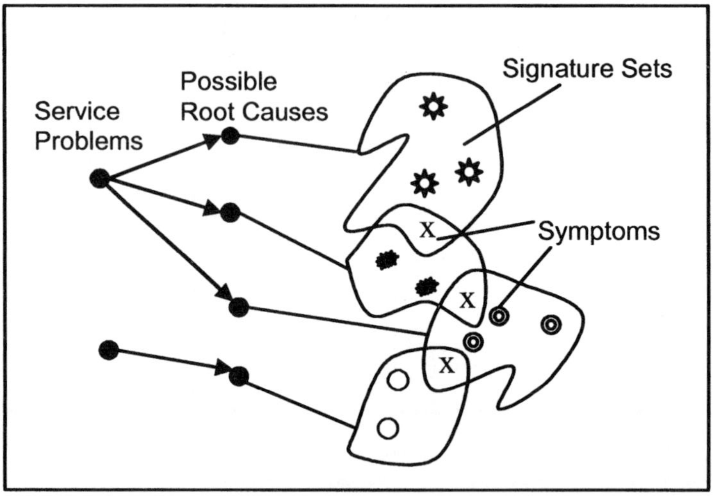

Figure 8.12 Signature sets and their relationship to root causes.

To illustrate the application of signature analysis, we use the same call-dropping example. In Figure 8.13, we highlight the observed symptoms in the signature set and show a statistical number associated with each symptom. This number represents the probability that the specific symptom is observed, given the root cause. In this case, the root cause for the high call-dropping rate is likely to be "SIP timer misconfigure," since its observed symptoms (in the signature set) weighted by their corresponding probabilities are the highest (total of 95%).

Once the possible root cause is identified, it is necessary to verify the hypothesis by removing the root cause and observing whether the service problem disappears.

8.4.2.5 Resolution

Once the root cause of the problem is identified and there is sufficient analysis and testing to support the conclusion, the resolution is to be implemented. Resolution may involve various steps, including:

- Equipment or software reset of reinitialization;
- Equipment configuration changes (routers, gateways, radio network parameters, SIP proxy);
- Network routing changes;
- Software upgrade;

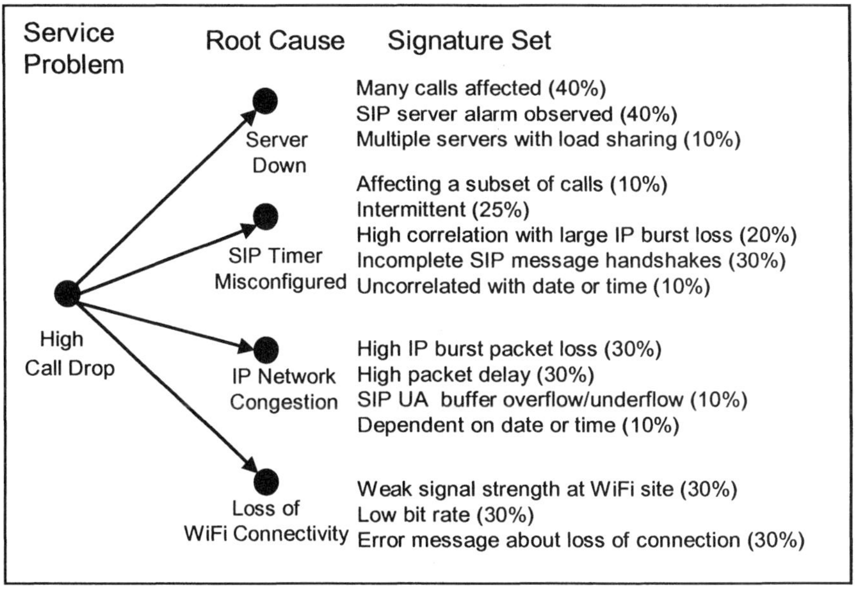

Figure 8.13 Root-cause and signature example.

- Replacement of faulty hardware;
- Change of policy (router, GGSN, SGSN, NAT, gateway).

After the implementation of the resolution and after the proper testing procedure is completed and logged, the diagnosis process is completed.

8.4.2.6 Mapping Diagnosis into Operator's Assurance Process

So far we have discussed the problem diagnostic functions and processes. In the next section, we will discuss how these functions are implemented in a VoWiFi/3G operator's operations process. We will examine issues including how the assurance process extends beyond a single operator's administration domain and how various functions are mapped into the traditional operations flows in a current mobile operator's network operations center. Finally, we will examine what new operations flows are necessary to achieve excellence in assurance operations.

8.5 OPERATOR'S ASSURANCE PROCESS FOR VOWIFI/3G SERVICE

Mobile operators have a well-defined process for maintaining and assuring service quality. The traditional assurance process is network centric and mainly focuses on maintaining and assuring proper operation of the radio access network. The addition of new IP-related services has prompted mobile operators to pay more attention to the management of a server farm similar to that of an Internet service provider. This section first examines the present mode of operations of a typical mobile operator's network. It then identifies the operations gaps as compared to industrial best practices and follows with a discussion of how the operations process can be improved with the introduction of service-centric management. We then describe a desirable targeted operations architecture based on the service model concept. A detailed description of how the VoWiFi/3G service model is used to capture the principles of operational best practices will serve as a culmination point for the introduction of service-model-based assurance in this book.

8.5.1 Traditional Operations Assurance Process

The traditional operator's service assurance process consists of a combination of fault management, performance management, trouble ticket management, testing, and field dispatch management. The overall process is illustrated in Figure 8.14. The service assurance process starts with the registration of trouble reports that come from the operator's staff, a peering provider's personnel, or the end customer. This first trouble reporting and registration is generally referred to as the "first tier," "first level," or simply the help desk. The help desk is usually staffed by nontechnical staff members whose main tasks are to speak to the customer and record complaints. For a large operator, there may be 100 to 200 help-desk staff members, and there may be over 100,000 trouble reports in a month. The first-tier help desk can usually answer the most commonly encountered questions where predesignated answers are provided. In practice, first-tier support is very effective, since roughly 70% to 80% of trouble calls can be resolved there.

When the first-tier staff is unable to resolve the customer problem, the problem is generally recorded in a trouble ticket system and sent to a second-tier (or tier 2) technical staff member. Second-tier support consists of technical members who have in-depth knowledge of the network and operations. They are much better equipped with testing tools and have access to alarm monitoring systems, performance reports, and even the design plans of the network. These technicians and engineers are located in a small number of NOCs, where the operator's entire network is monitored. An NOC operates on a 24 × 7 schedule. It can be considered the main control center of an operator's whole network. Inside an NOC, large screens showing thc critical status of the network are displayed, as shown in Figure 8.15. Large operators usually have multiple NOCs in their network, which are located in different geographical areas to avoid a single point of failure. Although tier 2 support deals with a smaller percentage of the trouble reports, the problems are usually more difficult and may require hours or days to resolve.

As shown in Figure 8.14, there are further hierarchies in the second-tier support. After receiving the trouble ticket, a second-tier engineer first examines the problem statement and decides whether it requires further escalation to experts in specific areas, such as the radio engineering group, the network facility group, or the IT department. If necessary, the trouble ticket will be forwarded to the appropriate department where specific maintenance personnel will take over the actual root-cause analysis. When the root cause of a trouble ticket is discovered, resolution may require dispatching field supporting staff to implement the solution. Examples include reconfiguring a terminal located in the outside plant, or sending an RF engineer to a base station transceiver to adjust the tilt of a cellular antenna. If such a dispatch is necessary, a work order will be issued.

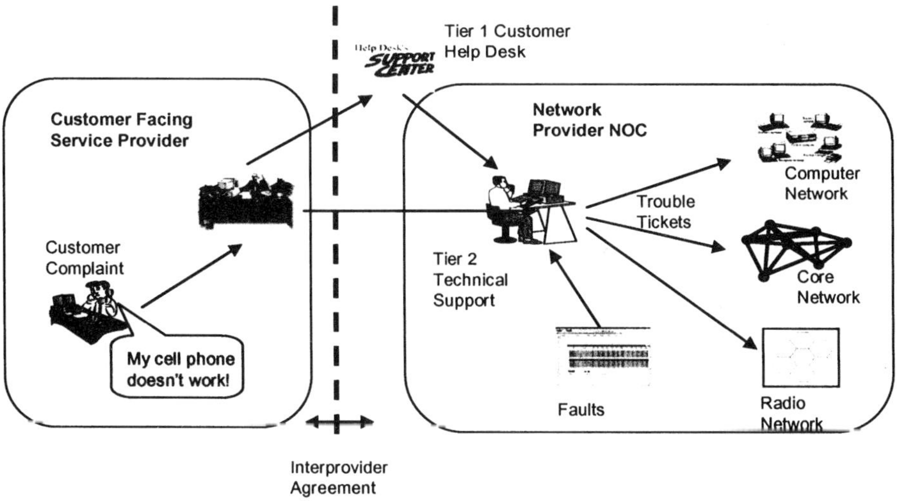

Figure 8.14 Traditional assurance process flow.

Figure 8.15 Typical network operations center.

Upon the completion of a trouble ticket resolution, the original second-tier engineer will need to verify and sign off on the completeness of the repair effort, thus certifying the recovery of normal service performance, then ensure appropriate measures will be taken to prevent similar resource troubles from occurring in the future. The last step is to close the trouble ticket, update the associated database, and inform the related parties according to the trouble ticket's specification.

The operator's tiered assurance structure typically deals with urgent performance and alarmed situations, where immediate actions are called for. In addition to the NOC, the operations staff also includes a team of performance and traffic engineers who are concerned with longer-term operations issues such as network and service performance, network and traffic planning, and network expansion. These engineers are also responsible for the instrumentation of collecting performance- and service-related data and performing analysis and data mining. Their work is used to provide critical statistics regarding service usage and to support market analysis. In addition, the reports can be used for identification of performance trouble spots, making recommendations on network expansion, and planning new services. Traditionally, these engineers belong to a different organization, although they may share a lot of the performance data and reports generated from the NOC personnel.

8.5.2 New Desirable Operations Features

Traditional operations procedures and processes are mainly optimized for circuit-switched voice service. When new services such as 3G and VoWiFi are deployed, a number of operations features and processes need to be modified or augmented to ensure smooth operation and guarantee customer satisfaction. Although the operations procedures required are different, the main goals are similar to those for traditional services. These include improving service quality and customer satisfaction and reducing churn and operations costs. The following is a list of desirable features and practices for the support of VoWiFi/3G operations:

- *Observability of overall voice quality within the domain of the operator:* Collect voice performance metrics from the infrastructure and conduct proper aggregation, statistical analysis, and correlation to produce a set of aggregated performance indicators. The metrics should be collected and categorized to fit different OSS applications, including revenue impact analysis, service assurance, and network planning. For new 3G networks supporting service portability and roaming of users, VoWiFi service and 3G services in general will traverse multiple operators' networks. It is therefore essential for a provider to be able to instrument enough service-monitoring capabilities that the necessary performance metrics are collected over the realm of its own network. This allows proper evaluation of customer perceived quality and proper optimization of its own network. Quantification of performance also allows the provider to write meaningful and enforceable SLAs with its peering providers or end customers.
- *Near-real-time access to critical service performance parameters:* In many scenarios, it is critical for the operator to be notified of the performance degradation in near real time, which is generally interpreted as within 5 minutes. This is important as performance metrics can be part of an SLA whose violation incurs a monetary penalty and results in disgruntled customers. Most operators still find supporting near-real-time performance feedback and alerting challenging, as many are still operating with a response time frame of hours or even days.
- *Labor and expertise reduction to generate SLA and performance reports:* Generation of SLA-compliance and performance reports is time-consuming and requires specialized expertise by multiple staff members. When the SLA is well defined and proper monitors are instrumented for collecting relevant data, it is beneficial to automate the process of report generation. This can significantly reduce the required labor and the level of expertise of the involved personnel. Moreover, there will be less chance of human error or miscommunication between organizations.
- *Integration of domain assurance functions including radio network, core network, server network, and backend systems:* Even within a single provider's domain, various technologies have resulted in separation of responsibility into different technical organizations for ease of management. Unfortunately, breaking an end-to-end assurance function into separate management domains discourages cooperation. Service assurance functions require proper coordination across technology domain boundaries and should be commonly accessible by different

organizations, including the radio and network department, the IT department, and network capacity and service planners.

- *A common service model that captures various relationships among components:* As the central idea brought forth in this book: a properly designed and implemented service model provides a framework suitable for supporting various new features necessary to optimize the operations process. It also provides a centralized structure for storing related parameters, including service identification, configuration parameters, equipment inventory, network logical connectivity, service-component relationship, and performance metrics. A service model framework provides modular design and allows incremental evolution towards a flow-through automated process. Commitments to building a service-model-driven OSS architecture will help to drive down operations cost in the long run, yet fulfill stringent service assurance requirements.
- *Avoidance of stand-alone OSS components:* In addition to an integrated view of end-to-end networks and services, operators should avoid deployment of isolated OSS components that do not have a logical representation in the centralized service model. This means the operational state of the stand-alone OSS is not connected to the centralized service model. While a stand-alone OSS, such as an IP performance system or a radio network debug tool, is easy to deploy and may solve some short-term problems, operations should strive to define the necessary "north-bound" management interfaces of these OSSs and to define a representation of the relevant managed objects in the service and network-management levels.
- *Reduction of training time for assurance operations expertise:* Service assurance operations are complicated and require much training to develop the necessary expertise. This is especially challenging when new technologies, such as 3G UMTS/IMS, are deployed. With proper design of operations flows and the automation of previously manual processes, operations can be carried out effectively, and the training time for assurance operations expertise can be drastically reduced.
- *Single log-on to OSS terminals for performance visualization:* Operations personnel should have on-demand access to customer information, critical performance data, SLA reports, and alarm logs. For troubleshooting, they should be able to initiate tracing of a session and retrieve correlated data at a single computer or workstation with a well-designed human interface (e.g., with a few mouse clicks).

This list contains desirable practices for general operations optimization. It will be shown next how a targeted VoWiFi/3G assurance operations architecture can fulfill these desirable features.

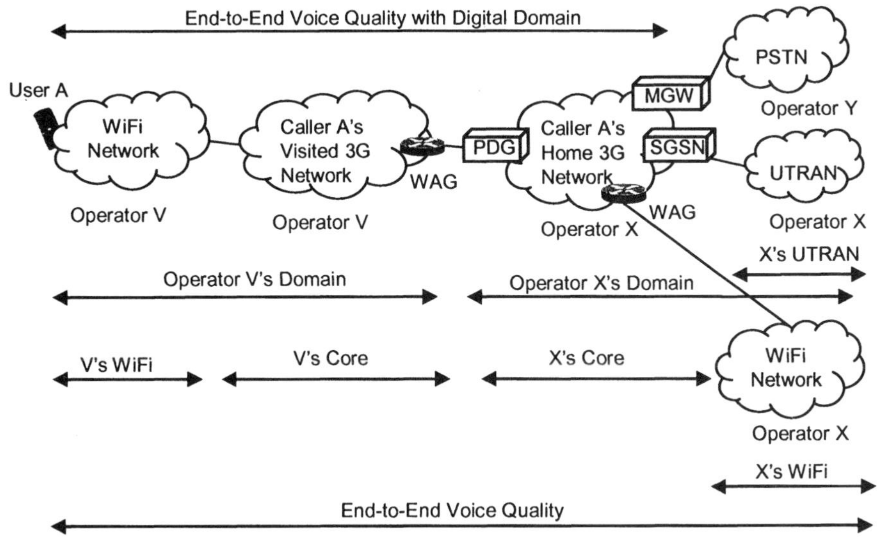

Figure 8.16 VoWiFi/3G service reference model.

8.6 TARGETED ARCHITECTURE SUPPORTING ADVANCED SERVICE ASSURANCE OPERATIONS

In this section, we focus on a targeted assurance operations architecture for the VoWiFi/3G service. We first lay out a network reference model, which highlights the high-level components and the involved providers. The reference model is then mapped to a service model, again focusing on illustrating the provider relationship and network components of the end-to-end voice path. The rest of the section describes how the service model is applied in conjunction with the targeted assurance operations process.

8.6.1 VoWiFi/3G Reference Scenario

Figure 8.16 shows the reference network connection of providing an integrated VoWiFi/3G. In the reference scenario, two service providers, X and V, are responsible for providing end-to-end packet telephony. For reference user A, provider X is considered to be the home service provider, which is the subscriber-facing provider, while provider V is the visiting network provider. Each provider may operate various access technologies, including WiFi hotspots, 3G UTRAN, or 2G cellular RAN. We assume the mobile station is a dual-mode handset that supports VoWiFi/3G and traditional 2G cellular circuit-switched service. Since we want to focus the following discussion on the operations process from operator X's perspective, we do not include the termination part of the call in Figure 8.16. Readers can refer to Section 7.2.3.4 for the complete end-to-end model description.

Referring to Figure 8.17, the service aspects include:

- WiFi-WiFi, both within a provider or across two providers;
- WiFi-UTRAN, in which either provider can have either access network;
- WiFi-2G cellular (or PSTN).

The high-level service model with respect to each of the above service cases is shown in Figure 8.17. Because every service case is supported via different access networks or different providers, the expected QoS and the need to perform monitoring and assurance may be different. KPIs need to be defined for each service case of the service model, and SLAs are defined across provider domains. From the provider's (say, X) service assurance viewpoint, the following performance metrics are important:

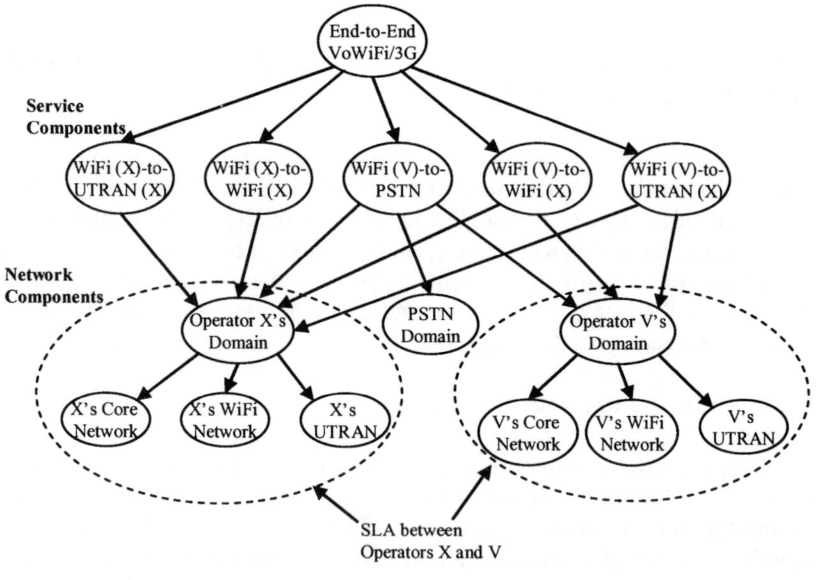

Figure 8.17 High-level VoWiFi /3G service model.

- Customer's overall perception of dual-mode VoWiFi/3G integrated service;
- Categorizing customer's perception of service quality for initiation of receiving calls from WiFi network, 3G UTRAN, and 2G cellular;
- Performance during handoffs between different access networks;
- Defining the SLA between X and V, such as the SLA including terms like IP-layer-performance KPIs, aggregate throughput, and other operations parameters, such as time to resolve a trouble ticket and the permission to test across provider's boundaries;
- Defining internal KPIs for each of the component networks.

8.6.2 Targeted Assurance OSS Architecture

Targeted assurance OSS architecture needs to address the following features, which are not included in the current practice of most cellular operators:

- A process to deal with the assurance problem;
- QoS planning;
- Integration of various technology-centric organizations;
- Improvement in response time to service-affecting events to near real time;
- A centralized data repository for easy access to information related to service- and network-related, real-time, and historic performance data;
- Easy access to logical and physical configuration data.

At a high level, the targeted assurance OSS flow is shown in Figure 8.18. Compared to the present mode of operation shown in Figure 8.14, we notice that a QoS alarm-handling process is added to the traditional fault alarms. This includes all the service- and network-monitoring capabilities discussed in Section 7.3.4, and the assurance procedures described in Section 8.2. In addition to providing performance information for the internal assurance process, performance data is also used to support the generation of QoS alerts and SLA reports to be exchanged between the two operators.

This automation reduces complexity and expedites QoS information exchange compared to the traditional manual process. Finally, there is an integrated QoS management process added between trouble ticket generation and distribution of the tickets to separate technical organizations. This integrated QoS process is most demanding in terms of its complexity and will take advantage of many features and capabilities of the service model for its implementation.

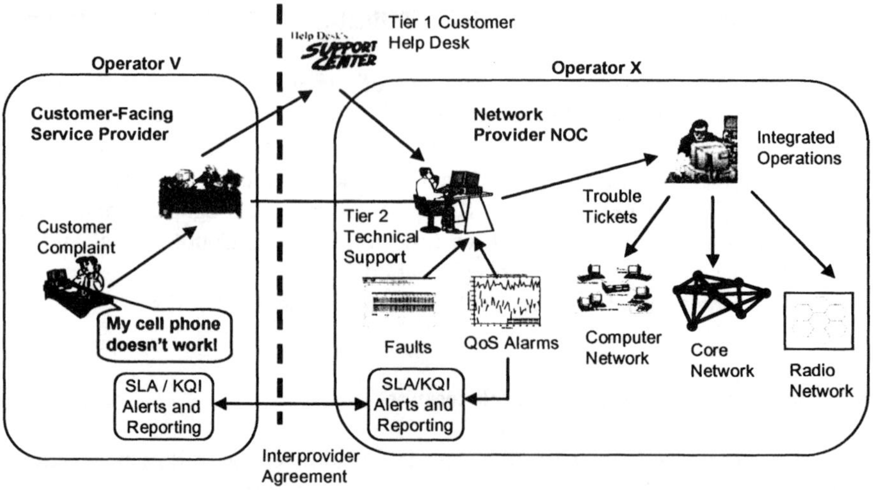

Figure 8.18 Targeted assurance OSS process.

Figure 8.19 shows the functional blocks supporting the targeted OSS flows of Figure 8.18. It shows the mapping between the high-level assurance process functions and the OSS flows. It also illustrates how the service model supplies relevant information to various assurance components. In addition, the targeted architecture includes a new, integrated, QoS-management functional block.

This new function is responsible for diagnosing the problem in an integrated manner. Rather than dispatching repair work orders to different organizations segmented by technologies, the integrated QoS manager performs a high-level diagnosis function before dispatching work orders to improve coordination and efficiency. It also has the overall responsibility of reporting the final results of the diagnosis.

8.6.2.1 Integrated QoS Management

Figure 8.20 provides a more detailed description of the integrated QoS management function. Its main components are the root-cause analysis (RCA) engine, various analysis and testing tools, and a self-learning and upgradeable knowledge database. In addition, it has easy access to end-to-end service and network configuration databases and performance data, as well as observability into the current status of the network (e.g., major network outage). The following further describes its key components.

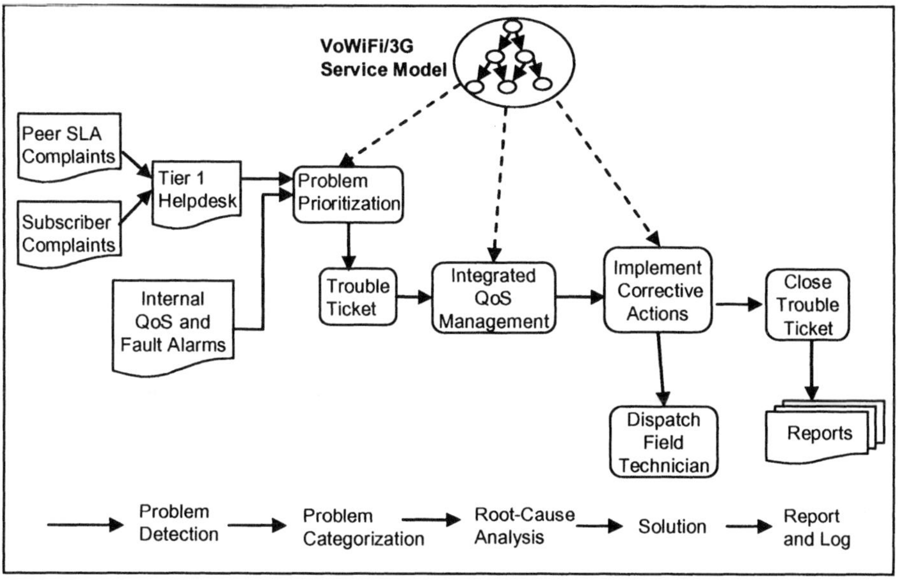

Figure 8.19 Functional block diagram of targeted assurance OSS flows.

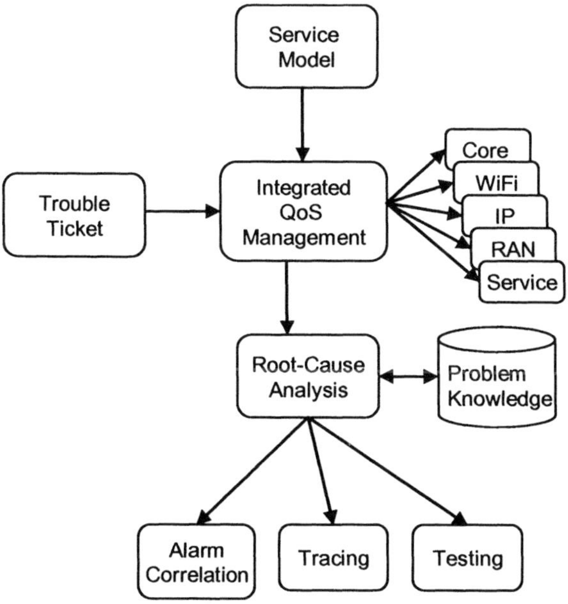

Figure 8.20 Detailed VoWiFi/3G assurance functions.

8.6.2.2 Root-Cause Analysis

As described in Section 8.4.2.4, the root-cause analysis engine performs the correlation between observed symptoms, signature sets, and the possible root causes. The root-cause analysis eliminates possible root causes by performing testing or drilling down to detailed performance data. The additional data may come from further analysis of historical performance data, testing results, or updated QoS or fault-alarm information.

Traditionally, root-cause analysis is performed by a small number of experts who know the end-to-end system in detail. As networks become more complex and new services emerge frequently, it is challenging for the technology experts to simply keep up with new operations guidelines. Problem diagnosis is much more demanding and requires deep knowledge beyond the normal operations procedures of the system. Also, as new technologies are introduced, it usually takes years for hidden network or software problems to be observed. In the first few years of mass deployment of the VoWiFi/3G service, many unfamiliar service problems are expected. It is therefore essential for an operator to be fully prepared with a well-defined OSS process in place, of which the root-cause engine is a key component.

8.6.2.3 Remote Diagnosis Capability

When an operator offers a VoWiFi/3G service with a dual-mode handset, the customer premises is required to install an 802.11 AP connected to a broadband network (DSL or cable). Normally, the AP will need to be configured with a number of parameters including:

- Type of WiFi AP (802.11a, b, g);
- Channel number (1–11 for 802.11b);
- SSID (default or user selected);
- Method to acquire an IP address (fixed assignment or via DHCP);
- Security options (WEP key, 802.1X, AP login and password);
- NAT/firewall configuration;
- Remote management access (a remote IP address needs to be specified by the vendor, and remote management access needs to be enabled).

These configurations are usually not difficult, and many of these parameters are set when the AP is first installed. However, when there are abnormal situations, such as a power outage, system reset, or forgotten WEP keys or login passwords, a nontechnical customer will likely need help from a help desk. In such a situation, it will be extremely useful if the provider has a remote diagnostic capability. This capability is illustrated in Figure 8.21, where the remote diagnostic workstation is located inside an NOC. The remote trouble-shooting work station has access to the customer's profiles and the corresponding AP's remote management access privilege. It will also be desirable that dynamic parameters such as DHCP-assigned IP addresses or WiFi channel numbers be available to the management workstation so that remote testing and reconfiguration is possible. This remote diagnosis may be one of the most useful features for saving operations cost, as dispatching of technicians to residential households is very expensive. This is also important from a service perspective, as the expectation of quality for telephony service is much higher than that for data services, and the consumer will be unlikely to tolerate the need to wait for hours or days to get help to reconfigure the AP for a dual-mode phone service.

It should be pointed out that, currently, remote management of APs is not generally available, although it can be a key differentiator.

8.6.2.4 Self-Learning and Upgrading

As mentioned before, many problems appearing in VoWIFI/3G networks are of the soft type (i.e., they occur intermittently, are statistical in nature, and are not easily reproducible). Therefore, once the root cause of a soft problem is identified with a level of confidence, the solution should be captured so that future references to it can be made. Due to the statistical nature of the problem, it is useful to have a mechanism that can predict the likelihood of a future match to a similar type of problem.

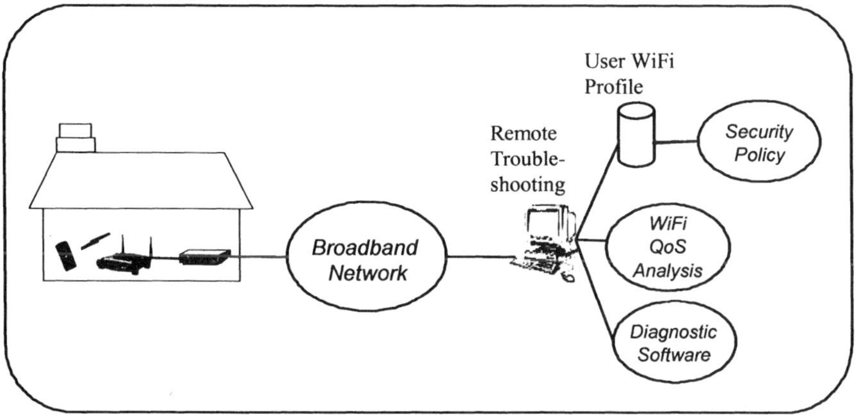

Figure 8.21 Remote diagnosis of a VoWiFi network.

For example, a set of KPI violations may be an indication that certain root causes are likely candidates. A table capturing these possible relationships will likely be very helpful to future diagnosis. As described in Section 8.4.2.4 (see Figure 8.13), self-learning with automatic database updates provides valuable statistics parameters for signature set analysis, which will result in higher diagnosis success rates as more experience is collected.

8.6.2.5 Tracing and Emulation Tools

When there are problems associated with the details of the signaling messages, such as SIP timer expiration or problems related to the handoff of a voice session, it is indispensable to have a detailed tracing capability that allows an expert to step through the messages recorded for one or more selected sessions. Tracing may include SIP signaling; the RTP data path; RTCP messages; Gi, Iu-PS, and Iu-b interface signaling; and SS7 messages (see Section 7.3.4). It is also desirable to have an emulated software agent that can be used to replicate problem scenarios. Most importantly, these tools need to be integrated into the operations process and readily accessible from different NOCs.

8.7 THE FUTURE OF VOWIFI/3G OPERATIONS

This chapter has addressed one of the most challenging tasks of mobile service providers: how to maintain and operate a large-scale VoWiFi/3G network and achieve operations excellence. Although the Internet has become popular in the last decade and wireline carriers have been migrating to VoIP networks in the past few years, the challenge of offering high-quality packet voice over a combination of WiFi and UMTS networks is still a daunting task. Not only has the Internet not proven its capability to support mission-critical QoS, but the uncertainty in the radio performance of WiFi and UTRAN, as well as the relatively new introduction of SIP technology, all contribute to the difficulty of assuring a service quality equivalent to that of the PSTN.

The goal of achieving superior quality will require something beyond a faster router or bandwidth over-provisioning or deploying new radio spectrum. A new and improved operations paradigm has to be part of the solution. The new operations will be different from the traditional circuit-switched and Internet operations. They will require an integrated service model where business goals, service management, network optimization and control, providers and end-customer relationships, and operations are tightly coupled. Operators will strive to automate the operations procedures and flows. Networks will operate with a focused business goal, and customer management itself will be a built-in component of the solution. The dependency provided by such a service model will guide the operations to focus on the area most essential for economic success.

Integrated operations cannot be achieved in a short time frame; nor can it become a reality via incremental and local optimization. Many critical components of the assurance process need to be planned out and implemented in the beginning of the deployment cycle for it to be effective. The planning and deployment of such an integrated process will best be enabled with an encompassing service model. Once the foundation of a service model is laid, the evolution of the operations process can take a much longer time frame, as the model can be adjusted to meet either immediate or long-term needs demanded by the services and business drivers.

Finally, we believe that in the competitive world of advanced mobile and Internet services, providers who are committed to delivering the best quality will earn a reputation that allows them to excel. Operators will see a much faster payoff for the advantages and the differentiation gained from a superior quality operations infrastructure. The benefits of such excellence are significant and far reaching as the new services are expected to be a growing area for the next decade!

References

[1] Gast, M. S., *802.11 Wireless Networks*, Sebastopol, CA: O'Reilly, 2002.

[2] Bims, H., "Enabling Voice over WLANs," White paper, Airflow Networks, Inc., September 2003.

[3] Alexander, T., Adaptive Signal Processing: Theory and Applications, New York: Springer-Verlag, 1986.

[4] ITU-T Recommendation G.168, "Digital Network Echo Cancellers," August 2004.

[5] Donovan, S., and J. Rosenberg, "Session Timers in the Session Initiation Protocol (SIP)," July 2004, Internet Draft, draft-ietf-sip-session-timer-15.

[6] Rosenberg, J., et al.,"SIP: Session Initiation Protocol," RFC 3261, June 2002.

[7] Boulton, C., and J. Rosenberg., "Best Current Practices for NAT Traversal for SIP," October 2004, Internet Draft, draft-ietf-sipping-nat-scenarios-02.

[8] Rosenberg, J., et al., "STUN—Simple Traversal of User Datagram Protocol through Network Address Translators (NATs)," RFC 3489, March 2003.

[9] Rosenberg, J., "Traversal Using Relay NAT (TURN)," October 2004, Internet Draft, draft-rosenberg-midcom-turn-06.

[10] Bhatia, M., et al., "SIP Session Border Control Requirements," January 2005, Internet Draft, draft-bhatia-sipping-sbc-00.

[11] UPnP Forum, http://www.upnp.org.

[12] Steindera, M., and A. Sthib, "A Survey of Fault Localization Techniques in Computer Networks," July 2004, http://www.sciencedirect.com.

Chapter 9

Conclusions

The introduction of new technologies, including 3G/IMS, broadband access, and disruptors such as WiFi or WiMax, are beginning to shape the next generation communications and entertainment system in a way that rivals the changes brought by the introduction of the Internet at the beginning of the 1990's. 3G/IMS is a logical evolution towards powerful core capabilities and enablers expected to be needed to support complicated business models and allow mobile operators to maintain control of advanced and valued-added applications. The IMS is also seriously considered by many wireline operators since the same types of capabilities and enablers are envisioned to be desired. At the same time, the advent and success of WiFi technology has prompted both cellular and wireline operators to factor its implications in their strategy. Although successful as a wireless extension of a LAN suitable for residential or enterprise deployment, it is marginally profitable when deployed in a hotspot (hotel or coffeeshop) environment. However, the maturity of the WiFi technology provides a number of options available to both cellular and wireline providers in such a way that it has to be considered as a basic building block in the technology and business value chain.

While the telecommunications industry is getting ready to embrace the new technology, it is important to understand the characteristics of a new breed of services and service enablers. Examples of these new services include any combination of VoIP, video telephony, instant messages, multimedia messaging service, push-to-X (where X can be talk, image, or video), click-to-dial, group-based messaging, digital TV, and many more. Driven by business needs, these services demand a new set of requirements, including fast-to-market, short life cycle, flexibility in mix and match of service components, and quantifiable support of fine-grained QoS. Each of these services may have different assurance requirements, depending on the type of service offering, agreement between providers, and the particular SLAs.

Although the long history of network-based management is entrenched in the world of traditional PSTN, it is becoming clear that a service-centric, model-based approach is necessary. This book brings forth a trend that is receiving a lot of attention in the industry: service assurance based on a formal service model to deal with the increasingly complex and demanding management of services. The service model as introduced in this book is constructed so that various technologies and business models are flexibly supported. A well-designed service model, with the right tools and automatic capability of collecting the right information, can properly prioritize the impact of service failures and recommend a correct decision, without much human intervention. An effective service model can embrace the experience and knowledge of many levels of experts in service assurance and automatically construct sensible plans to deal with tedious but necessary tasks.

As we were describing the VoWiFi/3G service model and assurance operations, we clearly recognized that a new set of assurance functions are required to deal with the demanding deployment support of a complex and emerging technology when SIP-based packet voice and wireless and mobility requirements are taken into consideration. These new problems are the results of a number of fundamental changes.

First, the fundamental philosophy of the Internet is that of "best effort." In the last decade, much research has been done aiming at evolving the Internet to be QoS-ready. In practice, the success of having a network where QoS is guaranteed is limited. Today, many premium services with guaranteed QoS are offered with over-provisioning of bandwidth. For a service in which QoS is critical, as in the case of VoWiFi/3G, guaranteeing and assuring quality are a challenge for providers, especially in capacity-scarce networks such as cellular networks. The service model and the emerging service assurance OSS can play a significant role in helping to ensure high service quality, which can be a significant differentiator.

Second, although SIP is emerging as the technology of choice in the session and control layer, and deployment of SIP in VoIP is also gaining significant momentum, many interoperability issues, such as compatibility across vendor equipment, consistent sets of SIP timers, SIP-related AAA structures, service features, and interactions, will need to be stabilized to achieve true interworking across administration boundaries. Many assurance issues can be a result of the configuration issues of various gateway and handset functions. Again, proper SIP-level monitoring and tracing capability will be indispensable when service-related problems are to be identified.

Third, the Internet and IP were originally designed to support peer-to-peer communications. Because of the popularity of the World Wide Web and the serious concern about security, firewalls emerged as a practical solution in the 1990s. Later, the lack of IP addresses in IPv4 also prompted the acceptance of NAT. Nowadays; the combination of NAT and firewalls is a common element in almost all the residential broadband access and enterprise networks. While the NAT/firewall works fine and provides a practical solution, it breaks the end-to-end IP model and creates significant problems for peer-to-peer traffic, such as SIP or H.323. There are many proposed methods to deal with NAT/firewall issues, but these patches all require either new network equipment or new software patches in the network or in the end devices. Moreover, due to the large variations among NAT/firewalls, it is hard to find a cure-all solution. The management of these fixes and new network elements will undoubtedly constitute a new set of assurance problems as explained in Chapter 8. Until the universal replacement of IPv4 to IPv6, the NAT/firewall and the associated assurance issues will remain a key problem that should be taken into account in assurance OSS flows.

Finally, the complicated business value chain demands that a new OSS interface be established among partnering providers. The SLA defined between providers is a critical element in end-to-end QoS. Once an SLA is agreed upon, it is up to the individual providers to define the proper monitoring agent and capabilities to ensure the SLA is enforced. However, most old-style SLAs cover only on Layer 2 metrics and lack proactive resolution for controlling violations related to network-management-specific views for customer-focused care. Other than the requirement for providers to know how the quality of service is, they also need an effective correlation mechanism in place to quantify revenue impact. The service assurance process, equipped with the model approach defined in this book, can provide real-time knowledge of dynamic IP and application-level SLAs with the appropriate level of metrics to support SLA negotiation and monitoring capabilities. It is even more important that an interoperator OSS mechanism be put in place so that management of service-affecting SLA violations can be quickly identified, prioritized, and finally resolved before end customers notice the degradation. A controlled view for customers to access their service levels for SLA verification is thus feasible.

The new service-model-based paradigm brought forth in this book is a starting point for a new assurance approach. It emphasizes modularity without loss of dependence, the proper level of abstraction to deal with complexity, and most importantly, measurable business impact and customer satisfaction. These new efforts are driven from business objectives such as new service development, system migration, business process reengineering, workflow automation, outsourcing, collaborative commerce, and others. In order to conduct business effectively in the future, these companies have to obtain the ability to exchange business information increasingly with strategic counterparts, vendors, and the infrastructure on a global basis. This vast external communication requires that the industry participants be able to exchange service-assurance-related intelligence that is fair and understandable among trusted community members. The increase in the use of real-time QoS indicators (also called real-time business intelligence) has driven the need for flexible applications that can reformat and reroute information with a business emphasis so that companies (e.g., service providers or enterprises) can easily focus their efforts on the appropriate service or operational resources (in terms of service assurance system environment). It is rather clear to us that not only telecommunications but also the info-entertainment industry have indeed reached the stage that modifying their general philosophy of service management to provide technological and operational leadership is an unavoidable direction.

We earlier defined business assurance as the implementation of processes and methodologies that allow service providers to measure and control the quality of operations. The operations and accountabilities within a business can be thought of as belonging to a kind of layered pyramid. Key to the economics of the solution is using existing business information systems rather than replacing them. The solution presented in this book illustrates a modular method allowing for a controlled and deliberate introduction to process monitoring and improvement. The scalability makes it suitable for monitoring service offerings or internal operations in both small (enterprise-level) and large (telecommunication service providers/operators) business environments.

Throughout the entire book, we have provided many useful ideas and methods that apply directly to the telecommunications industry. With the growing number of wireless applications in the enterprise space (e.g., wireless work force management and wireless CRM), the boundaries of many operational domains previously viewed as belonging exclusively to network carriers have become blurry. For instance, an enterprise using VoWiFi service at its premises can actually be an independent and isolated service domain. Even though this book does not cover a detailed service model for such an application, in this instance, we hope the examples presented for telecommunications operations can provide sufficient references for readers who might be interested in extending this model-based solution to their different business needs. Furthermore, the application of this service model-based paradigm is a technology-independent method. We see that our models and ideas have broader applicability outside the areas we have addressed so far.

Acronyms and Abbreviations

2G	Second Generation
3G	Third Generation
3GPP	Third Generation Partnership Project
3GPP2	Third Generation Partnership Project 2
AAA	Authentication, Authorization, Accounting
AAL2	ATM Adaptive Layer 2
AAL5	ATM Adaptive Layer 5
ACELP	Algebraic Code Excited Linear Prediction
ACK	Acknowledgment
ADPCM	Adaptive Differential Pulse Code Modulation
ADSL	Asymmetric Digital Subscriber Line
ALG	Application Level Gateway
AMR	Adaptive Multirate
AP	Access Point
ATA	Analog Telephone Adaptor
ATM	Asynchronous Transfer Mode
B2BUA	Back-to-Back User Agent
BER	Bit-Error Ratio
BICC	Bearer-Independent Call Control
BSC	Base Station Controller
BSS	Basic Service Set
BT	Bluetooth
BTS	Base Transceiver Station
CA	Collision Avoidance
CD	Collision Detection
CDR	Call Detail Record; Call Charging Record (used in 3GPP)
CLEC	Competitive Local Exchange Carrier
CMTS	Cable Modem Termination System
CN	Core Network; Corresponding Node (in MIP)
CNAME	Canonical Name
COPS	Common Open Policy Service
CORBA	Common Object Reference Broker Architecture
CSCF	Call State Control Function
CSI	Component Status Indicator
CSMA	Carrier Sense Multiple Access
CSRC	Contributing Source
CTS	Clear to Send
DCF	Distributed Coordination Function

DHCP	Dynamic Host Configuration Protocol
DIFS	Distributed Interframe Space
DNS	Domain Name Service
DOCSIS	Data over Cable Service Interface Specifications
DSL	Digital Subscriber Line
DSSS	Direct Sequence Spread Spectrum (IEEE 802.11)
DTMF	Dual-Tone Multifrequency
EAP	Extensible Authentication Protocol
EDCF	Enhanced Distributed Control Function
ENUM	E.164 number
ERL	Echo Return Loss
EMS	Element Management System
ESN	Electronic Serial Number
ESS	Extended Service Set
FCS	Frame Check Sequence
FER	FCS Error Rate; Frame-Error Rate
FHSS	Frequency-Hopping Spread Spectrum (IEEE 802.11)
GFSK	Gaussian Frequency Shift Keying
GGSN	Gateway GPRS Support Node
GPRS	General Packet Radio Service
GSM	Global System for Mobile Communications
HFC	Hybrid Fiber Coax
HLR	Home Location Register
HSS	Home Subscriber Server
ICMP	Internet Control Message Protocol
IDL	Interface Definition Language
IETF	Internet Engineering Task Force
IMEI	International Mobile Equipment Identity Number
IMS	IP Multimedia Subsystem
IMSI	International Mobile Station Identity
IN	Intelligent Network
IP	Internet Protocol
ISDN	Integrated Services Digital Network
ISIM	IMS Subscriber Identity Module
ISM	Industrial, Scientific, and Medical (unlicensed band)
ISUP	ISDN User Part
IT	Information Technology
ITU	International Telecommunications Union
KPI	Key Performance Indicator
KQI	Key Quality Indicator
LAN	Local Area Network
MAC	Medium Access Control (IEEE 802)
MAP	Mobile Application Part

MGC	Media Gateway Controller
MGCP	Media Gateway Control Protocol
MGW	Media Gateway
MIN	Mobile Identification Number
MIP	Mobile IP
M-IP	Managed IP (network)
MMS	Multimedia Message Service
MMSC	Multimedia Message Service Center
MN	Mobile Node
MNG	Measurement Navigation Graph
MOS	Mean Opinion Score
MPLS	Multiprotocol Label Switching
MRFC	Multimedia Resource Function Controller
MRFP	Multimedia Resource Function Processor
MS	Mobile Station
MTA	Media Terminal Adaptor
MU	Mobile Unit
NAT	Network Address Translation
NAV	Network Allocation Vector
NCS	Network-Based Call Signaling
NE	Network Element
NFS	Network File System
NIC	Network Interface Card
NOC	Network Operations Center
NTP	Network Time Protocol
OCS	Online Charging System
OFDM	Orthogonal Frequency Division Multiplexing
OSS	Operations Support System
PCF	Packet Control Function
PCM	Pulse Code Modulation
P-CSCF	Proxy Call State Control Function
PDD	Postdial Delay
PDF	Policy Decision Function
PDG	Packet-Data Gateway
PER	Packet-Error Rate
PLMN	Public Landline Mobile Network
PSTN	Public-Switched Telephone Network
QoS	Quality of Service
QPD	Quantitative Performance Diagnosis
QPSK	Quadrature Phase Shift Keying
PBX	Private Branch eXchange
RADIUS	Remote Authentication Dial-In User Service
RAN	Radio Access Network

RPE-LTP	Regular Pulsed Excited Long-Term Predication
RTCP	Real-Time Control Protocol
RTP	Real-Time Protocol
RTS	Request to Send
SBC	Session Border Controller
S-CSCF	Serving Call State Control Function
SDES	Source Description
SDP	Session Description Protocole
SGSN	Serving GPRS Support Node
SGW	Signaling Gateway
SI	Service Index
SIFS	Short Inter-Frame Space
SIGTRAN	Signaling Transport
SIM	Subscriber Identity Module
SIP	Session Initiation Protocol
SIP-T	SIP for Telephony
SLA	Service-Level Agreement
SMTP	Simple Mail Transfer Protocol
SNMP	Simple Network Management Protocol
SS7	Signaling System Number 7
SSID	Service Set Identification
SSRC	Synchronization Source
STUN	Simple Traversal of UDP through Network Address Translation
TCP	Transmission Control Protocol
TDM	Time Division Multiplexing
TDMA	Time Division Multiple Access
TRIP	Telephony Routing over IP
TURN	Traversal Using Relay NAT
UAC	User Agent Client (caller or mobile host)
UAS	User Agent Server (callee or correspondent host)
UDP	User Data Protocol
UMTS	Universal Mobile Telecommunications System
UPnP	Universal Plug and Play
USIM	UMTS Subscriber Identity Module
UTRA	Universal Terrestrial Radio Access
UTRAN	Universal Terrestrial Radio Access Network
VoIP	Voice over Internet Protocol
VoWiFi	Voice over WiFi
WAP	Wireless Application Protocol
WEP	Wired Equivalent Privacy
WiFi	Wireless Fidelity

About the Authors

Richard Lau is a chief scientist and fellow in applied research at Telcordia Technologies. During his 19 years of research in broadband and wireless technologies, he has pioneered fundamental work, including SONET survivable ring architecture, ATM timing recovery, and ATM self-synchronous scrambler ($1+x^{43}$). His recent work includes service modeling for wireless networks and 3G services, WiFi spectrum planning and network management, and quality assurance of packet voice networks.

Dr. Lau received his B.S., M.S., and Ph.D. in electrical engineering from the University of Pennsylvania, in 1983, 1984, and 1987, respectively. He holds five U.S. patents related to high-speed networking and has seven patents pending in the area of IP monitoring, ATM, ADSL, and IP service management. He received the New Jersey Inventor of the Year Award in 1997 for his pioneering work on SONET self-healing ring networks. In 1998 and 2003, he received the Telcordia Individual CEO Award for his long-term contribution to broadband technologies. He was selected to be Telcordia Fellow in 2004. He is a senior member of the IEEE and was an adjunct professor at Brooklyn Polytechnic University from 1987 to 1989.

Ram Khare is currently a principal systems engineer with the tactical systems and solutions business unit at Science Applications International Corporation (SAIC), where he has worked on networking and QoS architectures and systems engineering for future combat systems and next generation tactical radio systems. Previously, he was a director of mobile services control and management research at Telcordia Technologies, an assistant vice president of information technology at the enterprise and health care business unit of SAIC, a distinguished member of the technical staff at Bell Laboratories, and a senior engineer at NASA's Shuttle Avionics Integration Laboratory at the Johnson Space Center in Houston, Texas.

Dr. Khare holds an M.S. in electronics and radio physics from Indore University, Indore, India, and a Ph.D. in biophysics from the All-India Institute of Medical Sciences, New Delhi, India. Under a visitor exchange program, his multidisciplinary thesis work was done at Case Western, Kent State, and Akron Universities, in Ohio, followed by postdoctoral work at Carnegie-Mellon University in Pittsburgh, Pennsylvania. As a network and systems management professional, he has more than 25 years of in-depth experience in telecommunications, space and defense industries, and physics and life sciences.

His experience and interests are in developing systems architectures, developing business and IT requirements, assessing emerging technologies and standards, project management, network and systems management and integration, corporate strategic planning, and network architectures for domestic, international, and tactical services.

William Y. Chang is the cofounder of AcuMaestro, Inc., a system-development firm specializing in service design and management for information technology, financial services, and network service providers. He has 19 years of consulting, engineering, and development experience in major telecommunication and financial companies.

Mr. Chang's career combines leading-edge product/solution quality and system engineering, embracing deep information management, telecommunications technologies, mobile service creation and operation, and commercial software-product development. His affiliations include AcuMaestro, where he is the chief technology officer; Bell Communications Research (now Telcordia), where he was the principal systems engineer of the Next Generation Network OSS Engineering Group; Bell Laboratories, where he was a software architecture consultant for the AT&T Customer Network Management solution and federal government network management system (FTS2000); Tandem Computer Corporation, where he was an engineering consultant for a distributed banking solution; and Unisys Corporation, where he was a firmware engineer in the intelligent terminal development group. He holds an M.A. in computer science from the New Jersey Institute of Technology and has studied for two years in the doctor of engineering program in electrical engineering.

Index

Recent Titles in the Artech House Telecommunications Library

Vinton G. Cerf, Senior Series Editor

Access Networks: Technology and V5 Interfacing, Alex Gillespie

Achieving Global Information Networking, Eve L. Varma et al.

Advanced High-Frequency Radio Communications, Eric E. Johnson et al.

ATM Interworking in Broadband Wireless Applications, M. Sreetharan and S. Subramaniam

ATM Switches, Edwin R. Coover

ATM Switching Systems, Thomas M. Chen and Stephen S. Liu

Broadband Access Technology, Interfaces, and Management, Alex Gillespie

Broadband Local Loops for High-Speed Internet Access, Maurice Gagnaire

Broadband Networking: ATM, SDH, and SONET, Mike Sexton and Andy Reid

Broadband Telecommunications Technology, Second Edition, Byeong Lee, Minho Kang, and Jonghee Lee

The Business Case for Web-Based Training, Tammy Whalen and David Wright

Centrex or PBX: The Impact of IP, John R. Abrahams and Mauro Lollo

Chinese Telecommunications Policy, Xu Yan and Douglas Pitt

Communication and Computing for Distributed Multimedia Systems, Guojun Lu

Communications Technology Guide for Business, Richard Downey, Seán Boland, and Phillip Walsh

Community Networks: Lessons from Blacksburg, Virginia, Second Edition, Andrew M. Cohill and Andrea Kavanaugh, editors

Component-Based Network System Engineering, Mark Norris, Rob Davis, and Alan Pengelly

Computer Telephony Integration, Second Edition, Rob Walters

Customer-Centered Telecommunications Services Marketing,

Deploying and Managing IP over WDM Networks, Joan Serrat and Alex Galis, editors

Desktop Encyclopedia of the Internet, Nathan J. Muller

Digital Clocks for Synchronization and Communications, Masami Kihara, Sadayasu Ono, and Pekka Eskelinen

Digital Modulation Techniques, Fuqin Xiong

E-Commerce Systems Architecture and Applications, Wasim E. Rajput

Engineering Internet QoS, Sanjay Jha and Mahbub Hassan

Error-Control Block Codes for Communications Engineers, L. H. Charles Lee

Essentials of Modern Telecommunications Systems, Nihal Kularatna and Dileeka Dias

FAX: Facsimile Technology and Systems, Third Edition, Kenneth R. McConnell, Dennis Bodson, and Stephen Urban

Fundamentals of Network Security, John E. Canavan

Gigabit Ethernet Technology and Applications, Mark Norris

The Great Telecom Meltdown, Fred R. Goldstein

Guide to ATM Systems and Technology, Mohammad A. Rahman

A Guide to the TCP/IP Protocol Suite, Floyd Wilder

Home Networking Technologies and Standards, Theodore B. Zahariadis

Information Superhighways Revisited: The Economics of Multimedia, Bruce Egan

Installation and Maintenance of SDH/SONET, ATM, xDSL, and Synchronization Networks, José M. Caballero et al.

Integrated Broadband Networks: TCP/IP, ATM, SDH/SONET, and WDM/Optics, Byeong Gi Lee and Woojune Kim

Internet E-mail: Protocols, Standards, and Implementation, Lawrence Hughes

Introduction to Telecommunications Network Engineering, Second Edition, Tarmo Anttalainen

Introduction to Telephones and Telephone Systems, Third Edition, A. Michael Noll

An Introduction to U.S. Telecommunications Law, Second Edition, Charles H. Kennedy

LANs to WANs: The Complete Management Guide, Nathan J. Muller

The Law and Regulation of Telecommunications Carriers, Henk Brands and Evan T. Leo

Managing Internet-Driven Change in International Telecommunications, Rob Frieden

Marketing Telecommunications Services: New Approaches for a Changing Environment, Karen G. Strouse

Mission-Critical Network Planning, Matthew Liotine

Multimedia Communications Networks: Technologies and Services, Mallikarjun Tatipamula and Bhumip Khashnabish, editors

Next Generation Intelligent Networks, Johan Zuidweg

Open Source Software Law, Rod Dixon

Performance Evaluation of Communication Networks, Gary N. Higginbottom

Performance of TCP/IP over ATM Networks, Mahbub Hassan and Mohammed Atiquzzaman

Practical Guide for Implementing Secure Intranets and Extranets, Kaustubh M. Phaltankar

Practical Internet Law for Business, Kurt M. Saunders

Practical Multiservice LANs: ATM and RF Broadband, Ernest O. Tunmann

Principles of Modern Communications Technology, A. Michael Noll

A Professional's Guide to Data Communication in a TCP/IP World, E. Bryan Carne

Programmable Networks for IP Service Deployment, Alex Galis et al., editors

Protocol Management in Computer Networking, Philippe Byrnes

Pulse Code Modulation Systems Design, William N. Waggener

Security, Rights, and Liabilities in E-Commerce, Jeffrey H. Matsuura

Service Assurance for Voice over WiFi and 3G Networks, Richard Lau, Ram Khare, and William Y. Chang

Service Level Management for Enterprise Networks, Lundy Lewis

SIP: Understanding the Session Initiation Protocol, Second Edition, Alan B. Johnston

Smart Card Security and Applications, Second Edition, Mike Hendry

Strategic Management in Telecommunications, James K. Shaw

Strategies for Success in the New Telecommunications Marketplace, Karen G. Strouse

Successful Business Strategies Using Telecommunications Services, Martin F. Bartholomew

Telecommunications Cost Management, S. C. Strother

Telecommunications Department Management, Robert A. Gable

Telecommunications Deregulation and the Information Economy, Second Edition, James K. Shaw

Telecommunications Technology Handbook, Second Edition, Daniel Minoli

Telemetry Systems Engineering, Frank Carden, Russell Jedlicka, and Robert Henry

Telephone Switching Systems, Richard A. Thompson

Understanding Modern Telecommunications and the Information Superhighway, John G. Nellist and Elliott M. Gilbert

Understanding Networking Technology: Concepts, Terms, and Trends, Second Edition, Mark Norris

Videoconferencing and Videotelephony: Technology and Standards, Second Edition, Richard Schaphorst

Visual Telephony, Edward A. Daly and Kathleen J. Hansell

Wide-Area Data Network Performance Engineering, Robert G. Cole and Ravi Ramaswamy

Winning Telco Customers Using Marketing Databases, Rob Mattison

WLANs and WPANs towards 4G Wireless, Ramjee Prasad and Luis Muñoz

World-Class Telecommunications Service Development, Ellen P. Ward

For further information on these and other Artech House titles, including previously considered out-of-print books now available through our In-Print-Forever® (IPF®) program, contact:

Artech House
685 Canton Street
Norwood, MA 02062
Phone: 781-769-9750
Fax: 781-769-6334
e-mail: artech@artechhouse.com

Artech House
46 Gillingham Street
London SW1V 1AH UK
Phone: +44 (0)20 7596-8750
Fax: +44 (0)20 7630-0166
e-mail: artech-uk@artechhouse.com

Find us on the World Wide Web at:
www.artechhouse.com